Advances in Geophysical and Environmental Mechanics and Mathematics

Series Editor: Professor Kolumban Hutter

Board of Editors

Lukas Schneider · Kolumban Hutter

Solid-Fluid Mixtures of Frictional Materials in Geophysical and Geotechnical Context

Based on a Concise Thermodynamic Analysis

Springer

Lukas Schneider
TU Darmstadt
FG Energie- und
Kraftwerkstechnik
Petersenstr. 30
64287 Darmstadt
Germany
schneider@ekt.tu-darmstadt-de

Prof. Kolumban Hutter
Bergstrasse 5
8044 Zürich
Switzerland

ISSN 1866-8348 e-ISSN 1866-8356
ISBN 978-3-642-26934-9 e-ISBN 978-3-642-02968-4
DOI 10.1007/978-3-642-02968-4
Springer Dordrecht Heidelberg London New York

Cover design: deblik, Berlin

Printed on acid-free paper

Springer is part of Springer Science+Business Media (www.springer.com)

Preface

Mixture concepts are nowadays used in a great number of subjects of the biological, chemical, engineering, natural and physical sciences (to name these alphabetically) and the theory of mixtures has attained in all these disciplines a high level of expertise and specialisation. The digression in their development has on occasion led to differences in the denotation of special formulations as 'multi-phase systems' or 'non-classical mixtures', 'structured mixtures', etc., and their representatives or defenders often emphasise the differences of these rather than their common properties.

This monograph is an attempt to view theoretical formulations of processes which take place as interactions among various substances that are spatially intermixed and can be viewed to continuously fill the space which they occupy as mixtures. Moreover, we shall assume that the processes can be regarded to be characterised by variables which obey a certain degree of continuity in their evolution, so that the relevant processes can be described mathematically by balance laws, in global or local form, eventually leading to differential and/or integral equations, to which the usual techniques of theoretical and numerical analysis can be applied.

Mixtures are generally called non-classical, if, apart from the physical laws (e. g. balances of mass, momenta, energy and entropy), also further laws are postulated, which are less fundamental, but may describe some features of the micro-structure on the macroscopic level. In a mixture of fluids and solids – these are sometimes called particle laden systems – the fraction of the volume that is occupied by each constituent is a significant characterisation of the micro-structure that exerts some influence on the macro-level at which the equations governing the processes are formulated. For solid-fluid mixtures at high solids fraction where particle contact is essential, friction between the particles gives rise to internal stresses, which turn out to be best described by an internal symmetric tensor valued variable. Obviously each special application may give rise to its own such internal variable. The mixture is non-classical as a result that each such variable is described by its own dynamical equation.

Our own interest in mixtures has been their use in the description of the flow of debris, mud and slurry in various forms in the geophysical environment: avalanches of snow, gravel, soil, the catastrophic motion of debris as a solid-fluid compound, the motion of lahars in pyroclastic flows from volcanoes, sub-aquatic turbidity currents, catastrophic sediment transport in fluvial hydraulics, and the destabilisation of soil slopes and dams due to heavy rain fall, etc. Because of this background, this monograph is in many respects focussing on these applications, and certainly the geological, geophysical and geotechnical bias is apparent. Nevertheless, we believe that scientists from other fields might equally profit: process and chemical engineers interested in the transportation of products, mechanical engineers and chemists interested in fluidised and spouted beds, bubbly flows, sprays and combustion in flames, physical limnologists and oceanographers, atmospheric scientists etc. in subaquatic sediment transport and aerosol dispersion, etc.

The text uses methods of rational thermodynamics, which is an extensively developed field. We draw the readers' attention to TRUESDELLs writings [119], [121], to MÜLLER [97] and LIU [78], [80] for basic knowledge in continuum mechanics and thermodynamics. As for mathematics, we use the level of University calculus and analysis covering vector and Cartesian tensor calculus and some aspects of differential equations. Knowledge of the elements of exterior calculus may also be helpful, but is not absolutely necessary.

The work for this monograph commenced in early 2005 when LS was preparing his Diploma (M. Sc.) thesis 'A non-classical debris flow model, based on a concise thermodynamic analysis' under the supervision of KH. This thesis has been the basis for this monograph, but is founded on earlier work by SVENDSEN & HUTTER (1995) [115] and SVENDSEN, HUTTER and LALOUI (1999) [116]. These papers concentrated on the thermodynamic formulation of structured visco-elastic mixtures, on the one hand, and on constitutive modelling of dry granular materials when quasi-static frictional (plastic) behaviour is included, on the other hand. In early 2005 a thorough thermomechanical formulation of solid-fluid mixtures, in which the solid constituents would exhibit both dynamic viscous and quasi-static frictional effects, was still missing. However, there was some hope that the above works [115], [116] would provide a supporting guideline for the derivation of a structured thermodynamic mixture theory with frictional properties.

For several reasons, there is a necessity of such a fundamental analysis. Geotechnical engineers and engineering geologists are in need of mathematical models, which are capable of describing the motion of debris in unstable soil slopes, landslides, mud flows, etc., from initiation through their catastrophic advance down to their settlement in the run-out region. This embraces a huge range of mechanical behaviour, from quasi-static soil plasticity prior to rapid flow initiation through criticality, when shear banding initiates the ensuing catastrophic motion, which is basically viscous, to the strongly decelerating and likely compacting motion into the deposition area. This motion is often complicated by the presence of an additional fluid and possibly gas (air) and may be additionally affected by e. g. fluidization, reverse grading according to particle size, de-mixing and complete or partial saturation.

Geotechnical engineers encounter this complexity e. g. in dam break problems, artificial slope stability analyses and questions of safety of soil construction sites. In engineering geology, analogous problems arise in hurricane or typhoon generated mud flows and sturzstroms, in earthquake induced debris flows, sub-aquatic turbidity currents, lahars and related pyroclastic flows and dense as well as particle-laden snow avalanches (flow and powder snow avalanches).

To cover the entire flow regime from the quasi-static deformation of the water saturated soil prior to the formation of a strong shear zone and the associated transition to the dynamic behaviour of the relevant mixture, and finally into the deposition area, the dominant constitutive regimes change from elasto-visco-plastic, through primarily viscous to again consolidating elasto-visco-plastic behaviour. In a multi-constituent mixture, this may be coupled with segregation mechanisms due to particle size (Brazil nut effect) or a layering in essentially separated flows of a dense granular material, underlying a slurry, or in the deposit by a separation into a wet upper layer of solids underlain by a saturated mixture.

The development of a thermodynamically based set of field equations for a structured mixture is, however, desperately needed, even when only purely mechanical processes are in focus, since rather controversial opinions exist among scientists of different groups; these groups represent the *mixture* and *multi-phase* theorists. The controversy between these two groups centers around the question of 'which forms the solid and fluid stress tensors, and in turn, interaction force should be'. Claims are made about the structure of the flux terms in the momentum equations prior to any postulation of a constitutive model, a claim that is premature and pretty empty anyhow if stated prior to a complete exploitation of the Second Law of Thermodynamics with all its inferences.

It is the thesis of this small book that differences in these quantities are vacuous prior to a complete thermodynamic analysis, and indeed, it will be shown that exploitation of the Second Law of Thermodynamics will be the vehicle by which any such possible disagreement can be resolved. This is even so, when purely mechanical formulations are in focus, since the Second Law of Thermodynamics determines for a given constitutive class the equilibrium, among other things, values of the constituent stress tensors and interaction forces (and other field quantities). Differences of two distinct thermodynamic mixture formulations can then be clearly identified as differences in the constitutive postulates – and perhaps in differences of the exploitation of the Second Law of Thermodynamics and its underlain peculiarities.

This book, written during the last three years by both authors, is the product of a joint effort, with the analysis performed more by LS, and the conceptual activity and the physical motivation of the approach and the interpretation of the results primarily done by KH. He is also chiefly responsible for all errors which may still remain. In the process of development of the theory, we have been mostly working alone. Nevertheless, Dr. Ioana Luca, from Academia Sinica, Taiwan has closely followed the developments and used the results in her own work on avalanching solid-fluid avalanche flows. Her questions and critique has helped us in smoothing arguments here and there. Moreover, we thank Prof. Bob Svendsen, Institute of Mechanics, Department of Mechanical Engineering, Dortmund University of Technology, Germany and Prof. Leslie W. Morland, Department of Mathematics, University of East Anglia, UK for their reviews of an earlier draft. The criticism of these three people has led to improvements, which are now incorporated in the text.

Finally, we thank Springer Verlag and its personnel for the supportive help and advise in the production of the book.

November 2008

Academia Sinica, Nangang/Taipei (Taiwan), *K. Hutter*
Technische Universität Darmstadt (Germany), *L. Schneider*

Acknowledgments

KH: The work of this monograph is based on a precursory diploma thesis by LS, completed in early 2005 that was substantially extended in the following years with completion in late fall 2008. KH, who is retired from active duty at Darmstadt University of Technology, Darmstadt, Germany since April 2006, acknowledges the financial support received via the German Research Funding Agencies from 1987-2006, making such fundamental work possible. He now equally acknowledges the inspiring atmosphere at the Laboratory of Hydraulics, Hydrology and Glaciology at the Swiss Federal Institute of Technology (ETH) Zurich, where parts of the writings of this monograph have been done. He expresses, in particular, his thanks to Profs. E. MINOR, W. H. HAGER and M. FUNK for their steady support. In three visits of a total of more than six months duration in 2006/07/08 as Distinguished Visiting Professor at Academia Sinica, Nangang, Taipei, Taiwan, KH had the peace of mind to lay down large parts of the manuscript whose final form is now at the readers disposal. He considers his contribution to this book his scientific report to the authorities of Academia Sinica. KH is thankful to Prof. CHIEN. C. CHANG, Dr. CHIH-YU KUO and Dr. IOANA LUCA, (all Division of Mechanics, Research Center for Applied Sciences, Academia Sinica), and to Prof. YIH-CHIN TAI, from the National Chi Nan University, Puli, Taiwan, to all for the constructive cooperation and the friendly working atmosphere.

LS: An economist would probably regard this work as the first return of a long term investment. It is not only a financial investment – in particular from the German state and family Schneider – that was necessary for the success of this work but also time, good will and access to knowledge and experience. Of course, the excellent scientific infrastructure of Technische Universität Darmstadt (TUD) was crucial to this work but also the dedication of teachers like Mr. H. LANG (conservatoire in Heppenheim), Mr. W. GERECKE (Martin Luther Schule, Rimbach) and Prof. J. CASEY (University of California,

Berkeley) who laid down the mathematical basis that is indispensable for the work with rational thermodynamics.

The thesis 'A Non-classical Debris Flow Model – based on a concise thermodynamic analysis' [113] was the starting point of the work presented here. Under the supervision of KH it was written by LS in order to earn the Diploma degree in the studies of 'Applied Mechanics' at TUD. Without the motivating environment in the group 'AG III' of the former Department of Mechanics (TUD), which at that time was led by KH, this work would not have been possible.

LS also acknowledges the support from TUD, Prof. J. JANICKA, Prof. A. SADIKI and the group of 'Energy and Power Plant Technology', where LS is writing his Ph.D. thesis. The largest investment comes from his wife and family. He is very much indebted to them.

Contents

Acronyms, Symbols

Symbol	Name/Description	Page
$\mathrm{sym}(\cdot)$	Symmetry operator	17
$\mathrm{skw}(\cdot)$	Skew-symmetry operator	17
$[\cdot, \cdot]$	LIE-bracket	17
$\langle \cdot, \cdot \rangle$	JACOBI-bracket	17
$[\![\cdot]\!]$	Jump of the quantity in brackets	45
$\vert\mathbf{A}\vert$	Norm of $\mathbf{A}$; $\vert\mathbf{A}\vert := \sqrt{\mathrm{tr}(\mathbf{A}^2)}$	164
$\partial(\cdot)$	Partial time derivative	17
$\nabla(\cdot)$	Gradient	17
$\mathrm{div}(\cdot)$	Divergence	17
$\frac{\mathrm{d}^\alpha(\cdot)}{\mathrm{d}t}$	material time derivative following K_α	36
$\mathrm{d}(\cdot)$	Exterior derivative	82
$\bar{(\cdot)}$, $(\cdot)$	Density with respect to the mixture volume, constituent volume	42
$\dot{(\cdot)}$	Material time derivative following the mixture	61
$\overset{\circ}{(\cdot)}$	General objective time derivative	68
$\vec{(\cdot)}$	Represents the ordered collection of all constituent quantities of the quantity in brackets	61
$\overset{\diamond}{(\cdot)}$	Objective time derivative for binary, saturated mixtures with constant true mass densities	164

$\sum\limits_{\mathbf{A}}, \sum\limits_{\mathbf{v}}$	Sum over tensors **A**, Sum over vectors **v**	99
$(\cdot)\big\|_{\mathrm{E}}$	Equilibrium part of a quantity	127
$(\cdot)\big\|_{\mathrm{N}}$	Non-equilibrium part of a quantity	127
$(\cdot)^{\mathrm{N}}$	Non-linear part of a quantity	169
$(\mathbf{A})'$	Deviatoric part of **A**	172
$\overset{!}{=}$	'must be equal to'	129
$A(\mathcal{Q}_\alpha, \theta)$	Surface area of $\partial\Omega_{\alpha\theta}$	34
$a(\partial\mathcal{Q}_\alpha)$	Surface area of $\partial\omega_\alpha$	37
$\mathbf{a}_\alpha$	Constituent acceleration	36
$\mathbf{a}$	Vector for the mathematical formulation of the principle of objectivity for mixtures	71
$\mathbf{B}_\alpha$	Constituent left CAUCHY GREEN deformation tensor	37
$\mathcal{B}_\alpha$	Constituent material body	33
B_s	Inverse 'shear viscosity' of the solid constituent, 'shear fluidity'	173
$\mathbf{b}_\alpha$	Constituent external supply rate density for momentum	42
$\mathbf{b}$	Mixture external supply rate density for momentum	48
C	General constitutive variable	58
$\mathbf{C}^{\mathrm{D}}_{\alpha\beta}, \mathbf{C}^{\mathrm{W}}_{\alpha\beta}$	Tensor-valued coefficients of the isotropic expansion of c_α	138
c_α	Interaction supply rate density for mass	42
$c^{\dot\theta}_\alpha, c^{\theta\rho}_{\alpha\beta}, \ldots, c^{\mathrm{W}}_{\alpha\beta\gamma\delta}$	Scalar-valued coefficients of the isotropic expansion of c_α	139
$c^{\mathbf{N}}_\alpha$	Nonlinear remainder of the isotropic expansion of c_α	138
c	Vector-valued abbreviation for a combination of Δ^{*n}_D and ζ_n	126
c_β	Abbreviation for $\mathsf{c}_{,\mathbf{v}_\beta}$	145
$\mathbf{D}_\alpha$	Constituent stretch tensor	36
$\mathbf{D}$	Mixture stretch tensor	50
$\mathcal{E}^3$	EUCLIDian space, three dimensional	33
e_α	Interaction supply rate density for energy	42

e	Equilibrium variables	127
$\mathbf{F}_\alpha$	Constituent deformation gradient	36
$\boldsymbol{\mathcal{F}}$	Vector-valued one-form	82
$\mathcal{F}_{ij}^{x^J}$	Abbreviation for $(\boldsymbol{\mathcal{F}}_{\mathbf{x}^J})_{ij}$	105
$\mathcal{F}_{ijk}^{A}$	Abbreviation for $(\boldsymbol{\mathcal{F}}_{\mathbf{A}})_{ijk}$	105
f_1	Coefficient of barotropy in the hypo-plastic law	164
f_2	Coefficient of pyknotropy in the hypo-plastic law	164
$\mathcal{G}/\,\mathcal{G}_\alpha$	Physical variable characterising a particular aspect of the state of mixture/ constituent K_α	39
$\mathbf{I}$	Second order unit tensor	72
$\boldsymbol{\mathcal{I}}$	Fourth order unit tensor	73
I_A	First invariant of the second-rank tensor $\mathbf{A}$	160
II_A	Second invariant of the second-rank tensor $\mathbf{A}$	160
III_A	Third invariant of the second-rank tensor $\mathbf{A}$	160
$I,\ J,\ K,...$	Identifier for the constitutive variables	17
$i,\ j,\ k,...$	Identifier for the components	15
K_α	α^{th} constituent	33
K	Number of constitutive variables	17
K_s	Inverse 'bulk viscosity' of the solid constituent, 'bulk fluidity'	173
$\mathbf{k}$	Extra entropy flux vector	96
k	Abbreviation for $\frac{1}{2}\lambda^\varepsilon\rho_n(\mathbf{u}_n\cdot\mathbf{u}_n)$	115
k_α^v	Coefficient for the linear contribution of $\mathbf{v}_\alpha$ to $\mathbf{k}$	143
$\mathbf{k}^\mathbf{N}$	Vector-valued coefficient for the nonlinear contributions of $\mathbf{k}$	143
$\mathbf{L}_\alpha$	Constituent velocity gradient	36
L	Fourth order tensor for the modelling of hypo-plasticity	164
$\mathbf{L}$	Mixture velocity gradient	50
l_α^ρ	Linear combination of λ_α^ρ and $\boldsymbol{\lambda}_\alpha^v$	83
l_α^ν	Linear combination of l_α^ρ and λ_α^ν	83

${}_{\alpha}l^{\rho}_{\mathrm{I}}$	Extension of l^{ρ}_{α}	115
${}_{\alpha}l^{\nu}_{\mathrm{I}}$	Extension of l^{ν}_{α}	115
$\mathbf{M}_{\alpha}$	Interaction supply rate density for moment of momentum of constituent K_{α}	42
$\mathbf{M}_{s}$	Solid volume fraction tensor	170
M	Dimension of $\left\{\nabla\theta,\ \vec{\nabla}\rho,\ \vec{\nabla}\nu\right\}$	104
$\mathbf{m}_{\alpha}$	Interaction supply rate density for momentum of constituent K_{α}	42
$\mathbf{m}^{i}_{\alpha}$	EUCLIDian invariant part of $\mathbf{m}_{\alpha}$	57
m	$(n-m)$ constituents are density-preserving or m constituents are 'compressible'	68
$\mathbf{m}^{\mathrm{N}}_{\alpha}$	Vector-valued coefficient for the nonlinear contributions of $\mathbf{m}^{i}_{\alpha}$	144
$\mathbf{m}_{\mathrm{D}}$	Drag coefficient	172
$\mathbf{N}$	Second order tensor for the modelling of hypoplasticity	164
N	Dimension of $\left\{\nabla\theta,\ \vec{\nabla}\rho,\ \vec{\nabla}\nu,\ \mathbf{v}\right\}$	104
$\mathbf{N}^{\mathrm{D}}_{\alpha\beta},\ \mathbf{N}^{\mathrm{W}}_{\alpha\beta}$	Tensor-valued coefficients of the isotropic expansion of n_{α}	139
n	Number of constituents	55
n	Non-equilibrium variables	127
$\mathbf{n}$	Exterior normal unit vector to ω	44
$\mathbf{n}_{\sigma}$	Unit vector, perpendicular to the singular surface, σ	44
n_{α}	Volume (fraction) production rate density of constituent K_{α}	57
$n^{\dot{\theta}}_{\alpha},\ n^{\theta\rho}_{\alpha\beta},\ \ldots,\ n^{\mathrm{W}}_{\alpha\beta\gamma\delta}$	Scalar-valued coefficients of the isotropic expansion of n_{α}	139
n^{N}_{α}	Nonlinear remainder of the isotropic expansion of n_{α}	139
$\mathcal{P}$	Scalar-valued one-form	82
$\bar{p}^{G}_{\alpha}$	True thermodynamic pressure related to Ψ^{G}	117
$\mathbf{Q}$	Orthogonal time-dependent tensor	59
$\mathcal{Q}_{\alpha}$	An open set of elements X_{α}, with boundary $\partial\mathcal{Q}_{\alpha}$	33
$\boldsymbol{\mathcal{Q}}_{\mathbf{x}_{I}}$	Abbreviation for $\left(\mathbf{v}\otimes\mathcal{P}_{\mathbf{x}_{I}}-\boldsymbol{\mathcal{F}}_{\mathbf{x}_{I}}\right)$	88

$\mathbf{q}_\alpha$	Constituent heat flux vector	42
$\mathbf{q}$	Mixture heat flux vector	48
$\mathbf{q}^{\mathbf{N}}$	Vector-valued coefficient for the nonlinear contributions of $\mathbf{q}$	144
$\mathbf{R}_\alpha$	Rotational part of $\mathbf{F}_\alpha$	63
$\mathcal{R}_{\alpha 0}$	Region of constituent K_α in the reference configuration	33
$\mathcal{R}_t$	Region of the mixture in the present configuration	33
r_α	Constituent external supply rate density for the energy	42
r	Mixture external supply rate density for energy	48
$\mathbb{S}$	Set of constitutive variables	69
$\mathbb{S}_{\mathsf{n}}$	Set of non-equilibrium elements n	137
$\mathbb{S}_R$	Reduced set of constitutive variables	114
$\mathbb{S}_{\boldsymbol{\mathcal{Y}}}$	Set of elements $\boldsymbol{\mathcal{Y}}$	137
s	Saturation field	83
$\mathbf{s}$	Linear combination of s and $\boldsymbol{\Gamma}$	83
$\mathbf{s}^*$	Variant of $\mathbf{s}$	94
$\mathbf{T}_\alpha$	CAUCHY stress tensor of constituent K_α	42
$\mathbf{T}$	Mixture CAUCHY stress tensor	48
$\mathbf{T}_I$	'Inner' part of the CAUCHY stress tensor	51
$\mathbf{T}_D$	Diffusive part of the CAUCHY stress tensor	51
$\mathbf{T}_\beta^{\mathbf{N}}$	Tensor-valued coefficient for the nonlinear contributions of $\mathbf{T}_\beta$	144
$\bar{\mathbf{T}}_{\mathrm{cs}}$	Constraint part of the equilibrium solid CAUCHY stress tensor	158
$\bar{\mathbf{T}}_{\mathrm{es}}$	Elastic part of the equilibrium solid CAUCHY stress tensor	158
$\bar{\mathbf{T}}_{\mathrm{fric}}$	Frictional (hypo-plastic) part of the equilibrium solid CAUCHY stress tensor	158
t	Time	33
$\mathbf{u}_\alpha$	Diffusion velocity of constituent K_α	50
$V(\mathcal{Q}_\alpha, \theta)$	Volume of $\Omega_{\alpha\theta}$	34

$v(\mathcal{Q}_\alpha)$	Volume of ω_α	37
$\mathbf{v}_\alpha$	Velocity of constituent K_α	35
$\mathbf{v}$	Mixture velocity or barycentric velocity	49
$\mathbf{v}_{fs} := \mathbf{v}_f - \mathbf{v}_s$	Velocity difference of solid and fluid constituents	156
$\mathbf{v}_{vol} := \nu_s \mathbf{v}_s - \nu_f \mathbf{v}_f$	Volume weighted mixture velocity for a solid-fluid mixture	155
$\mathbf{W}_\alpha$	Vorticity tensor of constituent K_α	36
$\mathbf{W}$	Mixture vorticity tensor	50
$\mathbf{W}_{fs} := \mathbf{W}_f - \mathbf{W}_s$	Difference of solid and fluid vorticity tensors	156
X_α	Element of the material body $\mathcal{B}_\alpha$	33
$\mathbf{X}_{\alpha 0}$	Position vector of the element X_α in the reference configuration	33
$\mathbf{x}$	Position vector in the present configuration	34
$\mathfrak{Y}$	Subset of $\mathbb{S}$	137
$\mathbf{Z}_\alpha$	Partial internal variable for constituent K_α, necessary for hypo-plasticity	63
$\alpha,\ \beta,\ \gamma, \ldots$	Identifiers for the constituents	15
β_α^G	Configuration pressure of constituent K_α related to Ψ^G	117
β_α	Configuration pressure of constituent K_α related to Ψ	148
$\boldsymbol{\Gamma}$	Linear combination of $\mathbf{T}$ and Ψ^G	83
$\boldsymbol{\Gamma}^*$	Variant of $\boldsymbol{\Gamma}$	94
γ_α^Ψ	Constituent interaction supply rate density of ψ_α	46
$\dot{\gamma}$	Shear rate in a simple shear experiment	176
$\gamma_\alpha^{\rho\eta}$	Interaction supply rate density of constituent K_α for the entropy	42
$\Delta_D^{*\alpha}$	Combination of $\mathbf{T}_D$, Ψ_D, $\mathrm{sym}(\nabla\theta \otimes \mathcal{P}_{\nabla\theta})$ and $\mathbf{u}_\alpha \cdot \mathbf{u}_\alpha$	126
δ	Constant introduced along with [**A22**]	185
ε_α	Specific internal energy of constituent K_α	42
ε	Mixture specific internal energy	48

ε_I	'Inner' mixture specific internal energy	50
ε_D	Diffusive part of the mixture specific internal energy	50
ζ_α	Combination of configuration and saturation pressure	126
η_α	Specific entropy of constituent K_α	42
η	Mixture specific entropy	48
θ	(Absolute) temperature	58
$\boldsymbol{\iota}_\alpha$	Coefficient for the mass and volume-fraction interaction in the constitutive laws	119
$\boldsymbol{\kappa}^i_\alpha$	Configuration of constituent K_α at time t_i	33
κ^s_{1-6}	Coefficients in the isotropic representation of $\bar{\mathbf{T}}_s\big\|_{\mathrm{N}}$	170
κ^f_{1-3}	Coefficients in the isotropic representation of $\bar{\mathbf{T}}_f\big\|_{\mathrm{N}}$	170
κ_f, κ_s	'Bulk viscosity' of the fluid and solid constituent	172
λ^ε	LAGRANGE multiplier for the energy	79
λ^ρ_α	LAGRANGE multiplier for the mass	79
λ^ν_α	LAGRANGE multiplier for the volume fraction	79
$\boldsymbol{\lambda}^v_\alpha$	LAGRANGE multiplier for the momentum	79
$\boldsymbol{\lambda}^Z_\alpha$	LAGRANGE multiplier for hypo-plasticity	79
μ_α	Infinitesimal areal fraction of constituent K_α	38
$\mu^G_{\alpha\,\mathrm{I}}$	'Inner' parts of the constituent GIBBS-like free energies; similar to the GIBBS' free energies	118
$\mu_{\alpha\,\mathrm{I}}$	'Inner' parts of the constituent GIBBS' free energies	148
μ	Shear modules in the Neo-HOOKEian ansatz	162
μ_f	'Shear viscosity' of the fluid constituent	172
μ_s	'Shear viscosity' of the solid constituent	173
ν_α	Volume fraction and kinematic viscosity of constituent K_α	38
$\nu_{s\,\mathrm{crit}}$	Solid volume fraction at which the nominal particle distance is larger than, or equal to, the distance at which the particle contact ceases to exist	174
$\nu_{s\,\max}$	Maximum solid volume fraction	174

$\bar{\xi}_\alpha$	Mass fraction of constituent K_α	49
$\boldsymbol{\pi}^{\psi}_\alpha$	Constituent internal production rate density of $\boldsymbol{\psi}_\alpha$	41
$\pi^{\rho\eta}_\alpha$	Intrinsic entropy production rate density of constituent K_α	42
$\boldsymbol{\pi}^{\psi}$	Mixture internal production rate density of $\boldsymbol{\psi}$	48
$\pi^{\rho\eta}$	Mixture intrinsic entropy production rate density	48
π_f	Compressible part of the fluid pressure	157
π^{ρ_α}	Symbol for the balance equation of mass for constituent K_α	79
π^{ν_α}	Symbol for the balance equation of volume fraction for constituent K_α	79
$\pi^{\mathbf{Z}_\alpha}$	Symbol for the balance equation of the partial internal variable for constituent K_α	79
ϖ_α	Pressure term for the CAUCHY stress tensor of constituent K_α	146
ϖ_f	Pressure term for the CAUCHY stress tensor of the fluid constituent	182
ϖ_s	Pressure term for the CAUCHY stress tensor of the solid constituent	182
ϖ	'Total' pressure for the mixture CAUCHY stress tensor	147
$\rho_{\alpha 0}$	True mass density of constituent K_α in the reference configuration	38
ρ_α	True mass density of constituent K_α in the present configuration	38
ρ	Mixture mass density	49
$\sum$	Abbreviation for $\sum_{\alpha=1}^{n}$	49
σ	Singular surface, across which physical fields may suffer finite jump discontinuity	44
$\boldsymbol{\sigma}^{\Psi}_\alpha$	Constituent external supply rate density of $\boldsymbol{\psi}_\alpha$	46
$\sigma^{\rho\eta}_\alpha$	External supply rate density of constituent K_α for the entropy	42
$\boldsymbol{\sigma}^{\Psi}$	Mixture external supply rate density of $\boldsymbol{\psi}$	48
$\sigma^{\rho\eta}$	Mixture external supply rate density for the entropy	48
ς	Saturation pressure	117
ς_I	'Inner' part of ς	119
τ	Stress component in a simple shear experiment	176

$\boldsymbol{\Phi}_\alpha$	Constitutive quantity characterising the hypoplastic stress for constituent K_α	63
$\boldsymbol{\phi}^\psi$	Mixture flux density of $\boldsymbol{\psi}$	48
$\boldsymbol{\phi}^{\rho\eta}$	Mixture entropy flux vector	48
$\boldsymbol{\phi}_\alpha^\psi$	Constituent flux density of $\boldsymbol{\psi}_\alpha$	43
$\boldsymbol{\phi}_\alpha^{\rho\eta}$	Entropy flux vector of constituent K_α	42
$\boldsymbol{\chi}_\alpha$	Motion of the body $\mathcal{B}_\alpha$	33
Ψ^G	HELMHOLTZ-like free energy; similar to the HELMHOLTZ free energy	117
Ψ_I^G	'Inner' part of Ψ^G	117
Ψ_D^G	Diffusive part of Ψ^G	117
Ψ_{sf}^G	Non-elastic part of the 'inner' free energy, Ψ_I^G	163
Ψ_{es}^G	Elastic contribution of the solid constituent to the 'inner' free energy, Ψ_I^G	163
Ψ_{ef}^G	Elastic contribution of the fluid constituent to the 'inner' free energy Ψ_I^G	163
$\boldsymbol{\psi}_\alpha$	Physical field density per unit volume of constituent K_α	39
$\boldsymbol{\psi}$	Physical field density for the mixture	48
$\boldsymbol{\Omega}_\alpha$	General constituent spin tensor of constituent K_α	68
$\boldsymbol{\Omega}^*$	Tensor for the mathematical formulation of the principle of objectivity for mixtures	71
$\Omega_{\alpha\theta}$	Material region assigned to $\mathcal{Q}_\alpha$ in the reference configuration	34
$\partial\Omega_{\alpha\theta}$	Material surface assigned to $\partial\mathcal{Q}_\alpha$ in the reference configuration	34
ω_α	Material region assigned to $\mathcal{Q}_\alpha$ in the present configuration $\omega_\alpha = \Omega_{\alpha t}$	37
$\partial\Omega_\alpha$	Material surface assigned to $\partial\mathcal{Q}_\alpha$ in the reference configuration $\partial\omega_\alpha = \partial\Omega_{\alpha t}$	37

Assumptions

Number	Assumption	Page
[**A1**]	Areal and volume fractions are the same.	41
[**A2**]	The mixture exhibits only a single temperature and no constituent changes its aggregation state.	56
[**A3**]	$\bar{\mathbf{M}}_\alpha = \mathbf{x} \times \bar{\mathbf{m}}_\alpha$.	57
[**A4**]	All volume fractions, ν_α $(\alpha = 1, \ldots, n)$, are constitutive variables and $\bar{n}_\alpha = \partial\nu_\alpha + \operatorname{div}(\nu_\alpha \mathbf{v}_\alpha)$ holds.	63
[**A5**]	All internal variables $\bar{\mathbf{Z}}_\alpha$ $(\alpha = 1, \ldots, n)$ are constitutive variables and $\overset{\circ}{\bar{\mathbf{Z}}}_\alpha = \bar{\mathbf{\Phi}}_\alpha$ holds.	68
[**A6**]	Constituent density constraint.	69
[**A7**]	Saturation constraint.	69
[**A8**]	Constitutive law and set of constitutive variables $\mathbb{S}$.	71
[**A9**]	Existence of an ideal wall.	79
[**A10**]	$\sigma^{\rho\eta} = \sum \boldsymbol{\lambda}^v_\alpha \cdot \bar{\mathbf{b}}_\alpha + \lambda^\varepsilon r$.	80
[**A11**]	$\boldsymbol{\lambda}^v_\alpha = -\lambda^\varepsilon \mathbf{u}_\alpha$.	96
[**A12**]	Symmetry of $\boldsymbol{\mathcal{F}}_{\mathbf{v}_\alpha}$.	102
[**A13a**]	Constitutive assumption for $[\mathcal{P}_{\mathbf{B}_\alpha}, \mathbf{B}_\alpha]$ and $\left[\boldsymbol{\lambda}^Z_\alpha, \mathbf{Z}_\alpha\right]$ if $\boldsymbol{\Omega}_\alpha = \mathbf{W}_\alpha$.	102
[**A13b**]	Constitutive assumption for $[\mathcal{P}_{\mathbf{B}_\alpha}, \mathbf{B}_\alpha]$ and $\left[\boldsymbol{\lambda}^Z_\alpha, \mathbf{Z}_\alpha\right]$ if $\boldsymbol{\Omega}_\alpha$ is independent of $\mathbf{W}_\alpha$.	102

[**A14**]	$\boldsymbol{\mathcal{F}}_{\mathbf{x}^J}, \mathbf{x}^J \in \left\{\nabla\theta,\ \vec{\nabla\rho},\ \vec{\nabla\nu}\right\}$ are independent of $\nabla\rho_1,\ ...,\nabla\rho_m,\ \nabla\nu_1,\ ...,\nabla\nu_{n-1}$ and $\nabla\theta$.	103
[**A15**]	$\lambda^\varepsilon = \hat{\lambda}^\varepsilon(\theta,\ \dot{\theta})$.	113
[**A16**]	$\lambda^\varepsilon\big\|_{\mathrm{E}} = \theta^{-1}$.	134
[**A17**]	$\lambda^\varepsilon = \theta^{-1}$.	148
[**A18**]	Debris flows are isothermal processes.	155
[**A19**]	$\mathbf{q}^{\mathbf{N}}\big\|_{\mathrm{E}}$, $\bar{\mathbf{T}}^{\mathbf{N}}_s\big\|_{\mathrm{E}}$, and $\bar{\mathbf{m}}^{\mathbf{N}}_\beta\big\|_{\mathrm{E}}$ $(\beta = s, f)$ are omitted.	157
[**A20**]	'Principle of phase separation'.	159
[**A21**]	Decomposition of Ψ_I into the elastic and non-elastic parts.	159
[**A22**]	$\bar{\mathbf{T}}_{\mathrm{fric}} = \rho\delta\bar{\mathbf{Z}}_s, \quad \delta = \text{constant}$.	163
[**A23**]	$k^v_\alpha = 0 \qquad (\alpha = s, f)$.	185

Chapter 1
Introduction

Abstract A brief introduction to the aims and scopes of the monograph is given. In Section 1.1 a motivation for the application of the thermodynamic approach is provided. Starting with the balance laws for the masses and momenta of a binary continuous system, it is argued that differences in the apparent forms of these laws, which exist between the classical mixture and the multi-phase systems formulations, can only be resolved by use of the Second Law of Thermodynamics. We also make plausible that the relation between the partial densities, true densities and volume fractions for the constituents, necessarily requires one set of these variables to be internal and thus governed by evolution equations. That constitutive relations depend on the true densities and volume fractions must lead to thermodynamic and configuration pressures. Furthermore, the saturation condition gives rise to the saturation pressure that enters the mathematical formulation as a free field which replaces the lost variable due to the saturation condition.

In Section 1.2 we argue that debris flow models involving solid and fluid constituents must be described by at least a two-phase flow of solid and fluid components. Existing models in the literature employ either one-constituent descriptions or reduced, simplified binary mixture models. These are either reduced to classical DARCY models, in which the fluid accelerations are ignored, or models describing the influence of the fluid in an ad hoc fashion via a parameterisation of the pore pressure. Both ignore a justification of the equations by thermodynamic arguments.

Section 1.3 is concerned with frictional effects in avalanching flows of debris. Classical models incorporate these as viscous effects. To account for the destabilisation of a saturated soil heap in the pre- and post-critical phases of a catastrophic motion from initiation to deposition, plastic behaviour is introduced. This is suggested to be accounted for by symmetric tensorial variables that are related to the rate independent parts of the constituent stress tensors. Special choices of the production rate densities of these variables allow formulation of material laws of the class of hypo-plasticity.

L. Schneider, K. Hutter, *Solid-Fluid Mixtures of Frictional Materials in Geophysical and Geotechnical Context*, Advances in Geophysical and Environmental Mechanics and Mathematics, DOI 10.1007/978-3-642-02968-4_1,

1.1 Motivation

The literature on the dynamical interaction between a fluid and a solid or fluids and solids mixed to a heterogeneous body is immense and equally conflicting in several elements. The mechanical and thermal processes exhibited by such complex systems arise in many fields of the engineering and natural sciences. Examples are e. g. fluidized beds in chemical process engineering, particle laden flows in fluvial hydraulics and typhoon or hurricane and earthquake induced landslides in catastrophic avalanche flows, to name a few, for more details see HUTTER [61]. One characterisation of such flows is, that they often arise as interacting species flows, in which the different species are more or less continuously intermixed. It is then only a small step of abstraction to postulate, that (thermo)-dynamical models for them can be deduced by postulating that the various species continuously fill the space and form what is called a mixture. However, it is already at this level of the construction of a mathematical model that scientists tend to disagree in their opinion what a basic formulation of a mathematical model might be. According to a second alternative understanding the species interaction is achieved by an averaging or homogenisation operation of the effects exerted on one species element by all the other species elements that are present in the assemblage of the species.

The reader might have noticed that the terms 'mixture' and 'multi-phase system' were carefully selected above and avoided wherever possible. In fact, controversial views on mixtures and multi-phase systems start already here. However, our point is that there is no need for disagreement, because at last 'mixtures' and 'multi-phase systems' are structurally the same. Why? Well, in multi-phase systems the essential steps, done to reach structurally the same governing equations as in mixture theories (balances of mass, momenta, energy, entropy) from which models are deduced, are the motivation and formulation of interaction laws between the species elements and subsequently, the performance of a so-called 'phase averaging operation', by which the discrete distribution of the species elements is smeared or smoothed over a representative volume element. The emerging equations are balance laws as in mixture formulations, often, and inappropriately, called conservation laws for quantities typical of those arising in classical physics.[1] Expert presentations of the two approaches are by TRUESDELL [119] and MÜLLER [97] on the mixture side, and JACKSON [70], ANDERSON & JACKSON [6] and DREW &PASSMAN [34] on the multi-phase systems side. Of course, by postulating interaction laws between the species elements and performing the phase av-

[1] Differences may arise, if e.g. a non-polar structure of the mixture theory is postulated and the multi-phase approach suggests a theory with a polar structure. However, this is not what we have in mind. Such differences are of *fundamental nature* and are accepted by both groups as describing materials of different microphysical behaviour. What is meant here are differences in the postulation of flux and interspecies production terms within continuum formulations of the same 'class'.

eraging process, detailed knowledge and information has gone into the phase averaged equations, which is useful, but these equations still contain unspecified variables, e. g. stresses for which constitutive relations must be written down to arrive at the complete so-called closed system of equations. An illustrative example is given by PITMAN & LE [104].

We conclude this initial discussion with the assertion that the mixture and multi-phase systems approaches lead to systems of equations which have essentially the same balance structure for continuously differentiable fields, but both have at this level still unspecified fields which must be adequately parameterised to close the system of equations. Since these equations have the same structure, and their variables are obviously continuously occupying all points of the space of the body, it is not absolutely necessary to differentiate between 'mixtures' and 'multi-phase systems'. We shall treat these denotations as synonymous. This does not, however, remove all controversies. These start anew with the choice or postulation of the unspecified terms mentioned above. In fact, the disagreement between scientists seems often to be caused by performing ad-hoc closure statements at this level.

To explain the various issues, consider a binary mixture of a particle laden fluid. Let $\bar{\rho}_{f,s}$, $\bar{c}_{f,s}$, $\mathbf{v}_{f,s}$, $\bar{p}_{f,s}$, $\bar{\mathbf{T}}_{f,s}$, $\bar{\mathbf{m}}_{f,s}$ and $\bar{\rho}_{f,s}\mathbf{g}$ be the partial densities, the constituent mass production rate densities, the constituent velocities, the constituent pressures, the constituent stress tensors, the specific interaction forces and the external body forces of the fluid (f) and solid (s) constituents, respectively. Moreover, let $\nu_{f,s}$ be the volume fractions and $\rho_{f,s}$ the true densities[2] of the fluid and the solid constituents, respectively. Then, accept the fact (we shall explain this in detail later) that $\bar{\mathbf{T}}_{f,s}$ are symmetric tensors, the balances of mass and linear momentum can be written down for the solid and the fluid as follows:

$$
\begin{aligned}
&\frac{\partial \bar{\rho}_f}{\partial t} + \nabla \cdot (\bar{\rho}_f \mathbf{v}_f) = \bar{c}_f \ , \\
&\frac{\partial \bar{\rho}_s}{\partial t} + \nabla \cdot (\bar{\rho}_s \mathbf{v}_s) = \bar{c}_s \ , \\
&\frac{\partial (\bar{\rho}_f \mathbf{v}_f)}{\partial t} + \nabla \cdot (\bar{\rho}_f \mathbf{v}_f \otimes \mathbf{v}_f) = -\nabla \bar{p}_f + \nabla \cdot \bar{\mathbf{T}}_f + \bar{\mathbf{m}}_f + \bar{\rho}_f \mathbf{g} \ , \\
&\frac{\partial (\bar{\rho}_s \mathbf{v}_s)}{\partial t} + \nabla \cdot (\bar{\rho}_s \mathbf{v}_s \otimes \mathbf{v}_s) = -\nabla \bar{p}_s + \nabla \cdot \bar{\mathbf{T}}_s + \bar{\mathbf{m}}_s + \bar{\rho}_s \mathbf{g} \ ,
\end{aligned}
\tag{1.1}
$$

in which

$$
\bar{c}_f = -\bar{c}_s, \qquad \bar{\mathbf{m}}_f = -\bar{\mathbf{m}}_s, \tag{1.2}
$$

since mass and momentum of the mixture body are conserved (see main text for justification). Moreover, irrespective of whether the solid and the fluid

[2] The 'true' density is the mass of a constituent per volume of that constituent.

constituents are compressible or not, one has

$$\bar{\rho}_{f,s} = \nu_{f,s}\rho_{f,s} \ . \tag{1.3}$$

Assume now that we wish to derive a model for this binary mixture that is only based on the partial differential equations (1.1) (i. e. not involving the variables $\nu_{f,s}$ and $\rho_{f,s}$). Then, equations (1.1) and (1.2) comprise 8 equations for 26 unknowns, namely $\bar{\rho}_{f,s}$ (2), $\mathbf{v}_{f,s}$ (6), $\bar{c}_{f,s}$ (1), $\bar{p}_{f,s}$ (2), $\bar{\mathbf{m}}_{f,s}$ (3) [see (1.2)] and $\bar{\mathbf{T}}_{f,s}$ (12). Treating $\bar{\rho}_f$, $\bar{\rho}_s$, $\mathbf{v}_f$ and $\mathbf{v}_s$, as the basic fields, closure relations are needed for all quantities arising on the right-hand side of (1.1). For the most simple case of vanishing constituent mass production $\bar{c}_f = \bar{c}_s = 0$, we then need closure statements for $\bar{p}_{f,s}$, $\bar{\mathbf{T}}_{f,s}$ and $\bar{\mathbf{m}}_f$. Taking a very naive view, we may set $\bar{\mathbf{T}}_f = \mathbf{0}$ on grounds that the fluid is ideal, and we may conjecture Newtonian viscous behaviour for the solid stress $\bar{\mathbf{T}}_s$. This still leaves us with closure relations for $\bar{p}_{f,s}$ and $\bar{\mathbf{m}}_f$. For the latter we may write down a DARCY type relation: $\bar{\mathbf{m}}_s = \gamma(\mathbf{v}_f - \mathbf{v}_s)$ where γ is a permeability, but this ignores the fact that there may be a buoyancy contribution to $\bar{\mathbf{m}}_f$ of the fluid exercising a static force on the solid, see JACKSON [70] or PITMAN & LE [104]. Leaving this question aside for the moment, we are still left with equations of state for $\bar{p}_{f,s}$. For a single fluid the thermal equation of state follows from thermodynamics via the second law, and there it is a relation involving the true density (which is the same as the partial density for a single constituent material). Here, however, since according to (1.3) partial mass densities may change by changes of the true mass densities or the volume fractions or both, we must conclude that (i) introduction of single constituent pressures $\bar{p}_f$ and $\bar{p}_s$ is insufficient and (ii) a model based on equations (1.1) and (1.2) alone is equally defective. Furthermore, we suspect the existence of *thermodynamic pressures* $p^{\mathrm{th}}_{f,s}$, which must be governed by changes of the true mass densities, and so-called *configuration pressures* $p^{\mathrm{conf}}_{f,s}$, that are then governed by variations of the volume fractions, $p^{\mathrm{conf}}_{f,s} = \beta_{f,s}$. Obviously, this calls for a rigorous thermodynamic derivation which delivers explicit formulae for the pressures. This requires that the partial densities $\bar{\rho}_{f,s}$ are not basic field variables, but instead $\rho_{f,s}$ and $\nu_{f,s}$ are, and that the volume fractions are treated as internal variables for which also evolution equations are required, because $\nu_{f,s}$ obviously describe the configuration change of the mass distribution. We write these equations in the form

$$\begin{aligned} &\frac{\partial \nu_f}{\partial t} + \nabla \cdot (\nu_f \mathbf{v}_f) = n_f \ , \\ &\frac{\partial \nu_s}{\partial t} + \nabla \cdot (\nu_s \mathbf{v}_s) = n_s \ , \end{aligned} \tag{1.4}$$

where $n_{f,s}$ are production rate densities.

There are further compelling demands for a thermodynamic foundation of the theory. To see this, assume that the fluid and solid are both density-preserving. This means that the true mass densities $\rho_{f,s}$ are constant and that the mass balance equations $(1.1)_{1,2}$ reduce to equations $(1.4)_{1,2}$ with

$$n_f = \frac{c_f}{\rho_f}, \qquad n_s = \frac{c_s}{\rho_s} \tag{1.5}$$

for consistency. If we then, for some reason, would assume that the constitutive relations do not depend on $\nu_{f,s}$ either, the pressure terms $p^{\text{th}}_{f,s}$, $p^{\text{conf}}_{f,s}$ would both vanish, a strange behaviour, in particular for equilibrium conditions. So, dropping $\nu_{f,s}$ as independent constitutive variables will in this case most likely lead to singular behaviour.

There is still a further case which gives rise to concern. Suppose that we want to use equations (1.1) for the situation that the mixture is saturated. This means that

$$\nu_f + \nu_s = 1 \qquad \text{(saturation)} . \tag{1.6}$$

This equation expresses that the fluid constituent fills the entire pore space. Its requirement amounts to loosing an independent variable. Equation (1.6) may thus be interpreted as a constraint condition and as such gives therefore rise to a constraint pressure, just as the pressure in incompressible fluids. This pressure is called *saturation pressure* ς, so that in this case

$$\bar{p}_{f,s} = \bar{p}^{\text{th}}_{f,s} + \bar{p}^{\text{conf}}_{f,s} + \bar{p}^{\text{sat}}_{f,s} , \tag{1.7}$$

with

$$\bar{p}^{\text{sat}}_f = \lambda\varsigma , \quad \bar{p}^{\text{sat}}_s = (1-\lambda)\varsigma , \tag{1.8}$$

distributing this pressure via a parameter λ among the constituents. In the usual approaches this distribution is postulated in an ad hoc manner, just as now, since the effect of a single variable of the nature of stress must somehow be distributed among the constituents in a reasonable manner. 'Reasonable' means, that the sum over the constituent pressures ought to be equal to the total saturation pressure. But is this correct and what is the functional form for λ? It can take any finite value, but is, in the literature, unanimously postulated to have the form

$$\lambda = \nu_f , \tag{1.9}$$

see IVERSON [66], IVERSON & DENLINGER [69], DENLINGER & IVERSON [33], PITMAN & LE [104], JACKSON [70] and many others. Assumption (1.8) or its generalisation

$$\bar{p}^{\text{sat}}_\alpha = \nu_\alpha\varsigma \qquad (\alpha = 1, \dots, n) , \qquad \sum_{\alpha=1}^{n} \nu_\alpha = 1 \tag{1.10}$$

for saturated mixtures of n constituents is called the 'assumption of pressure equilibrium'.[3] It is only correct for special cases. Its validity or replacement must and can be proved by thermodynamic arguments.

In the above discussion nothing specific was yet said about the interaction forces $\bar{\mathbf{m}}_{f,s}$ except that their sum must vanish, an obvious statement in view of NEWTON's third law. However, experts in multi-phase systems emphasise that the interaction forces can take different forms, see ANDERSON & JACKSON [6], JACKSON [70]. Conceptually, this is quite obvious if $\bar{\mathbf{m}}_f$ is thought to be additively decomposed as follows:

$$\bar{\mathbf{m}}_f = \bar{\mathbf{m}}_f^{(1)} + \operatorname{div} \bar{\mathbf{T}}_f^{(1)} . \tag{1.11}$$

For instance, from an a priori estimate[4], one might have a first guess of the contribution of $\bar{\mathbf{m}}_f$ and it so happens that it arises as a divergence term. Then, the right-hand sides of $(1.1)_{3,4}$ could be written in the form

$$\begin{aligned} \text{RHS}(1.1)_3 &= -\nabla \bar{p}_f + \nabla \cdot ({}^{(1)}\bar{\mathbf{T}}_f^{\text{New}}) + \bar{\mathbf{m}}_f^{(1)} + \bar{\rho}_f \mathbf{g} , \\ \text{RHS}(1.1)_4 &= -\nabla \bar{p}_s + \nabla \cdot ({}^{(1)}\bar{\mathbf{T}}_s^{\text{New}}) - \bar{\mathbf{m}}_f^{(1)} + \bar{\rho}_s \mathbf{g} , \end{aligned} \tag{1.12a}$$

or even as

$$\begin{aligned} \text{RHS}(1.1)_3 &= -\nabla \bar{p}_f + \nabla \cdot ({}^{(2)}\bar{\mathbf{T}}_f^{\text{New}}) + \bar{\mathbf{m}}_f^{(1)} + \bar{\rho}_f \mathbf{g} , \\ \text{RHS}(1.1)_4 &= -\nabla \bar{p}_s + \nabla \cdot ({}^{(2)}\bar{\mathbf{T}}_s^{\text{New}} - {}^{(2)}\bar{\mathbf{T}}_f^{\text{New}}) - \bar{\mathbf{m}}_f^{(1)} + \bar{\rho}_s \mathbf{g} , \end{aligned} \tag{1.12b}$$

with

$${}^{(1)}\bar{\mathbf{T}}_f^{\text{New}} = \bar{\mathbf{T}}_f + \bar{\mathbf{T}}_f^{(1)} , \qquad {}^{(1)}\bar{\mathbf{T}}_s^{\text{New}} = \bar{\mathbf{T}}_s - \bar{\mathbf{T}}_f^{(1)} , \tag{1.13a}$$

and

$${}^{(2)}\bar{\mathbf{T}}_f^{\text{New}} = \bar{\mathbf{T}}_f + \bar{\mathbf{T}}_f^{(1)} , \qquad {}^{(2)}\bar{\mathbf{T}}_s^{\text{New}} = \bar{\mathbf{T}}_s + \bar{\mathbf{T}}_f , \tag{1.13b}$$

respectively. In these equations, $\bar{\mathbf{T}}_{f,s}^{\text{New}}$ and $\bar{\mathbf{m}}_f^{(1)}$ are still left arbitrary. Obviously, in a constitutive postulate these terms must be determined in thermodynamic formulations. Needless to say, that the representations $(1.1)_{3,4}$ and (1.12), (1.13) are equivalent to one another if indeed a decomposition of the form (1.11) exists.

Looking at (1.13), one might be tempted to conclude that ${}^{(1)}\bar{\mathbf{T}}_f^{\text{New}}$ exhibits only fluid properties, whilst ${}^{(1)}\bar{\mathbf{T}}_s^{\text{New}}$ expresses both fluid and solid properties. However, such an interpretation is not compelling, for if one writes (1.11) as

[3] In the literature 'pressure equilibrium' seems to be applied to the total pressures not just to the saturation pressure.

[4] In multi-phase systems of ANDERSON & JACKSON [6, 70] such *a priori* estimates are explicitly made.

$$\bar{\mathbf{m}}_s = \bar{\mathbf{m}}_s^{(1)} + \operatorname{div} \bar{\mathbf{T}}_s^{(1)} , \tag{1.14}$$

which expresses the same as (1.11), since $\bar{\mathbf{m}}_s + \bar{\mathbf{m}}_f = \mathbf{0}$, (1.12a) is still obtained, but (1.13a) now takes the form

$$^{(1)}\bar{\mathbf{T}}_f^{\text{New}} = \bar{\mathbf{T}}_f - \bar{\mathbf{T}}_s^{(1)} , \qquad ^{(1)}\bar{\mathbf{T}}_s^{\text{New}} = \bar{\mathbf{T}}_s + \bar{\mathbf{T}}_s^{(1)} , \tag{1.15}$$

and the new interaction-force-sum relation, $\bar{\mathbf{m}}_s^{(1)} + \bar{\mathbf{m}}_f^{(1)} = \mathbf{0}$, is preserved. Now, the new solid stress seems to exhibit pure solid properties, whilst the fluid stress is 'mixed'.

To present an even further alternative, one may incorporate the divergence term in (1.11) into the fluid stress, but refrain from doing the analogous step in the momentum equation for the solid. In this case the interaction forces of the new formulation do no longer sum up to zero, a requirement, which is generally left unquestioned. However, in all cases demonstrated so far, the fluid and solid momentum equations remain unchanged.

Obviously, even other decompositions are thinkable. A popular one is due to the multi-phase systems defenders, see e. g. JACKSON [70]. This author demonstrates by scrutinising the forces that are exerted by a fluid flow on suspended particles that

$$\bar{\mathbf{m}}_f = -\bar{\mathbf{m}}_s = -\nu_s \nabla \cdot \bar{\mathbf{T}}_f + \bar{\mathbf{m}}_f^{(2)} . \tag{1.16}$$

With this choice, we may write for the right hand sides of $(1.1)_{3,4}$

$$\begin{aligned} \text{RHS}(1.1)_3 &= -\nabla \bar{p}_f + \nu_f \nabla \cdot \bar{\mathbf{T}}_f + \bar{\mathbf{m}}_f^{(2)} + \bar{\rho}_f \mathbf{g} , \\ \text{RHS}(1.1)_4 &= -\nabla \bar{p}_s + \nabla \cdot \bar{\mathbf{T}}_s + \nu_s \nabla \cdot \bar{\mathbf{T}}_f - \bar{\mathbf{m}}_f^{(2)} + \bar{\rho}_s \mathbf{g} . \end{aligned} \tag{1.17}$$

These equations have lost the divergence property for the stresses. They are yet a different version of the constituent momentum equations, but they are obviously analogous to $(1.1)_{3,4}$ provided that the decomposition (1.16) holds.[5]

It is now understandable that it must be very difficult to bring such disparate formulations into coincidence, or at least to establish a certain degree of harmony. We claim that rational thermodynamics is the vehicle to achieve this. In so doing, one must select a formulation, either equations $(1.1)_{3,4}$, (1.12) or (1.17) for the balance laws of constituent momenta, postulate constitutive relations for a desired material class and subsequently exploit the entropy principle to reduce the constitutive relations to the adequate 'minimal' form. In principle, this can be done for all three of the above presented systems of balance laws or any other one that is available. If for different

5 Readers with advanced knowledge in structured mixture theory will realise that momentum equations based on (1.17) cannot become equivalent to the equations with the original variables, if for the constitutive theory the principle of phase separation is employed.

formulations final results can be brought into coincidence, then the formally different theories describe indeed the same material behaviour of the mixture in question, if not, the decompositions (1.11) or (1.16) are doubtful. This holds for a system that can be derived from (1.1). We regard equations (1.1) to be the most primitive ones, because they have the structure of balance laws, which possess clearly defined global, integrated forms with the usual mathematical properties.

In the above, two constituent mixtures were the basis for us to explain the crucial issues that are encountered with such continuous complex systems, but it is quite clear that nothing essential changes, if the mixture or the multi-phase system consists of $n > 2$ components. We shall develop the theory for arbitrary n but are aware that $n = 2$ and perhaps $n = 3$ or $n = 4$ are the most significant cases. The analysis, not easy to grasp, will demonstrate that the above mentioned uncertainties or open holes in the derivation of the models will all be resolved. There is no reason for scientists to disagree and debate on different reasonable ad hoc assumptions on some parts of the stresses or interaction forces in different formulations; the different theories may well agree with one another, but *agreement or disagreement of two formulations can only be judged after both have been subjected to a complete thermodynamic analysis*.

1.2 Mixtures and Debris Flow Models

Our own interest in fluid-solid mixtures is guided by their application to catastrophic movement of snow, ice and debris avalanches, by mud flows in fluvial hydraulics, e. g. as a result of heavy rain fall when soil on mountain sides or dams break off or when large river discharge triggers sediment erosion and generates its transport and deposition further down in the river basin, often giving rise to devastating effects to life and property by erosion and sedimentation. Such catastrophic events occur nowadays worldwide and can almost daily be found in the news media. Many avalanche models have been developed for dry granular fluids (for a collection of references, see HARBITZ [50], HUTTER [59], PUDASAINI & HUTTER [105], ANCEY [5] and others). Such models are adequate for dry dense snow avalanches, for dry debris flows and landslides that may be triggered by an earthquake. For flow of water-soaked soil or debris, a large number of models for catastrophic motion are based on single constituent models, in which the constitutive relation for the stress tensor is motivated by non-Newtonian rheology. These models employ a closure relation, for which the stress is proportional to the strain rate tensor with scalar factor – the effective viscosity – which itself depends on the second invariant of the stress tensor, sometimes in a rather complex way.

There is an extensive literature on this subject in the rheological sciences, mostly dealing with polymeric and other complex fluids and using stress-stretching relations of the class of visco-plastic materials. Rheologists differentiate between fluids exhibiting a yield stress (BINGHAM, HERSCHEL BULKLEY, DE KEE-TURCOTTE fluids) and fluids without a yield stress (Newtonian, power law fluids with shear thinning and shear thickening behaviour). ANCEY [5] reviews the recent literature, emphasising the viscous and plastic nature of such models. Because of the singularities that arise in some of these fluid models (e. g. infinite viscosity at zero stretching for shear thinning fluids, an essential singularity for all plastic models at yield), regularizations have been introduced which remove these singularities. Such a regularization, making the behaviour mathematically viscous has been suggested by ZHU et al. [130] for the De Kee-Turcotte fluid. LUCA et al. [82] proposed such a regularisation for shear thinning and shear thickening fluids with Newtonian behaviour at small stretching and power law behaviour at large stretching. They derive thin layer approximations on arbitrary topography for these kinds of fluid models and show that at least two classes of avalanche models must be differentiated, the first class being applicable for fluids with zero or small yield stress and the second class for materials exhibiting large yield stress.

These models are likely adequate for flows of particle laden systems up to moderately high solids concentration. They are however, likely inappropriate for high solids concentration when frictional contact between the grains is frequent. In the debris flow community mixture concepts have so far only been incorporated in a broad fashion. Geotechnical engineers and engineering geologists (MCDOUGALL & HUNGR [85, 86], IVERSON [66, 67], IVERSON & DENLINGER [69]) have early recognised that the role of the fluid, or more precisely the pore pressure, is decisive for the run-out of a granular mass subjected to an interstitial fluid. In a first attempt to quantify the role played by the fluid, IVERSON assumed the saturated binary mixture to consist of a fluid and a solid constituent, but imposed the simplifying assumption that both constituents move with the same velocity, $\mathbf{v}_f = \mathbf{v}_s$. This implied that the interaction force (the DARCY term) disappeared from the formulation, but the total pressure needed to be distributed between the solid and the fluid pressures according to

$$p_f = \lambda p \, , \qquad p_s = (1-\lambda) p \, . \tag{1.18}$$

To close the system, an ad hoc closure relation is needed for the pore fluid pressure p or for λ. A similar procedure was also taken up by PUDASAINI et al. [106] in a slightly different avalanche formulation, but using the same physics. Computations have shown that the pore (fluid) pressure exercises a significant effect on granular mass run-out distances. The major disadvantage of this model is the ad hoc nature of the distribution of the pore pressure among the constituents according to (1.18). This equation has no theoretical basis other than to close the system of equations. Computations have also shown that results depend chiefly on the choice of the parameterisation of

λ, an equation for which there is no rational background. In an attempt to introduce some notion of rationalism in such a simplistic approach a diffusion equation for the pore pressure was derived from an independent principle; see, IVERSON [68]. However, such an approach does still not remove the fact that the pore pressure equation is not rationally connected to the mixture equations.

A mixture model, deserving the qualification 'mixture' and serving the purpose of describing water saturated debris flows, was presented by PITMAN & LE [104] using the mass balances $(1.1)_{1,2}$ and the momentum balances that are based on equations (1.16), $(1.1)_{3,4}$. Density preserving and saturation assumptions were also used. The constitutive theory uses a plastic description which is essentially a MOHR-COULOMB behaviour due to RANKINE [108]. In a reduced model the fluid acceleration terms are dropped, so that the fluid equation reduces to a classical DARCY equation. Only preliminary computations for this reduced model have been conducted, and they demonstrate that the fluid component enhances the run-out distances. Interestingly, this theory is complementary to that of IVERSON; it emphasises the role of the dynamic interaction force, whilst IVERSON ignores it, and it ignores the pore pressure, whilst IVERSON accounts for it. Both, however, predict run-out distances that are increased by the presence of the fluid.

The above mixture models are the only ones we know of, which have been proposed to be applied to debris flows and landslides, etc. in the mentioned geological applications. They are of ad hoc nature, i. e. closure relations have been suggested by physical reasoning and plausibility arguments, without any detailed thermodynamic analysis. It is our belief that for a constitutive class, which acceptably embraces possibly successful debris flow models, a thorough thermodynamic analysis ought to be performed to achieve a high degree of certainty of the model equations from which fluid-solid avalanche models in the geophysical context can be further deduced.

1.3 Mixtures with Frictional Components

Landslides and debris flows are often initiated from slopes that have been artificially consolidated and exhibit a high degree of 'stability'. Such flows can also develop from artificial dams. Soil slumps arise in such cases often in inhabited areas (e. g. Hong Kong as a well known site!). In these circumstances the geotechnical engineer or engineering geologist wishes to have a theoretical model for the behaviour of the soil prior to any catastrophic movement when the soil structure is still stable under quasi-static loads, which allows also (i) the prediction of failure by identifying shear bands and other possible localisation features and (ii) the determination of the catastrophic avalanching motion from this initial post-critical state through all phases of the motion to the deposition. The model should predict the pre-critical state of stress, the onset of the localisation, the initiation of the ensuing motion, the speeds and the mass distribution in motion as well as in the final deposition. The models mentioned in the preceding sub-section lack the potential to describe all these phases. A stress 'component' capable of predicting the localisation to identify the breaking soil mass is missing. There are several options to incorporate this component into the theory. We have decided to do this in a form that generalizes the well known constitutive models of hypo-plasticity. These models were developed originally in the Geotechnical Institute of the University of Karlsruhe by D. Kolymbas in his Ph. D. dissertation [74] and independently in the Geomechanics Institute of the University of Grenoble, see CHAMBON [23, 24], CHAMBON et al. [25, 26] by F. Darve [29, 30] and their associates. Since then, they have been generalized and applied by many geotechnical engineers. A review will be given in Chapter 4. Here we wish to mention that, (i) even though the models are likely in conformity with the Second Law of Thermodynamics, this has not been demonstrated, and (ii) that the hypo-plastic models have so far not been set into a clear mixture concept with a number of fluid and solid constituents. It is not clear how this can be done. When interstitial water plays a role, the mixture context shows up in the classical hypo-plastic models only indirectly via the somewhat 'cryptic' statement, that effective stresses are looked at. A thermodynamic setting of their derivation is, however, useful per se, since it will pave the route for extending hypo-plasticity to mixtures with several solid constituents.

In the present work, incorporation of the frictional effects into the theoretical formulation is done in much greater generality than just for hypo-plastic materials. However, this approach allows us to prove how and under what conditions hypo-plasticity fulfils the Second Law of Thermodynamics. We shall show this for the situation that every constituent of the mixture entails hypo-plastic material behaviour, but allows for the possibility that it can be dropped for particular constituents, e. g. fluid components. The demonstration of the thermodynamic consistency has, however, not been easy, at least not in the initial stages. The reason is, that unlike the customary hypo-plastic approach, where an evolution equation for the CAUCHY stress tensor is postulated, we introduce the hypo-plastic behaviour indirectly by adding an evolution equation for an objective symmetric second rank tensor valued internal variable, to the physical balance laws, and postulating its production rate density in plausible form. This has first been demonstrated by SVENDSEN et al. [116] for single phase dry granular materials on the basis of an idea by SVENDSEN and is here generalized in the context of a general mixture theory. This is a necessary step, because it is not at all clear how one should postulate hypo-plastic contributions to the interaction forces or to the partial stress tensors of the solid constituents, if there are more than one. The method can describe frictional behaviour more generally than just by hypo-plasticity, but our interest is primarily in the latter. We claim this work forms the basis for a general material theory of a mixture of solid and fluid constituents that should in principle be able to describe the thermodynamical response of fully saturated soils from their quasi-static behaviour through a phase of destabilisation via the formation of localisation features, to the post-critical behaviour in catastrophic avalanching motions, from their initiation and while the material is subjected to large deformations, down to the deposition. There is certainly still a long way to go before this goal will be fully achieved, but this analysis makes a start at an acceptable level of rigour.

1.4 Objectives, Methods and Structure of the Present Work

As outlined in the previous sections, the objective of the present work is a *thermodynamically consistent derivation* of a general non-classical mixture model that, in a reduced form, allows the reproduction of the behaviour of debris flows. This model shall, on the one hand, be able to describe the above properties of flowing water-filled sediments and, on the other hand, be sufficiently simple to allow its numerical treatment with reasonable effort.

The backbone of this model is a mixture theory that includes both, SVENDSEN's balance equation for the volume fractions (cf. SVENDSEN & HUTTER [115]) and a variant of his frictional theory (cf. SVENDSEN et al. [116]). These extensions of the classical mixture theory are necessary, because debris material, as defined above, is an *immiscible mixture* (to be specified), in which structures at small scales evolve independent of the overall motion.[6] We employ and extend the approach to the concept of the frictional resistance of SVENDSEN [116], because we think that the frictional effects between the grains, which are influenced by the fluid can be described with it. We even hope that the model to be developed has the capacity to appropriately reproduce the resistance of the reposed debris material to shear stresses at and close to equilibrium. At this early stage, the model is designed to treat frictional effects of a rather general kind, more general than hypo-plasticity, which turns out to be one particular application. In fact, in the thermodynamic setting of the theory, as outlined in Chaps. 5 to 7, no explicit connection of the results with hypo-plasticity is apparent. The identification of hypo-plasticity will be achieved in Chap. 8 by specially choosing a representation for the production rate density of a tensorial internal variable. So, when hypo-plasticity is mentioned in lieu of frictional behaviour, we have a specialisation in mind that is important for the geotechnical-geophysical application. The developments of Chaps. 5 to 7 hold for a material with more general or more specialized frictional behaviour.

For the general mixture model it will be assumed that (i) the mixture is *saturated*, (ii) $(n-m)$ of the n constituents are *density-preserving* whilst

---

[6] An example for this evolution is the abrupt outset and termination of fluidisation.

the remaining m components are compressible, and (iii) the stress tensors are *symmetric*, i. e. no *polar effects* are considered.[7] It may be appropriate at this juncture to draw the reader's attention to the fact that in mixture theories the definitions of 'saturation' and 'incompressibility' must be taken with care. In this work, saturation means that the mixture constituents fill the entire space; there is no empty space which may be envisaged to consist of mass-less voids. The word 'incompressibility' should be avoided entirely and only the notions 'volume and density preserving' should be used. Looking at a single grain, this grain can be made of a one-constituent material which is 'volume-preserving'. This then means that the grain does not change its volume. The grain by itself is then also 'density-preserving', because it is well known that in one-constituent materials 'volume-preserving' and 'density-preserving' describe the same behaviour.[8] Not so for the mixture! And the crucial relation is equation (1.3). The partial density of a constituent can vary because the true density varies or its volume varies, or both. It follows immediately that the mixture density may vary, even if all constituents are density-preserving; that is, the mixture is not density-preserving. It is clear that replacing the notions of 'density-preserving' and 'volume-preserving' by 'incompressible' is not sensible.

The thermodynamic analysis which is presented in the sequel is based on the MÜLLER-LIU approach of exploiting the entropy principle (cf. MÜLLER [97]). Also, use will be made of methods and definitions of SVENDSEN & HUTTER [115], LIU [80] and SVENDSEN et al. [116].

The present work commences with some mathematical preliminaries and notation issues (Chapter 2), followed by an explanation of the main ideas (i. e. premises, balance equations, definitions) of a mixture theory that accounts for the internal structure of the mixture (Chapter 3). Following the lines of material modelling in Chapter 4, constitutive assumptions are postulated, first, to introduce the specific material behaviour of water-soaked debris and second, to close the system of field equations. With the help of the entropy principle additional restrictions for the constitutive quantities are found in Chapter 5. This chapter presents the formulation of the entropy principle

---

[7] In the recent literature on granular materials (cf. HUANG & BAUER [56]) micro-polar effects have been considered, specifically to account for the rotation of the grains in order to accurately describe, besides other things, the formation of shear bands and with it the effect of fluidisation. Theories using this refinement are not yet well tested, and not established to model granular materials. To avoid additional complexity, we assume a priori all stress tensors to be symmetric.

[8] This follows from the mass balance, $\dot{\rho} + \rho \operatorname{div} \mathbf{v} = 0$. Indeed, 'density-preserving' means $\dot{\rho} = 0$, and this implies $\operatorname{div} \mathbf{v} = 0$. Considering a fixed volume $\mathcal{V}$ with boundary $\partial\mathcal{V}$ we thus have $\int_{\mathcal{V}} \operatorname{div} \mathbf{v}\, dV = 0$ or $\int_{\partial\mathcal{V}} \mathbf{v} \cdot \mathbf{n}\, da = 0$, if $\mathbf{n}$ is the unit normal vector on $\partial\mathcal{V}$ exterior to $\mathcal{V}$. The last expression says that the volume $\mathcal{V}$ does not change with time. Thus:

$$\underset{\text{'density-preserving'}}{\dot{\rho} = 0} \quad \Longleftrightarrow \quad \underset{\text{'volume-preserving'}}{\operatorname{div} \mathbf{v} = 0}$$

in the form as postulated by MÜLLER [95, 96, 97] and transformed by LIU into a variational principle and deals with its mathematical transformation that allows direct identification of the applicability of LIU's lemma [78]. In Chapter 6 first inferences are drawn, i. e., the LIU identities and the residual entropy inequality are identified. Exploitation of the isotropy of the one-form suggested by the former paired with a number of ad-hoc assumptions allows derivation of a substantially simplified thermodynamic potential from which thermodynamic and configuration pressures can be derived, including all LAGRANGE multipliers except that corresponding to saturation and defining the saturation pressure. In Chapter 7 thermodynamic equilibrium is defined and equilibrium forms of the stress tensor, interaction-force and heat flux vector are derived. It is in this chapter where the dependences of these fields are unravelled and the structure of the theory becomes transparent. Chapter 8 is devoted to the reduction and simplification of the derived, general mixture theory towards a manageable debris flow model, where besides other things, the '*principle of phase separation*' (to be specified) is applied. In the same chapter, we compare the unphysical assumption of '*pressure equilibrium*' (to be specified), which, despite its unnecessary strong implications, is still in use, (cf. PITMAN & LE [104], IVERSON and others [33, 68, 69]) with a much weaker assumption proposed by HUTTER et al. [63]. Moreover, we give a detailed account of the non-equilibrium solid stress parameterisation, which is based on typical viscometric experiments of non-Newtonian rheology of elasto-visco-plastic fluids. In Chap. 9 a brief review of the modelling concepts is given and the achievements and limitations of the taken approach are discussed. At the end of the book, Appendix A presents concepts of exterior calculus that are employed in Chaps. 5 to 7. A number of the more complex computational steps that are encountered in the developments are explained in detail in Appendix B.

Chapter 2
Mathematical Preliminaries and Notations

Abstract In the first part of this chapter we present the symbolic and the Cartesian tensor notations and show how these are applied in this book. Tensor calculus is presumed known to the reader; so, only specifics and peculiarities pertinent to the work are discussed. In the second part the elements of exterior calculus are explained, but only to the extent as they are used in the thermodynamic approach treated later on, in particular in Chap. 5.

2.1 Tensors

It is assumed that the reader is familiar with the elements of tensor algebra, analysis and calculus. There are many books which present this subject, among them e. g. Bowen & Wang [15, 16] or Chadwick [22] or Klingbeil [73].

Subsequently, not only symbolic but also index notation will be used, because often proofs and auxiliary results are easier to derive that way. Notation is a crucial issue and has to be treated with care. In particular, this is true for mixture theory. In the symbolic notation we choose Greek letters, $(\alpha,\ \beta,\ \gamma, \ldots)$, to identify the constituents of the mixture and place them in the right lower corner of a quantity. In index notation, the Greek letters for the constituents are moved to the right upper corner. Indices identifying the Cartesian components of tensors are written in small Latin minuscules, $(i,\ j,\ k,\ l,\ \ldots)$, in the lower right corner of a quantity. As usual, the Einstein summation convention is used for the component indices but not for the Greek constituent indices. Consequently, summations over the constituents are always written out explicitly.

In ensuing calculations, we think of vectors and tensors over $\mathbb{R}^3$ as quantities that consist of components and an associated basis. Thus, we write

$$\mathbf{v} = \mathrm{v}_i\ \mathbf{e}_i, \qquad \mathbf{A} = \mathrm{A}_{ij}\ \mathbf{e}_i \otimes \mathbf{e}_j, \qquad \mathbf{B} = \mathrm{B}_{ijk}\ \mathbf{e}_i \otimes \mathbf{e}_j \otimes \mathbf{e}_k\ , \tag{2.1}$$

L. Schneider, K. Hutter, *Solid-Fluid Mixtures of Frictional Materials in Geophysical and Geotechnical Context*, Advances in Geophysical and Environmental Mechanics and Mathematics, DOI 10.1007/978-3-642-02968-4_2,

where an orthonormal basis $\mathbf{e}_i$ $(i = 1, 2, 3)$ is used that spans $\mathcal{W}$ which is a three-dimensional vector space over $\mathbb{R}$. $\mathbf{e}_i \otimes \mathbf{e}_j$, $\mathbf{e}_i \otimes \mathbf{e}_j \otimes \mathbf{e}_k$, etc. represent *dyadic products* of these basis vectors. It follows that $\mathbf{v}$ is an element of $\mathcal{W}$ and the second rank tensor $\mathbf{A}$ can be understood as a linear mapping of a vector from $\mathcal{W}$ to $\mathcal{W}$. This statement can be written as

$$\mathbf{A}\mathbf{v} = \mathrm{A}_{ij}\,(\mathbf{e}_i \otimes \mathbf{e}_j)\mathrm{v}_k\mathbf{e}_k \;=\; \mathrm{A}_{ij}\mathrm{v}_k\,\delta_{jk}\,\mathbf{e}_i \;=\; \mathrm{A}_{ij}\mathrm{v}_j\,\mathbf{e}_i =: \mathrm{y}_i\,\mathbf{e}_i \;. \tag{2.2}$$

Analogously, higher order tensors can be understood as multi-linear forms, for details see e. g. BOWEN & WANG [15]. In (2.2) the usual KRONECKER delta,

$$\delta_{ij} = \begin{cases} 1 & \text{for } i = j \;, \\ 0 & \text{for } i \neq j \;, \end{cases} \tag{2.3}$$

and the definition of the dyadic product

$$(\mathbf{a} \otimes \mathbf{b})\;\mathbf{c} := \mathbf{a}\;(\mathbf{b} \cdot \mathbf{c}) \;, \tag{2.4}$$

have been used, where $\mathbf{a}$, $\mathbf{b}$ and $\mathbf{c}$ are any vectors in the vector space $\mathcal{W}$. The operation $\mathbf{a} \cdot \mathbf{b}$ of $\mathbf{a}$ and $\mathbf{b}$ is called the *scalar product* and reveals a scalar. Henceforth, the *dot product* $\mathbf{A} \cdot \mathbf{B}$ of two tensors $\mathbf{A}$ and $\mathbf{B}$ of the same, but arbitrary rank results in a scalar. For second, $\mathbf{A}$, $\mathbf{B}$, and third, $\mathbf{C}$, $\mathbf{D}$, rank tensors we define this product as

$$\mathbf{A} \cdot \mathbf{B} := \mathrm{A}_{ij}\;\mathrm{B}_{ij} \;, \qquad \mathbf{C} \cdot \mathbf{D} := \mathrm{C}_{ijk}\;\mathrm{D}_{ijk} \;. \tag{2.5}$$

One can think of several other products, e. g. in $\mathbb{R}^3$ the *cross product* of the two vectors $\mathbf{a}$ and $\mathbf{b}$

$$\mathbf{a} \times \mathbf{b} := \mathrm{e}_{ijk}\;\mathrm{a}_i\mathrm{b}_j\;\mathbf{e}_k \;, \tag{2.6}$$

where e_{ijk} stands for the alternator,

$$\mathrm{e}_{ijk} := \begin{cases} 1 & \text{if } i, j, k \text{ are an even permutation of 1,2,3} \;, \\ -1 & \text{if } i, j, k \text{ are an odd permutation of 1,2,3} \;, \\ 0 & \text{else} \;. \end{cases} \tag{2.7}$$

Later in this chapter the *trace operator*

$$\mathrm{tr}(\mathbf{A}^{\mathrm{T}}\mathbf{B}) \;=\; \mathrm{tr}(\mathbf{A}\mathbf{B}^{\mathrm{T}}) := \mathbf{A} \cdot \mathbf{B} \tag{2.8}$$

will also be applied, where $\mathbf{A}$ and $\mathbf{B}$ are second rank tensors. It can be seen from its definition that the trace operator transforms $\mathbf{A}\mathbf{B}^{\mathrm{T}}$ into a scalar. The *transpose* $\mathbf{A}^{\mathrm{T}}$ of the tensor $\mathbf{A}$ is defined as follows

$$\mathbf{a} \cdot \left(\mathbf{A}^{\mathrm{T}}\mathbf{b}\right) \;=\; \mathbf{b} \cdot (\mathbf{A}\mathbf{a}) \,, \qquad \forall\; \mathbf{a}, \mathbf{b} \in \mathcal{W} \;. \tag{2.9}$$

As done in the literature, for calculations in index notation the bases of vector- and tensor-valued quantities are occasionally omitted. Tacitly assuming that we are always dealing with an orthonormal basis, we will follow the same line. Therefore, in place of

- $v_i \mathbf{e}_i$ we shall write v_i ,
- $A_{ij}\mathbf{e}_i \otimes \mathbf{e}_j$ we shall write A_{ij} ,
- $(B_{ik}\mathbf{e}_i \otimes \mathbf{e}_k)(a_j \mathbf{e}_j)$ we shall write $B_{ij}a_j$,

and we shall call v_i and $B_{ij}a_j$ vectors and A_{ij} a second rank tensor even though this is, strictly, not correct.

To make calculations easier we define

$$\begin{aligned} \mathrm{sym}(\mathbf{A}) &:= \tfrac{1}{2}\left(\mathbf{A} + \mathbf{A}^{\mathrm{T}}\right) , \quad & [\mathbf{A},\ \mathbf{B}] &:= \mathbf{AB} - \mathbf{BA} , \\ \mathrm{skw}(\mathbf{A}) &:= \tfrac{1}{2}\left(\mathbf{A} - \mathbf{A}^{\mathrm{T}}\right) , \quad & \langle \mathbf{A},\ \mathbf{B} \rangle &:= \mathbf{AB} + \mathbf{BA} , \end{aligned} \tag{2.10}$$

where the operators $\mathrm{sym}(\cdot)$ and $\mathrm{skw}(\cdot)$ extract the symmetric and the skew-symmetric parts of $\mathbf{A}$, respectively. The latter two definitions specify the LIE and JACOBI-brackets, respectively[1].

If we follow the notation of SVENDSEN & HUTTER [115] the temporal (or partial time) derivative of a general quantity φ (it can be a scalar-, vector- or tensor-valued function) is denoted by $\partial\varphi$ and its spatial (or partial space) derivative is given by $\nabla\varphi = \partial\varphi/\partial\mathbf{x} = \varphi_{i\ldots j,k}\ \mathbf{e}_i \otimes \ldots \otimes \mathbf{e}_j \otimes \mathbf{e}_k$.

In the thermodynamic analysis we will be dealing with dependent constitutive quantities $\mathbf{f}$ ($\mathbf{f}$ stands e. g. for the CAUCHY stress tensor or the heat flux vector, ...) and independent (constitutive) variables

$$\vec{\mathrm{x}} = (\mathrm{x}_1,\ \ldots, \mathrm{x}_K) .$$

Examples for x_s $(s = 1,\ \ldots, K)$ are the temperature field, the velocity of a constituent or its gradient etc. The dependence of $\mathbf{f}$ on $\vec{\mathrm{x}}$ is written as

$$\mathbf{f} = \hat{\mathbf{f}} \circ \vec{\mathrm{x}} = \hat{\mathbf{f}}(\vec{\mathrm{x}}) .$$

Due to the chain rule, the temporal and spatial derivatives of $\mathbf{f}$ take the forms

$$\begin{aligned} \partial\mathbf{f} &= \sum_{I=1}^{K} \hat{\mathbf{f}}_{,\mathrm{x}_I} (\partial\mathrm{x}_I) , \\ \nabla\mathbf{f} &= \sum_{I=1}^{K} \hat{\mathbf{f}}_{,\mathrm{x}_I} (\nabla\mathrm{x}_I) , \end{aligned} \tag{2.11a}$$

[1] In the sequel, the brackets $[\cdot,\ \cdot]$ and $\langle\cdot,\ \cdot\rangle$ will exclusively be used for the LIE and JACOBI operations, $(2.10)_{2,4}$.

$$\nabla \cdot \mathbf{f} := (\mathbf{f})_{i(\ldots)jk,k}\, \mathbf{e}_i \otimes \ldots \otimes \mathbf{e}_j$$
$$= \sum_{I=1}^{K} \frac{\partial(\hat{\mathbf{f}})_{i(\ldots)jk}}{\partial \mathrm{x}_I} \frac{\partial \mathrm{x}_I}{\partial x_k}\, \mathbf{e}_i \otimes \ldots \otimes \mathbf{e}_j \,. \tag{2.11b}$$

The partial derivative $\mathbf{f}_{,x_I}$ which occurs in equations (2.11) can be defined according to FRÉCHET. For a detailed definition of this type of partial derivative the reader is referred e. g. to MARSDEN & HUGHES [83], EDELEN [35], or CASEY [20], where explicit definitions and calculations of some important derivatives can also be found.

2.2 Results from Exterior Calculus

The mathematically complete introduction to exterior calculus can be found in the book 'Applied Exterior Calculus', by D.G.B. EDELEN [35]. Its formal treatment goes beyond the mathematical knowledge that is commonly absorbed by geophysicists and engineers; so, the intention here is to present those results established in this special mathematical field which are useful in the ensuing developments and facilitate the algebraic manipulations in the calculations of the thermodynamic analysis in Chapters 5 to 7. In this book only those aspects are of significance which concern so-called *differential* or *Pfaffian forms* and inferences which can be drawn from them when these forms are *total* or *perfect.* Alternative presentations of exterior calculus to [35] are by CARTAN [18, 19] and HEIL [53]. Here we follow mainly the beautiful 'down to earth' presentation by BAUER [10] in Chapter 4 to his Ph.D. dissertation 'Thermodynamische Betrachtung einer gesättigten Mischung', which we present here in our own English version, with additions and alterations where felt necessary. A formal exposition of the Exterior Calculus, presenting the ground work of what follows in the summary below is given in Appendix A

2.2.1 What is integrability?

Let

$$dF = \sum_{i=1}^{n} X_i(x_j)dx_i \tag{2.12}$$

be a *differential form* dF, which is expressed as a linear combination of differentials dx_i with coefficient functions X_i $(i = 1, \ldots, n)$ which depend on some or all of the x_j $(j = 1, \ldots, n)$. Equation (2.12) is also called a *Pfaffian form.* Under what conditions is the denotation dF on the left-hand side of (2.12) justified in the sense that the expression (2.12) represents a total differential? In other words, under what conditions does integration over the right-hand side of (2.12) deliver a value that is independent of the path of integration in 'configuration space' of the 'independent' variables x_i ($i = 1, \ldots, n$)?[2] If this is true, this value will only depend upon the initial and final points of the integration. If this should not be the case and $\sum_i X_i dx_i$ does not represent a total differential, there still remains the question: can we alter this situation by multiplying the right-hand side of (2.12) with an adequate function? Under those situations this function is called an *integrating factor*.[3] In order to geometrically interpret the roles played by the variables X_i and the differential forms dx_i, it is advantageous to write (2.12) in vectorial form

$$dF = \mathbf{X} \cdot d\mathbf{x} \; , \tag{2.13}$$

where $\mathbf{X} \in \mathbb{R}^n$ and $\mathbf{x} \in \mathbb{R}^n$ are ordered arrays $\mathbf{X} = (X_1, \ldots, X_n)$, $\mathbf{x} = (x_1, \ldots, x_n)$ and the dot denotes the scalar product over $\mathbb{R}^n$. If one writes (2.13) in the homogeneous form, $dF = 0$, it becomes clear that $\mathbf{X}$ defines a normal field which is orthogonal to the hypersurfaces on which the value of F does not change. Solutions of the equation $dF = \mathbf{X} \cdot d\mathbf{x} = 0$ are surfaces (or curves according to dimension) $\mathbf{x} = \mathbf{x}(\sigma, \tau, \ldots)$ on which F is constant. Locally such a solution can always be constructed, however, this surface may possibly not have the largest dimension $(n-1)$. If the local solutions possess the maximal possible dimension, one says that equation (2.12) is *completely integrable* (see also HEIL [53]). In this case it is possible, starting at a particular point, to construct a hypersurface – the mentioned manifold of dimension $(n-1)$ – within which any arbitrary integration of the right-hand side of (2.12) delivers the result zero. In this way one achieves the result to fill the entire phase space with 'onion shells' on which the equation $dF = 0$ holds and which never touch or intersect each other. If one imagines that the phase space is 'partitioned' in this way, there still remains the problem to assign to each 'onion shell' a value for the potential F and to guarantee that an integration of the differential between the various shells delivers always the same difference between these values, irrespective of where this integration is performed. For even if the construction of the surfaces of constant F-values is successful, this does not yet guarantee that the 'distance' between the surfaces does not depend on the position at the surface. However, once the 'onion shells' are constructed and appropriate potential values assigned to them, these facts then define in a unique way a scalar valued function – the above mentioned *integrating factor* by which the right-hand side of (2.12) must be multiplied to create everywhere the correct 'distance' between the

[2] Configuration space is the space of the independent variables x_j ($j = 1, \ldots, n$).

[3] In the classical thermodynamic literature, authors often use a different notation for a differential form depending on whether it is total (dF) or not ($đF$). So, dF is total, but $đF$ is not. The modern mathematical literature does not distinguish the two cases and omits the differential symbol on the left-hand side of (2.12) (see (5.13) and (5.14)).

potential surfaces, respectively to create the desired connection between the differential and potential, provided it is not a priori given.

As we will see, a certain arbitrariness or possibility of choice remains unresolved because the 'labeling' of the 'onion shells' with potential values is not unique. Except for this freedom, it is, however, possible in this way to construct an integrating function for a vector field or a differential which locally allows in each point in phase space the construction of an equi-potential surface. With the aid of this function the vector field can be derived from a scalar potential or, alternatively, the differential form becomes total so that integrals between two points along arbitrary paths have all the same value. The above qualification of such a differential as being completely integrable is to be understood in this way.

In the following the conditions will be studied which formally must be satisfied in order that a differential form which by itself is not total can be made total by multiplying it with a scalar function, respectively to see whether a differential form is total already ab initio. Generally, the advantage of such a reduction of a vector valued function to a single scalar valued function is that mathematical operations are generally easier to perform with scalars than with vectors or tensors.

In what follows the differential dF will define the entropy (and in a second case the entropy flux) which must in all circumstances be a potential. This requirement allows inferences to be deduced for the coefficients of the differential form, X_i, which must be compatible with the potential properties.

2.2.2 Requirements to be imposed on the 'normal fields'

Recall that a vector field $\mathbf{v}(x, y, z)$ over $\mathbb{R}^3$ is a gradient field of a scalar potential field P, $\mathbf{v} = \operatorname{grad} P$, if the vecor field $\mathbf{v}(x, y, z)$ is irrotational,

$$\nabla \times \mathbf{v} = \mathbf{0} \, . \tag{2.14}$$

If this property is not fulfilled, one may try to enforce it by multiplication of $\mathbf{v}$ with a scalar function $f(x, y, z)$. This would make f an integrating factor. Instead of requiring the vanishing of $\nabla \times \mathbf{v}$, one will then request

$$\nabla \times (f\mathbf{v}) = f\nabla \times \mathbf{v} + (\nabla f) \times \mathbf{v} = \mathbf{0} \, . \tag{2.15}$$

Scalar multiplication of this equation with $\mathbf{v}$ yields, since only $f \neq 0$ is reasonable,

$$(\nabla \times \mathbf{v}) \cdot \mathbf{v} = 0 \, , \tag{2.16}$$

which is a necessary condition that a non-trivial function can exist by which (2.15) can be fulfilled. However, that (2.16) is also a sufficient condition for the existence of a non-trivial f is not easy to prove. Rather than to pursue this restricted case for $\mathbf{v} \in \mathbb{R}^3$ it is advantageous here to address the generalization of this theorem of differential forms for arbitrary dimensions. This proposition is known in the theory of differential forms or in exterior calculus as the *Frobenius condition* and is well known. Its derivation is somewhat complicated and requires algebraic techniques of exterior calculus, see CARTAN [18], [19], EDELEN [35]. Just as the FROBENIUS condition generalizes equation (2.16), so condition (2.14), which is a statement restricted to a vector field over $\mathbb{R}^3$ to make it derivable from a potential, can be generalized to POINCARÉ's *theorem*, valid in a space of arbitrary dimension, see CARTAN [18], [19], EDELEN [35]. We shall state these propositions without proof.

a) POINCARÉ's theorem:

The formal mathematical statement is as follows:[4]
To a differential form ω of given order p there exists a differential form Ω of order $p-1$, from which ω ensues via an exterior derivative according to $d\Omega = \omega$, provided ω is closed, (that is, if $d\omega = 0$).
If this statement is translated into the common language of this book, it means that a differential form $dF = \sum_i X_i dx_i$ is total or exact and therefore derivable from a potential, if and only if after a further differentiation the coefficients are crosswise equal, viz.,

$$\frac{\partial X_i}{\partial x_j} = \frac{\partial X_j}{\partial x_i} \,. \tag{2.17}$$

When the vector space is $\mathbb{R}^3$, (2.17) states that the vector field over $\mathbb{R}^3$ must be irrotational in order to be derivable from a potential. If the conditions of POINCARÉ's theorem are not fulfilled, then one is confronted with the question whether introduction of an integrating factor may lead to a success. In this regard the theory of exterior calculus or differential forms makes the following statement.

b) The condition of FROBENIUS:

Let the differential form ω not be closed, that is, let $d\omega \neq 0$. Under such a condition the differential form is completely integrable, if $\omega \wedge d\omega = 0$, where '$\wedge$' is the exterior or 'veck' product defined as

$$\mathbf{w}_1 \wedge \mathbf{w}_2 := \mathbf{w}_1 \otimes \mathbf{w}_2 - \mathbf{w}_2 \otimes \mathbf{w}_1, \qquad \mathbf{w}_1, \mathbf{w}_2 \in \mathbb{R}^3 \tag{2.18}$$

[4] For the formal presentation of the terminology used in this theorem, see Appendix A.

In the simpler notation used in this book the FROBENIUS condition means the following: consider that for the differential form $dF = \sum_i X_i dx_i$ the condition (2.17) is not satisfied. Then, this differential can be transformed with an integrating factor into a total differential, if the condition

$$\sum_{ijk} \mathrm{e}_{ijk} \left(\frac{\partial X_i}{\partial x_j} \right) X_k = 0 \tag{2.19}$$

holds, in which the sum stretches over all possible combinations of the indices i, j, k; moreover, e_{ijk} is the alternating symbol defined in (2.7). The indices $\{i, j, k\}$ can be arbitrarily selected from the set of available indices in any initial order. This is so since all permutations of an initially selected order are contained in the sum (2.19). In $\mathbb{R}^3$ the FROBENIUS condition is equivalent to the satisfaction of the requirement (2.16) that the curl of a vector field must be perpendicular to the field itself. In $\mathbb{R}^2$ the curl of a vector field, interpreted as a field in $\mathbb{R}^3$ is trivially perpendicular to the field (if the vector field lies in the $x-y$-plane, the curl points into the z-direction), and in $\mathbb{R}^1$, there is only a single route along which a function can be integrated between two points, making every differential a total one. In spaces $\mathbb{R}^n, n > 3$, the condition of FROBENIUS can be interpreted as follows:
If the mixed derivatives of a differential form with respect to x_i and x_j with different sequences differ from one another (and this only holds for this single pair of variables) i. e., if

$$\frac{\partial X_i}{\partial x_j} \neq \frac{\partial X_j}{\partial x_i} \tag{2.20}$$

then in points where (2.20) is valid all other coefficient functions X_k of the differential dF with $k \notin (i, j)$ must vanish. The FROBENIUS condition (2.19) then reduces to

$$\sum_{k} \mathrm{e}_{ijk} \left(\frac{\partial X_i}{\partial x_j} \right) X_k = 0, \quad \text{for fixed} \quad i \neq j \,. \tag{2.21}$$

In the geometric language of $\mathbf{X}$ as vector in R^n $(n > 3)$ this means, if two mixed derivatives are not equal (as in (2.21)), that the normal vector $\mathbf{X}$, formed by the components X_i $(i = 1, \ldots, n)$ must lie in the plane spanned by the coordinates belonging to these derivatives (otherwise (2.21) does not hold). Only in this case a hypersurface can locally and consistently be defined, which is perpendicular to the normal vectors and thereby guarantees the existence of an adequate integrating factor. Hence, the problem must essentially be 'locally two-dimensional'. The larger the dimension of the configuration space is, the more restrictive will be the constraints which correspond to the conditions of FROBENIUS.

Let us briefly summarize what the POINCARÉ theorem and condition of FROBENIUS imply for different dimensions n.

- $n = 1$ In the one-dimensional case each differential is total. There is no choice between different integration paths to reach point b from point a; every integral along a 'closed path' vanishes trivially.
- $n = 2$ In two dimensions, there are infinitely many non-trivial possibilities to vary the path of integration between two given points. Not every field satisfies by itself the condition that the result of this integration will be independent of the choice of the path of integration. Where this is a priori not the case, an integrating factor can always be found which establishes this property and makes the differential form of the vector field a total one.
- $n \geq 3$ Not every vector field is so structured that it could be derived from a potential. Neither can it be guaranteed that for such an 'unpleasant' field an integrating factor could always be found that makes the corresponding differential a total one. Thus, there are 'pathological' differentials which are neither total nor would be transformable with an integrating factor into total differentials. The question under which circumstances a differential form can be made total, is equivalent to the question whether the FROBENIUS condition is satisfied. In $\mathbb{R}^3$ satisfaction of the FROBENIUS condition is tantamount to stating that the vector field is perpendicular to its vorticity field. In higher dimensions a good interpretation in terms of geometry is not available. In these cases one is left with the algebraic requirement (2.19) that must formally be verified or required for vector fields to be potential fields.

In this book the differential forms which are encountered are those of the entropy and its flux and arise first in (5.13) and (5.14) as scalar and vector valued one-forms. The variables x_i ($i = 1, \ldots, n$) here are the independent constitutive variables there. Moreover, dF here is written as $\mathcal{P}$ and $\mathcal{F}$, depending on whether the entropy $\mathcal{P}$ or the entropy flux $\mathcal{F}$ is in focus. The explicit forms of the coefficient functions X_i ($i = 1, \ldots, n$) follow from the exploitation of the entropy principle (Second Law of Thermodynamics). Since the number of independent constitutive variables for the mixture theory of this book is much greater than three, the question of $\mathcal{P}$ and $\mathcal{F}$ to be total or not is crucial. The requirement that the entropy is meaningfully defined as a potential then corresponds to the requirement that the FROBENIUS condition is satisfied. This then implies restrictions to the constitutive variables which constitute necessary constraints for the satisfaction of the Second Law of Thermodynamics.

2.2.3 On the non-uniqueness of the integrating factors

As already explained, the single requirement that a differential form be total does not lead to the determination of a unique integrating factor. As an example, simply imagine that a successfully determined integrating function is globally multiplied with an arbitrary non-vanishing constant factor, then it is clear that a new integrating function is obtained; however, this factor f stretches ($f > 1$) or compresses ($0 < f < 1$) or mirrors ($f = -1$) the scale of the assumed potential values relative to the first function with the chosen factor. A mirroring operation will generally be excluded because with it a significant different interpretation of the related quantity would go along with such a change; alternatively, stretches or compressions are relatively harmless and only correspond to a change in the employed unit for the potential. More precisely, a given integrating factor g (here written as an integrating denominator) of an arbitrary non-total differential df,

$$dF = \frac{df}{g} , \tag{2.22}$$

can always be multiplied with an arbitrary non-trivial differentiable function G of F, $G(F)$, without destroying the integrability properties. Indeed, by multiplication with G a new differential form dH is obtained which is given by

$$dH = \frac{G(F)}{g} df = \frac{G(F)}{g} \sum_i \frac{\partial f}{\partial x_i} dx_i \, . \tag{2.23}$$

POINCARÉ's theorem can now be employed to verify whether the conditions for a total differential are also fulfilled for H: With condition (2.17) one obtains

$$\text{for } k = i \quad : \quad \frac{\partial^2 H}{\partial x_j \partial x_i} = \frac{dG}{dF} \frac{\partial F}{\partial x_j} \frac{1}{g} \frac{\partial f}{\partial x_i} + G \frac{\partial}{\partial x_j} \left(\frac{1}{g} \frac{\partial f}{\partial x_i} \right) , \tag{2.24}$$

$$\text{for } k = j \quad : \quad \frac{\partial^2 H}{\partial x_i \partial x_j} = \frac{dG}{dF} \frac{\partial F}{\partial x_i} \frac{1}{g} \frac{\partial f}{\partial x_j} + G \frac{\partial}{\partial x_i} \left(\frac{1}{g} \frac{\partial f}{\partial x_j} \right) \tag{2.25}$$

for $\{i, j\} \in (1, \dots, n)$. In (2.24) and (2.25) the last terms on the right-hand sides are equal, since g is an integrating factor by assumption; the first terms on the right-hand side of (2.24) and (2.25) are also identical, since $F = \int dF/g$. Indeed, the two expressions

$$\begin{aligned} \frac{dG}{dF} \frac{\partial F}{\partial x_j} \left(\frac{1}{g} \frac{\partial f}{\partial x_i} \right) &= \frac{dG}{dF} \frac{\partial F}{\partial x_j} \frac{\partial F}{\partial x_i} , \\ \frac{dG}{dF} \frac{\partial F}{\partial x_i} \left(\frac{1}{g} \frac{\partial f}{\partial x_j} \right) &= \frac{dG}{dF} \frac{\partial F}{\partial x_i} \frac{\partial F}{\partial x_j} , \end{aligned} \tag{2.26}$$

are equal. Therefore, it is ascertained that

$$dH = G(F)\frac{df}{g} \tag{2.27}$$

is a total differential of a function H, which, however, is not identical to F. Rather, if $\Delta_{a\to b}F$ denotes the value of the integral $\int_a^b dF$ between the states a and b in phase space of the considered system, then

$$\Delta_{a\to b}H = \Delta_{a\to b}(GF) - \int_a^b F\left(\frac{dG}{dF}\right) dF \ . \tag{2.28}$$

Depending upon the properties of G, different total differentials can be formed from the original differential df. This arbitrariness holds for each integrating factor of any differential form and is not particularly surprising either. Such an operation, as it leads here from F to another potential, H, is also somewhat irrelevant. True, the values of the original potential are stretched and moved, but the equi-potential surfaces remain unchanged thereby. This can easily be seen by looking at equation (2.13) or its homogeneous variant

$$dF = \mathbf{X} \cdot d\mathbf{x} = 0 \tag{2.29}$$

which defines the hypersurfaces of constant values of the potential. If the normal vector $\mathbf{X}$ is stretched by a certain factor, then this process does not change any property of the surface whatsoever that is defined by (2.29). Since the vector can not vanish if both F and H are well defined and the integrating factor must also be continuous, the function G must be only of one sign; avoiding mirror transformation this requires G to be positive valued. Applied to the entropy, this requirement guarantees that all entropies which can be defined this way maintain the 'ordering' of their values. Indeed, if equation (2.28) is written for an infinitesimal process, it takes the form

$$\Delta_{a\to b}H = \Delta_{a\to b}(GF) - F\Delta_{a\to b}G = G\Delta_{a\to b}F \tag{2.30}$$

A positive entropy difference remains in such a transformation positive, if G is selected according to the above description. If this holds true for every infinitesimal partial transformation, so it will hold also for the entire finite process. This then also guarantees that a configuration which in one formulation possesses minimum entropy and thus corresponds to a thermodynamic equilibrium state also possesses minimum entropy in every other such formulation. There remains the question whether with the functions $G(F)$ of the potential F all possible transformations have been found. That this is so can be seen, if one recalls that neighbouring equi-potential surfaces must have 'the same distance' everywhere. This, alternatively, implies that the normal vectors X on such a surface must everywhere on this surface be stretched with the same value. Consequently, the transformation factor G must at most be

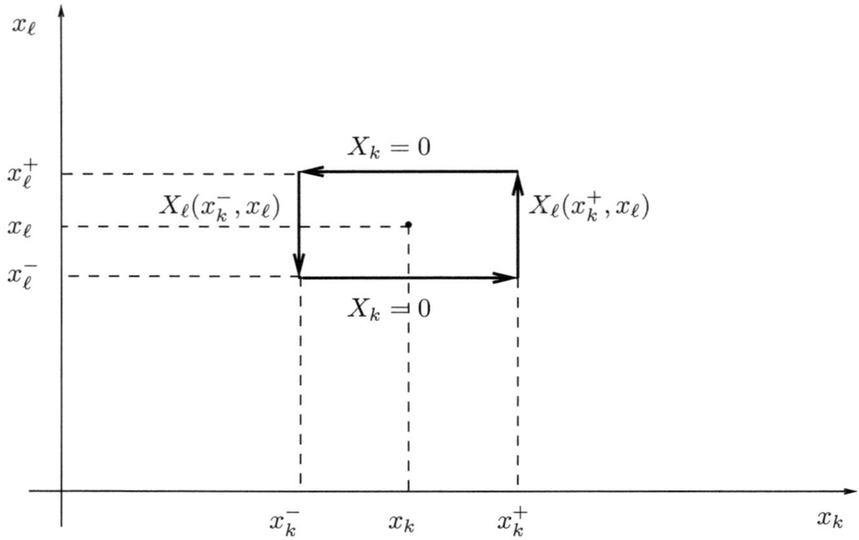

Fig. 2.1 Infinitesimal rectangle centered at $(x_k,\ x_\ell)$ with coefficient functions X_k and X_ℓ along the four sides of the infinitesimal rectangle.

a function of F for this is the only quantity which does not change on the equi-potential surfaces.

A further property in the context of functional dependences of integrating functions and coefficients in differential forms ensues if a given total differential only involves some but not all of the variables of the configuration space, i. e., if

$$dF = \sum_{i=1}^{n} X_i dx_i \qquad \text{with} \qquad X_k = 0 \quad \forall k \in (k_1, k_2, \ldots, k_m), \quad m < n \ . \tag{2.31}$$

In this case one can prove the following

Proposition regarding the dependence of the coefficients X_i on the x_j: If a total differential of the form (2.31) has coefficients which vanish identically, then those coefficients which do not vanish equally only depend functionally on those variables x_i which belong to them (that is, they do not depend on x_k, $k \in (k_1, k_2, \ldots, k_m)$, $m < n$ in (2.31)).

The proof follows by contradiction of the opposite assumption. So, let X_ℓ with $\ell \neq k$ depend on x_k, and assume X_k to vanish identically. Consider, moreover, the closed infinitesimal rectangular integration path shown in Fig.

2.1, centered at x_k, x_ℓ. It is now clear that the integrals $X_k dx_k$ along the horizontal edges vanish because $X_k = 0$; analogously, the integrals along the left and right sides of the rectangle of $X_\ell dx_\ell$ differ in absolute value from one another since by assumption $\partial X_\ell / \partial x_k \neq 0$. Therefore, the integral along the closed path composed of all four contributions, $\oint dF$, does not vanish. This result is in conflict with the assumption that dF is a total differential; consequently, X_ℓ cannot depend on x_k as was assumed above. This statement holds locally or globally, depending upon whether the respective coefficients of the differential form vanish locally or globally.

This result is particularly important because, later, it will imply non-trivial consequences for the differential of the entropy via the so-called GIBBS relation, which connects it with the internal energy and additional variables with an integrating denominator which in classical thermostatics of CARATHEODORY [17] is proved to agree with the KELVIN temperature. Ananalogous result should also hold here. As we shall see, it will follow in this mixture theory from a judicious application of the POINCARÉ theorem, the FROBENIUS condition and ad hoc assumptions which are plausible on the basis of physical or mathematical arguments.

Chapter 3
Introduction to Mixture Theory

Abstract After a general description of mixtures and multi-phase systems and their difference, reasons are given why their distinctions are premature prior to a complete thermodynamic exploitation of postulated constitutive relations by the Second Law of Thermodynamics. Consequently, both systems are here denoted as mixtures. Kinematics is treated first. Then, the general balance laws and their specializations for constituent mass, momenta, energy and entropy are discussed in global and local forms as well as jump conditions across singular surfaces. Based on TRUESDELL's metaphysical principles the sum relations define the corresponding mixture quantities which obey the physical balance laws for the mixture as a whole.

3.1 Basic Principles of Mixture Theory

Consider a glass of water and a tea spoon of sodium chloride crystals (NaCl = salt). If we insert the salt into the water and stir the compound with the spoon, we obtain salt water. The salt has gone into solution, i. e., we have the water molecules and Na^+ and Cl^--ions (which are charged particles) between them. These have nearly the same or only slightly different velocities. We say that the Na^+ and Cl^--ions diffuse through the water. However, we feel comfortable that, at a particular spatial point at time t, water and the ions coexist together. Moreover, since to each Na^+-ion there belongs a Cl^--ion that form the salt crystal we think of them as single salt entities that diffuse through the water. The continuous distribution of (Na^+, Cl^-) ions within the water is an abstraction that is actually never questioned but automatically postulated, and the number density of them at any particular point defines the salt concentration or salinity of the water. This continuous description of the pure water and the ions (Na^+, Cl^-) is called a mixture.

On the other hand, if we add to the pure water in the glass a powder with a certain colour and stir, we obtain a coloured fluid, uniform in its darkness, if the stirring has been sufficient to uniformly distribute the particles of the

L. Schneider, K. Hutter, *Solid-Fluid Mixtures of Frictional Materials in Geophysical and Geotechnical Context*, Advances in Geophysical and Environmental Mechanics and Mathematics, DOI 10.1007/978-3-642-02968-4_3,

powder (we may think of the powder to be premixed with water at very high concentrations to form ink). The ink particles are here no longer just molecules but clusters of such and so small that we have no difficulty in assuming that at every spatial point there are water and ink particles. The two substances are miscible and the conglomerate can again be called a mixture of water and ink particles.

If, however, we try to generate a homogeneous compound of water and sand particles (of clay or silt or grains of larger size), then a homogeneous, uniform distribution can still be obtained or nearly obtained by stirring, but the concept that each spatial point may be occupied by water and sand particles can no longer be maintained. Such a compound is called *heterogeneous*. In chemical process engineering spouted and fluidized beds belong to such bodies, in river hydraulics, sediment transport in the form of bed load, suspended load and wash load[1] characterize such compounds and in geophysics and geology debris and mud flows, sub-aquatic turbidity currents and avalanches belong to them. In the chemical literature such systems are often called 'multi-phase systems' and, as outlined in the Introduction, it is emphasised that such systems are different from mixtures. The idea is that a differential element, i. e. a Representative Volume Element (RVE) consists of many fluid and solid particles with sub-element properties which must be brought up to the macroscopic level by some homogenization procedure.

It is our thesis that differences in the formulations of such heterogeneous materials can only be identified once a full (thermodynamic) analysis has been performed, but not at an intermediate stage. We, therefore, do not semantically differentiate between 'mixtures' and 'multi-phase systems' and will henceforth call materials that are composed of several different constituents *mixtures*. For our purpose a *constituent* consists of a group of particles that have the same mechanical and thermodynamical properties. Thus, we can distinguish constituents according e. g. to their chemical composition, their state of aggregation, their size or other significant criteria. In general it is assumed that the constituents are distributed 'equally', i. e. continuously over the region of the mixture.

Alternative terms for constituents that are equivalent are *components* or *phases*. The terms 'constituents' and 'components' are used in a general context, but often 'phases' denote different states of aggregation of the same material, say ice, water and water vapour. The latter example also leads naturally to the characterization of mixtures of *miscible constituents* or *immiscible constituents*. A specimen of ice from an Alpine glacier often consists

[1] The solid matter that is transported by the fluid in rivers is occasionally divided into three different components. The largest particles form the sediment bed and are transported within a thin basal layer, the moving sediment bed. The suspended material is divided into two fractions: the larger particles are distributed in the water with a concentration that varies with position within the depth of the river, the smallest particles are, roughly, uniformly distributed over the river depth and make the river water a slurry. This is the so-called wash load.

of several ice crystals and water embedded in inclusions. This is an immiscible mixture of ice and water, because, strictly, the ice and the water occupy different positions. Similarly, in a mud flow the water and the suspended solid particles form an immiscible mixture of solid and fluid constituents but in contrast to the above example they are made of different materials. Moreover, the water in the atmosphere arises in two forms, as vapor and as clouds or fog. Vapour is the gas phase of H_2O and in air it is in *solution* within the other components O_2, O_3, N_2, CO_2, etc., and it is invisible by eye, but in the clouds the water appears as droplets suspended in the remaining components and it is visible by eye. These droplets are in *suspension* and, strictly, this mixture is immiscible. In what follows we shall mostly be concerned with immiscible mixtures, but the differences of the various kinds of are not apparent in the mathematical formulation. This is partly due to the continuity assumption of the distribution of the constituents which we shall now address.

Consider a mixture of n arbitrary constituents. To describe its mechanical and thermodynamical behaviour it would be advantageous if we could extend the principles of continuum mechanics for bodies of a single constituent (e. g. balance laws, principle of determinism, principle of local action, etc.) to that of many constituents, i. e. mixtures. This is possible indeed, but certain simplifications and suppositions are to be made. The suppositions proposed by TRUESDELL [118, 119] first in 1957 and then again in 1968 (he called them '*metaphysical principles*') achieved wide acceptance and constitute now the basic principles of mixture theory. His suppositions read as follows:

1. Each spatial point of the mixture is simultaneously occupied by material of all constituents.
2. 'All properties of the mixture must be mathematical consequences of properties of the constituents' (first metaphysical principle).
3. 'So as to describe the motion of a constituent, we may in imagination isolate it from the rest of the mixture, provided we allow properly for the actions of the other components upon it' (second metaphysical principle).
4. 'The motion of the mixture is governed by the same equations as is a single body' (third metaphysical principle).

Obviously, this 'mixture continuum hypothesis' is physically not correct, but it is made whenever mathematical formulae are laid down to describe the physical processes that take place within the mixture, be it miscible or immiscible. Of course, the atomistic structure of matter makes this assumption always dubious at sufficiently small scales. In fact, the mixture continuum hypothesis should always be viewed as a certain homogenization process over a RVE, which is sufficiently small to be able to describe processes that vary over length scales which are larger than the side lengths of the RVE, but equally also sufficiently large to smear over rapid variations in the interior of the RVE. Figure 3.1 gives a picture of this idea. If the side length of the RVE is large (regime I in Fig. 3.1), then variations of variables in the sub-

RVE regime can be ignored, and the continuous distribution of mass is an adequate assumption. When they are so small that sub-RVE heterogeneities gain influence on the larger scales (regime II in Fig. 3.1), then these effects may have to be accounted for by additional variables. For instance in a very fine porous material the pores may not be 'visible' except by the numerical values of the phenomenological coefficients. In a soil specimen the pore space is not negligible as it is known that it exerts a quantitative influence on the soil behaviour. In regime III of Fig. 3.1 the RVE scale is comparable to the sub-RVE elements, in a soil this is the diameter of the individual grains. At this scale the continuum description obviously breaks down. TRUESDELL's second supposition can be restated as: *'The whole is no more than the sum of its parts'* ([119]) and the fourth can be expressed by the words: *'In its motion as a whole a body does not know whether it is a mixture or not'*([119]). The third supposition allows for the application of the balance laws (e. g. balance laws of mass, momentum, moment of momentum, energy and entropy) for every constituent, but in contrast to the balance laws for single-material bodies, these equations are no longer *conservative.*[2] One can also think of the third supposition as cutting free a constituent and introducing the correct reaction quantities which counteract the cutting operation.

The fields that will be defined according to the above suppositions are called *macroscopic*, as they can not resolve the exact microscopic fields, e. g. position, velocity etc. of the constituents. This shortcoming of TRUESDELL's mixture theory becomes considerable if we draw attention to immiscible mixtures (e. g. soil). On a submacroscopic scale of such materials, structures do exist that are not measured by the macroscopic mixture quantities introduced before. We have given examples above.

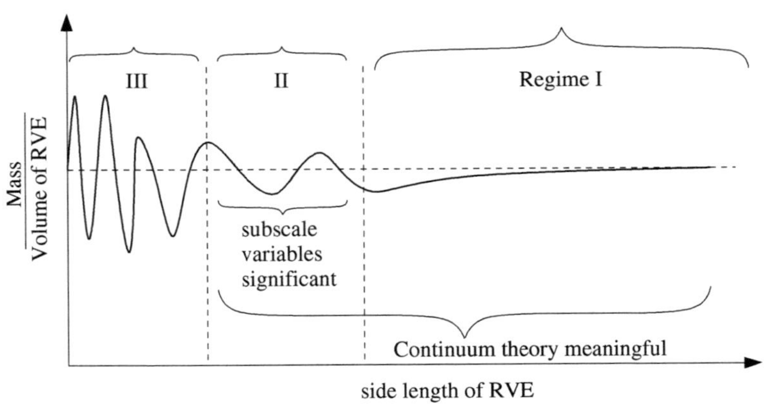

Fig. 3.1 Density, i. e. mass per unit volume of the RVE, plotted against the side length of the RVE (principal sketch).

[2] This means that interaction (production) terms may be non-zero and then express the effect of all other constituents upon one particular constituent and vice versa.

To circumvent this drawback researchers incorporate internal variables like the constituent volume fractions or the porosity or tortuosity of the pore space into the set of *independent variables*[3]. They also postulate balance equations for these variables, but there is no agreement on what form these equations should take. Examples for volume fraction balance equations are the second-order balance equation of GOODMAN & COWIN (cf. PASSMAN et al. [103]) or the first order balance equation of SVENDSEN & HUTTER [115] which does not include a volume fraction flux term. Another example for a porosity balance equation can be found in WILMAŃSKI [124]. FANG et al. [41, 42] follow essentially GOODMAN & COWIN but introduce an internal length scale of the grains or pore space and write down a balance relation for it. In what follows the volume fraction balance equations of SVENDSEN & HUTTER will be used, as they simplify the thermodynamic analysis considerably.

For some more references and more detailed discussions of the basic principles the reader is referred e. g. to ERINGEN & INGRAM [38], BOWEN [14], TRUESDELL [119], MÜLLER [97], RAJAGOPAL & TAO [107], DREW & PASSMAN [34] or HUTTER & JÖHNK [62].

3.2 Kinematics of Multi-phase Mixtures

In this section our aim is to take a closer look at the kinematics of multi-phase mixtures.[4] Thus, we start, as in every continuum theory, from configurations and motions of a three-dimensional non-empty continuum $\mathcal{B}$, the so-called material body.

As we are dealing with mixtures, we must consider configurations one for each constituent and motions of a collection of n non-empty continua $\mathcal{B}_\alpha$ (for each constituent one), that constitute the body $\mathcal{B}$.

An element, X_α, of continuum $\mathcal{B}_\alpha$ can be understood as a name tag for one specific material particle of constituent K_α. An open set of these elements is called $\mathcal{Q}_\alpha$, and its surface $\partial\mathcal{Q}_\alpha$. Every $\mathcal{B}_\alpha$ has a set of configurations $\{\boldsymbol{\kappa}_{\alpha\theta}\}_{\theta\in\mathrm{I}}$ with $\mathrm{I}\subset\mathbb{R}$ that are bijective[5] mappings from $\mathcal{B}_\alpha$ into connected and compact regions, $\{\mathcal{R}_{\alpha\theta}\}_{\theta\in\mathrm{I}}$, in the Euclidian space, $\mathcal{E}^3$. In other words, one specific $\boldsymbol{\kappa}_{\alpha\theta}$ assigns a vector $\mathbf{X}_{\alpha\theta}$ to the material particle X_α at a fixed *time* θ, i. e.,

[3] It is not always so that such internal variables have a clear physical meaning. In an extreme situation they may be defined by no other specification than the equations laid down for them. In these situations their effect is only recognisable by the effects they exert on observable physical quantities. Internal variables have always some degree of inexplicability and are therefore also called 'hidden' variables.

[4] This derivation has primarily been influenced by writings of TRUESDELL [118, 119], MÜLLER [94], but has also profited from those of ERINGEN & INGRAM [38], MÜLLER [94], HUTTER & JÖHNK [62] and 'NAGHDI's Notes on Continuum Mechanics' (not published).

[5] A mapping is bijective if it is one-to-one and onto.

$$\begin{aligned} \boldsymbol{\kappa}_{\alpha\theta} &: \mathcal{B}_\alpha \rightarrow \mathcal{E}^3 , \\ \mathbf{X}_{\alpha\theta} &= \boldsymbol{\kappa}_{\alpha\theta}(X_\alpha) \in \mathcal{R}_{\alpha\theta} , \end{aligned} \qquad \text{for } \theta \text{ fixed} , \tag{3.1}$$

$\mathbf{X}_{\alpha\theta}$ is called the position vector of the corresponding material particle and $\mathcal{R}_{\alpha\theta}$ is the region occupied by constituent K_α. In this spirit we assign to the open set $\mathcal{Q}_\alpha$ and its surface $\partial\mathcal{Q}_\alpha$ a *material* region $\Omega_{\alpha\theta}$ and a material surface $\partial\Omega_{\alpha\theta}$ within $\mathcal{R}_{\alpha\theta}$. $\Omega_{\alpha\theta}$ and $\partial\Omega_{\alpha\theta}$ are called material, because they are tied to the sets of material particles $\mathcal{Q}_\alpha$ and $\partial\mathcal{Q}_\alpha$. Without going into details of integration theory[6] we equip the region $\mathcal{R}_{\alpha\theta}$ with volume and surface measures. We employ the following notation

$$V(\mathcal{Q}_\alpha, \theta) = \int_{\Omega_{\alpha\theta}} dV, \qquad A(\partial\mathcal{Q}_\alpha, \theta) = \int_{\partial\Omega_{\alpha\theta}} dA \tag{3.2}$$

for the volume of $\Omega_{\alpha\theta}$ and the surface area of $\partial\Omega_{\alpha\theta}$, respectively.

Now, let us choose any element from the set of configurations and call it a reference configuration, $\boldsymbol{\kappa}_{\alpha 0}$. $\mathcal{R}_{\alpha\theta}$ and $\mathbf{X}_{\alpha\theta}$ corresponding to $\boldsymbol{\kappa}_{\alpha 0}$ are then $\mathcal{R}_{\alpha 0}$ and $\mathbf{X}_{\alpha 0}$, respectively. As already observed by TRUESDELL [119], $\boldsymbol{\kappa}_{\alpha 0}$ can be different for every constituent. The present configuration can be obtained by sequentially applying the mappings $\boldsymbol{\kappa}_{\alpha 0}^{-1}$ and $\boldsymbol{\kappa}_{\alpha t}$ (see Fig. 3.2), where $\boldsymbol{\kappa}_{\alpha t}$ is the mapping from the continuum $\mathcal{B}_\alpha$ to the present configuration. However, the first supposition of TRUESDELL (see Section 3.1) requires that all constituents are located at the same present configuration $\mathcal{R}_t$, i. e. all $\mathcal{R}_{\alpha t}$ fall together to one $\mathcal{R}_t$ and thus, in the present configuration we have only one *position vector* $\mathbf{x}$. In mathematical formulas $\mathbf{x}$ can be expressed as

$$\mathbf{x} = \left(\boldsymbol{\kappa}_{\alpha t} \circ \boldsymbol{\kappa}_{\alpha 0}^{-1}\right)(\mathbf{X}_{\alpha 0}) , \tag{3.3}$$

or

$$\mathbf{x} = \chi_\alpha(\mathbf{X}_{\alpha 0}, t) , \tag{3.4}$$

where $\boldsymbol{\kappa}_{\alpha t} \circ \boldsymbol{\kappa}_{\alpha 0}^{-1}$ denotes a mapping that results from sequentially applying first $\boldsymbol{\kappa}_{\alpha 0}^{-1}$ and then $\boldsymbol{\kappa}_{\alpha t}$. The symbol χ_α denotes a vector-valued function of location $\mathbf{X}_{\alpha 0}$ and time t of which the value is $\mathbf{x}$. If we assume that χ_α is continuously differentiable in the neighbourhood of a material point $\mathbf{x}_\alpha$ except possibly at some singular points, lines and surfaces, it is also invertible there (cf. HUTTER & JÖHNK [62] and ERINGEN & INGRAM [38]). The function χ_α is called *the motion* of the constituent material body $\mathcal{B}_\alpha$. With these definitions in mind we can now specify the *constituent velocity* $\mathbf{v}_\alpha$, the *material derivative* $\mathrm{d}^\alpha(\cdot)/\mathrm{d}t$ following the motion of constituent K_α, the *constituent acceleration* $\mathbf{a}_\alpha$, the *constituent velocity gradient* $\mathbf{L}_\alpha$ and the *constituent deformation gradient* $\mathbf{F}_\alpha$. They are defined as

[6] As the rigorous definition of the integrals would exceed the scope of this treatise we refer the reader to the respective literature on integration theory, e. g. BAUER [11].

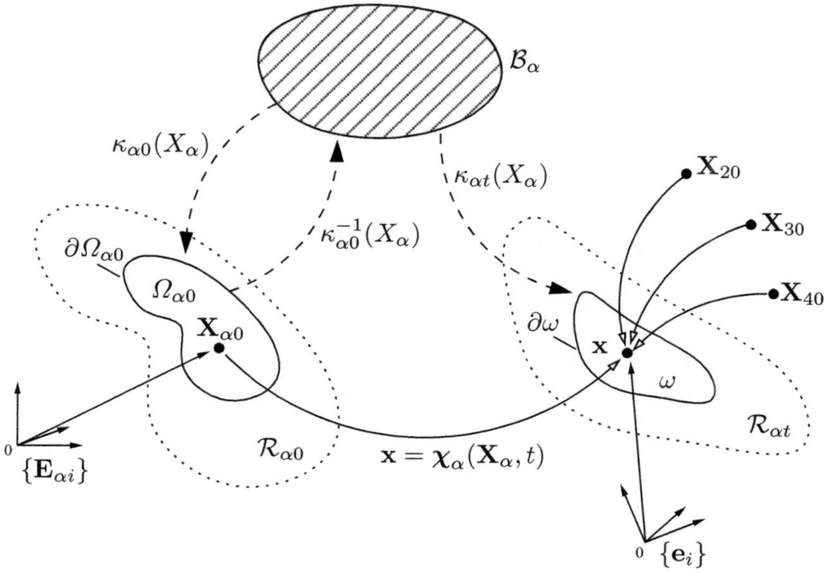

Fig. 3.2 Material body $\mathcal{B}_\alpha$ of constituent K_α. An open set $\mathcal{Q}_\alpha \in \mathcal{B}_\alpha$ becomes in the reference configuration the material region $\Omega_{\alpha 0} \in \mathcal{R}_{\alpha 0} \subset \mathcal{E}^3$ with boundary $\partial\Omega_{\alpha 0}$, and the material particle X_α is mapped onto $\mathbf{X}_{\alpha 0} \in \mathcal{R}_{\alpha 0}$. Similarly, in the present configuration, $\mathcal{Q}_\alpha$ is mapped into the open set $\Omega_{\alpha t} := \omega_\alpha \in \mathcal{R}_{\alpha t} \subset \mathcal{E}^3$ with boundary $\partial\Omega_{\alpha t} := \partial\omega_\alpha$, and the material particle X_α is mapped onto $\mathbf{x} \in \mathcal{R}_{\alpha t}$. $\{\mathbf{E}_i\}$ $(i = 1, 2, 3)$ is a basis for K_α in the reference configuration, and $\{\mathbf{e}_i\}$ $(i = 1, 2, 3)$ is a basis for K_α in the present configuration. Note all different $\mathbf{X}_{\alpha 0}$ $(\alpha = 1, 2, \ldots, n)$ are mapped in the present configuration onto the same point $\mathbf{x}$. So ω_α and $\partial\omega_\alpha$ are the same region ω and boundary $\partial\omega$ for all α.

$$
\begin{aligned}
\hat{\mathbf{v}}_\alpha(\mathbf{X}_{\alpha 0}, t) :&= \frac{\partial\, \chi_\alpha(\mathbf{X}_{\alpha 0}, t)}{\partial t} = \frac{\mathrm{d}}{\mathrm{d}t}\, \chi_\alpha(\mathbf{X}_{\alpha 0}, t)\Big|_{\mathbf{X}_{\alpha 0}} \\
&= \frac{\mathrm{d}^\alpha \chi_\alpha\left(\chi_\alpha^{-1}(\mathbf{x}, t), t\right)}{\mathrm{d}t} =: \tilde{\mathbf{v}}_\alpha(\mathbf{x}, t) =: \acute{\mathbf{x}}_\alpha(\mathbf{x}, t)\ ,
\end{aligned}
\tag{3.5}
$$

where $\acute{(\cdot)}_\alpha$ is an abbreviation for $\mathrm{d}^\alpha(\cdot)/\mathrm{d}t$, the material time derivative, following the motion of constituent K_α. Moreover, $|_{\mathbf{X}_{\alpha 0}}$ indicates that the indexed quantity is held fixed among the variables that are indexed; the functions $\hat{\mathbf{v}}_\alpha$ and $\tilde{\mathbf{v}}_\alpha$ take the same values for the same $\mathbf{X}_{\alpha 0}$, t and $\mathbf{x}$ evaluated from (3.4), but they are different functions, as they depend on different position vectors. The same is true for the accelerations

$$\hat{\mathbf{a}}_\alpha\left(\mathbf{X}_{\alpha 0},\ t\right) := \frac{\partial^2\ \chi_\alpha\left(\mathbf{X}_{\alpha 0},\ t\right)}{\partial t^2} = \frac{\mathrm{d}^2}{\mathrm{d}t^2}\ \chi_\alpha\left(\mathbf{X}_{\alpha 0},\ t\right)\Big|_{\mathbf{X}_{\alpha 0}}$$

$$= \frac{\mathrm{d}^{\alpha 2}\chi_\alpha\left(\ \chi_\alpha^{-1}\left(\mathbf{x},\ t\right),\ t\right)}{\mathrm{d}^2 t} =: \tilde{\mathbf{a}}_\alpha\left(\mathbf{x},\ t\right)\ . \tag{3.6}$$

Let us, for a moment, look at a general field φ_α (scalar-, vector- or tensor-valued) of constituent K_α that can be written in the forms

$$\varphi_\alpha = \hat{\varphi}_\alpha\left(\mathbf{X}_{\alpha 0},\ t\right) = \tilde{\varphi}_\alpha\left(\mathbf{x},\ t\right)\ . \tag{3.7}$$

The first description is called Lagrangean (or material) and the second Eulerian (or spatial). For the sake of completeness, we also mention $\varphi = \breve{\varphi}_\alpha\left(X_\alpha,\ t\right)$, which is only meaningful for philosophical considerations, but is of no practical use. If we now take the material derivative $\mathrm{d}^\alpha(\cdot)/\mathrm{d}t$ of φ_α in the spatial representation we obtain[7]

$$\frac{\mathrm{d}^\alpha\ \varphi_\alpha\left(\mathbf{x},\ t\right)}{\mathrm{d}t} := \acute{\varphi}_\alpha = \frac{\partial\tilde{\varphi}_\alpha}{\partial t} + \nabla\tilde{\varphi}_\alpha\ \tilde{\mathbf{v}}_\alpha\ , \tag{3.8}$$

via the chain rule of differentiation. The first term on the right-hand side of equation (3.8) describes local effects at a fixed position $\mathbf{x}$ and the second term is associated with convective effects. Thus, $\mathrm{d}^\alpha(\cdot)/\mathrm{d}t$ can be understood as a material derivative following the motion of constituent K_α. In the sequel, we will omit the tilde-sign on top of the spatial quantities, because we are mainly operating in the Eulerian description.

Next, we introduce the *constituent velocity gradient*

$$\begin{aligned} \mathbf{L}_\alpha &:= \nabla\mathbf{v}_\alpha = \mathbf{D}_\alpha + \mathbf{W}_\alpha\ , \\ \mathbf{D}_\alpha &:= \mathrm{sym}\left(\nabla\mathbf{v}_\alpha\right)\ , \quad \mathbf{W}_\alpha := \mathrm{skw}\left(\nabla\mathbf{v}_\alpha\right)\ , \end{aligned} \tag{3.9}$$

where we have decomposed $\mathbf{L}_\alpha$ into a symmetric, $\mathbf{D}_\alpha$, and a skew-symmetric part, $\mathbf{W}_\alpha$, that represent the *stretching* or *rate of deformation tensor* and the *vorticity* or *spin tensor* of constituent K_α, respectively.

On the basis of the presented kinematics the *constituent deformation gradient*, $\mathbf{F}_\alpha$, is defined as

$$\mathbf{F}_\alpha := \frac{\partial\ \chi_\alpha}{\partial\mathbf{X}_{\alpha 0}}\ \left(\mathbf{X}_{\alpha 0}, t\right)\ . \tag{3.10}$$

Thus, $\mathbf{F}_\alpha$ is a second rank tensor that maps a unit vector $\mathbf{M}_\alpha$ in the reference configuration of constituent K_α into a vector $\lambda_\alpha\mathbf{m}$ in the present configuration which can be viewed as a stretched and tilted image of $\mathbf{M}_\alpha$; λ_α is called the *stretch factor* and $\mathbf{m}$ is a unit vector in the present configuration, i. e.

[7] If $\tilde{\varphi}_\alpha$ in (3.8) is a scalar then $\nabla\tilde{\varphi}_\alpha\tilde{\mathbf{v}}_\alpha$ must be interpreted as the dot product $(\nabla\tilde{\varphi}_\alpha)\cdot\tilde{\mathbf{v}}_\alpha$; if $\tilde{\varphi}_\alpha$ is a tensor, then $\nabla\tilde{\varphi}_\alpha\tilde{\mathbf{v}}_\alpha$ is to be interpreted as $\tilde{\varphi}^\alpha_{ijk\ldots,l}\tilde{v}^\alpha_l$.

$$\mathbf{F}_\alpha \mathbf{M}_\alpha = \lambda_\alpha \mathbf{m} \; . \tag{3.11}$$

With the above assumption on $\boldsymbol{\chi}_\alpha$, $\mathbf{F}_\alpha$ satisfies $\det(\mathbf{F}_\alpha) \neq 0$ and therefore it is invertible.

We then define the symmetric *constituent left* CAUCHY-GREEN *deformation tensor* by

$$\mathbf{B}_\alpha := \mathbf{F}_\alpha \mathbf{F}_\alpha^{\mathrm{T}}, \qquad \mathbf{B}_\alpha = \mathbf{B}_\alpha^{\mathrm{T}} \; . \tag{3.12}$$

In treatises on kinematics, temporal and spatial derivatives of the latter quantities are given. We omit them here and cite the appropriate literature, e. g. BOWEN [14], HAUPT [51], GREVE [46] or HUTTER & JÖHNK [62].

For the derivation of the constituent balance equations we need to define volume and surface measures for the present configuration see Fig. 3.3. The constituent volume and surface measures in the present configration are defined as

$$v(\mathcal{Q}_\alpha) = \int_{\omega_\alpha} \mathbf{1}_\alpha^{(3)} dv, \qquad a(\partial\mathcal{Q}_\alpha) = \int_{\partial\omega_\alpha} \mathbf{1}_\alpha^{(2)} da \; , \tag{3.13}$$

where ω_α and $\partial\omega_\alpha$ are the mappings of the material regions $\Omega_{\alpha 0}$ and their surfaces $\partial\Omega_{\alpha 0}$ in the present configuration, and $\mathbf{1}_\alpha^{(3)}$ and $\mathbf{1}_\alpha^{(2)}$ are the characteristic functions of K_α in ω_α and $\partial\omega_\alpha$, respectively.[8] At time t, ω_α and $\partial\omega_\alpha$

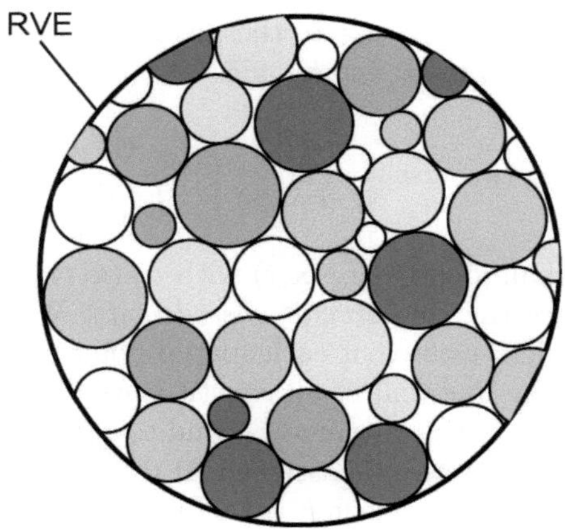

Fig. 3.3 A two-dimensional picture of an RVE-circle filled with constituents indentified by different shadings. It is evident, that the total area of the RVE is larger than the sum of the areas of any particular shading. The size of the RVE is assumed to be typical for a differential length on the macro-scale.

[8] To be precise, it is assumed that the constituents occupy disjoint regions in ω_α and $\partial\omega_\alpha$, respectively. In other words $\mathbf{1}_\alpha^{(3)} \cap \mathbf{1}_\beta^{(3)} = \emptyset$, if $\alpha \neq \beta$ and similarly $\mathbf{1}_\alpha^{(2)} \cap \mathbf{1}_\beta^{(2)} = \emptyset$. So, $|v(\mathcal{Q}_\alpha)| \leq \left|v\left(\bigcup_\beta \mathcal{Q}_\beta\right)\right|$ and $|a(\partial\mathcal{Q}_\alpha)| \leq \left|a\left(\bigcup_\beta \partial\mathcal{Q}_\beta\right)\right|$

are coincident with the region ω and boundary $\partial\omega$ of the mixture region. This is a consequence of the first supposition of TRUESDELL. The overall volume of the mixture in the region ω is denoted by

$$v(t) = \int_{\omega} dv \, . \tag{3.14}$$

It is the volume which the mixture would occupy if we would regard it as a continuum composed of only one material. We also define

$$a(t) = \int_{\partial\omega} da \tag{3.15}$$

as the surface measure of the mixture region. With these definitions at hand we introduce the (*local*) *volume fraction* or simply *volume density* of constituent K_α

$$\nu_\alpha(\mathbf{x}, t) := \lim_{\omega \to \mathrm{RVE}(\mathbf{x})} \left\{ \frac{\int_\omega \mathbf{1}_\alpha^{(3)} dv}{\int_\omega dv} \right\} \tag{3.16}$$

as the limit of the ratio of the solid volume when ω approaches the RVE-volume centred at $\mathbf{x}$. This implies that in the context of Fig. 3.1 regimes II and III are ignored and regime I is formally extrapolated to the RVE-scale equal to zero. The volume fraction quantity plays a key role in mixture theories for granular flows as it forms a measure to represent the fine grain structures. For completeness, we also mention that the above introduction of surface measures gives also rise to the definition of *local areal fraction* or simply *area density* of constituent K_α via

$$\mu_\alpha(\mathbf{x}, t) := \lim_{\partial\omega \to \mathrm{RAE}(\mathbf{x})} \left\{ \frac{\int_{\partial\omega} \mathbf{1}_\alpha^{(2)} dv}{\int_{\partial\omega} dv} \right\} , \tag{3.17}$$

however, later we shall identify $\mu_\alpha(\mathbf{x}, t)$ with $\nu_\alpha(\mathbf{x}, t)$.

We now introduce the *constituent mass density*. We recall TRUESDELL's first supposition which states that each spatial point of the mixture is simultaneously occupied by material of all constituents. In view of this, it seems reasonable to go back to the continua $\mathcal{B}_\alpha$ and to require that to every particle X_α a positive *mass density* is assigned. If this density is defined as mass per unit volume of the constituent K_α, there are two mass densities, one for the reference configuration $\rho_{\alpha 0}(\boldsymbol{X}_{\alpha 0}, t)$ and one for the present configuration $\rho_\alpha(\mathbf{x}, t)$.

3.3 Balance Equations and Sum Relations

It is the objective of any mixture theory to 'calculate' the motion of each continuous bodies $\mathcal{B}_\alpha$, and along with it, the evolution of the associated fields, e. g. mass densities, temperatures etc. The equations which must be established to accomplish this task are of functional differential and algebraic character and have to be completed by sufficient initial and boundary conditions such that a *well posed* mathematical problem is obtained. The *balance equations* for the masses, momenta, energies and entropies of the constituents describe the general behaviour of the bodies $\mathcal{B}_\alpha$ without containing material specific information. In contrast, the *constitutive relations*, which are the main objective of this book, supply exactly this information.

In the Introduction we have already mentioned that different paths exist to derive balance equations for a system that consists of more than one constituent. Irrespective of the type of derivation, e. g. mixture or multi-phase approach, they all obey balance equations of basically the same structure. Modern mixture theories, see TRUESDELL [118, 119], MÜLLER [94], ERINGEN & INGRAM [38], HUTTER & JÖHNK [62] or RAJAGOPAL & TAO [107] employ the following

Postulate. Let ω and $\partial\omega$ be a region and its boundary in the three-dimensional Euclidian space $\mathcal{E}^3$ continuously filled with material of the constituents K_α, $\alpha = 1, 2, \ldots, n$ of a mixture. When a particular constituent K_α is considered, we shall write ω_α and $\partial\omega_\alpha$ for all $\alpha \in [1, n]$, even though $\omega = \omega_\alpha$ and $\partial\omega = \partial\omega_\alpha$. Let $\mathcal{G}_\alpha$ be a physical variable characterising a particular aspect of the state of constituent K_α in ω_α. Let, moreover, $\mathcal{G}$ be the corresponding physical variable characterising the same particular aspect of the state of the mixture. Then we request the following equations to hold:

(*i*) for the constituents

$$\frac{d\mathcal{G}_\alpha}{\mathrm{d}t} = \mathcal{P}_\alpha + \mathcal{S}_\alpha + \mathcal{F}_\alpha\,, \tag{3.18}$$

(*ii*) for the mixture

$$\frac{d\mathcal{G}}{\mathrm{d}t} = \mathcal{P} + \mathcal{S} + \mathcal{F}\,. \tag{3.19}$$

(*iii*) The constituent quantities $\mathcal{G}_\alpha$, $\mathcal{P}_\alpha$, $\mathcal{S}_\alpha$ and $\mathcal{F}_\alpha$ and the mixture quantities $\mathcal{G}$, $\mathcal{P}$, $\mathcal{S}$ and $\mathcal{F}$ are related to one another by

$$\mathcal{G} := \sum_{\alpha=1}^{n} \mathcal{G}_\alpha, \quad \mathcal{P} := \sum_{\alpha=1}^{n} \mathcal{P}_\alpha, \quad \mathcal{S} := \sum_{\alpha=1}^{n} \mathcal{S}_\alpha, \quad \mathcal{F} := \sum_{\alpha=1}^{n} \mathcal{F}_\alpha. \tag{3.20}$$

Equations (3.18) and (3.19) are global balance laws, in which the time rate of change $d\mathcal{G}_\alpha/dt$ ($d\mathcal{G}/dt$) is assumed to be a consequence of a *production* $\mathcal{P}_\alpha$ ($\mathcal{P}$) in the body, an *external supply* $\mathcal{S}_\alpha$ ($\mathcal{S}$) from outside the body and a *flux* $\mathcal{F}_\alpha$ ($\mathcal{F}$) through the surface of the body.

All quantities $\mathcal{G}_\alpha$, $\mathcal{G}$, ..., $\mathcal{F}_\alpha$, $\mathcal{F}$ in equations (3.18) and (3.19) depend only on time and on ω_α, ω and $\partial\omega_\alpha$, $\partial\omega$, respectively. So, we should write $\mathcal{G}_\alpha(t, \omega_\alpha)$, $\mathcal{G}(t, \omega)$, ..., $\mathcal{F}_\alpha(t, \partial\omega_\alpha)$, $\mathcal{F}(t, \partial\omega)$, to identify the dependences on ω_α, ω and $\partial\omega_\alpha$, $\partial\omega$, respectively. For simplicity, we will, however, only write $\mathcal{G}_\alpha = \mathcal{G}_\alpha(t)$, $\mathcal{G} = \mathcal{G}(t)$, ..., $\mathcal{F}_\alpha = \mathcal{F}_\alpha(t)$, $\mathcal{F} = \mathcal{F}(t)$.

As we assume the constituent body $\mathcal{B}_\alpha$ to fill the space ω_α continuously, we are free to cut this body into smaller parts, e. g. infinitesimal bodies as long as we properly account for the interactions of the respective parts on each other. If we now equip these bodies with the same physical properties as $\mathcal{B}_\alpha$ and map them into the present configuration we are in the position to define the *physical field density* ψ_α per unit volume of constituent K_α at every point within the region ω_α.[9]

The general *true* density, ψ_α, stands, in particular, for the mass density, momentum 'density', moment of momentum 'density', energy 'density' and entropy 'density' per unit volume of constituent K_α (for their mathematical formulation see second column in Table 3.1).

This density must be differentiated from $\bar{\psi}_\alpha$, which is defined as the quantity $\mathcal{G}_\alpha$ per unit mixture volume. In the sequel we shall denote densities with an overbar as *partial densities* and those without as *true densities*. Whilst the former are related to the mixture volume v, the latter are associated with the constituent volume $v(\mathcal{Q}_\alpha)$. The two are related to one another according to

$$\bar{\psi}_\alpha = \nu_\alpha \psi_\alpha \ , \tag{3.21}$$

in which ν_α is the volume fraction defined in (3.16).

With $\bar{\psi}_\alpha$ defined as the amount of $\mathcal{G}_\alpha$ per unit mixture volume, $\mathcal{G}_\alpha$ is given by the sum relation

$$\mathcal{G}_\alpha(t) = \int_{\omega_\alpha} \bar{\psi}_\alpha(\mathbf{x}, t) dv \ . \tag{3.22}$$

In the same spirit we now also write

[9] The assumption that any of the four field quantities in (3.18) and (3.19) can be expressed as integrals of field densities over the space at which they are defined, is known as *additivity assumption*. It is postulated to hold here for $\mathcal{G}_\alpha$, $\mathcal{P}_\alpha$, $\mathcal{S}_\alpha$ and $\mathcal{F}_\alpha$ and in view of (3.20) carries automatically over to $\mathcal{G}$, $\mathcal{P}$, $\mathcal{S}$ and $\mathcal{F}$.

$$\mathcal{P}_\alpha(t) = \int_{\omega_\alpha} \Big(\bar{\boldsymbol{\pi}}^\psi_\alpha(\mathbf{x},t) + \bar{\gamma}^\psi_\alpha(\mathbf{x},t)\Big) dv \ , \tag{3.23}$$

$$\mathcal{S}_\alpha(t) = \int_{\omega_\alpha} \bar{\boldsymbol{\sigma}}^\psi_\alpha(\mathbf{x},t) dv \ , \tag{3.24}$$

$$\mathcal{F}_\alpha(t) = \int_{\partial\omega_\alpha} {}^*\bar{\boldsymbol{\phi}}^\psi_\alpha(\mathbf{x},t,\mathbf{n}) da \ , \tag{3.25}$$

in which $\bar{\boldsymbol{\pi}}^\psi_\alpha$, $\bar{\gamma}^\psi_\alpha$, $\bar{\boldsymbol{\sigma}}^\psi_\alpha$ and ${}^*\bar{\boldsymbol{\phi}}^\psi_\alpha$ are partial densities per unit mixture volume and unit mixture area, respectively. The partial production density has been divided into two contributions. We interprete henceforth $\bar{\boldsymbol{\pi}}^\psi_\alpha$ as the (self) production rate of $\bar{\boldsymbol{\psi}}_\alpha$ by constituent K_α, whilst $\bar{\gamma}^\psi_\alpha$ is the production rate of $\bar{\boldsymbol{\psi}}_\alpha$ by all constituents other than K_α. This can be interpreted as production by interaction. For mass, linear and angular momentum, energy and entropy these quantities are defined in columns 3 and 6 of Table 3.1. Analogously, $\bar{\boldsymbol{\sigma}}^\psi_\alpha$ is the partial density of the supply rate of $\bar{\boldsymbol{\psi}}_\alpha$, a source of $\bar{\boldsymbol{\psi}}_\alpha$ outside the body volume ω_α but affecting material points within ω_α.

The partial density of the flux of $\bar{\boldsymbol{\psi}}_\alpha$ through the boundary of the body is defined per unit area on the surface $\partial\omega_\alpha$ and is denoted by ${}^*\bar{\boldsymbol{\phi}}^\psi_\alpha$. Thus, the total flux through the boundary $\partial\omega_\alpha$ into the body volume ω_α is the surface integral (3.25). As opposed to $\bar{\boldsymbol{\psi}}^\psi_\alpha$, $\bar{\boldsymbol{\pi}}^\psi_\alpha$ and $\bar{\gamma}^\psi_\alpha$, which are only functions of $\mathbf{x}$ and t, ${}^*\bar{\boldsymbol{\phi}}^\psi_\alpha$ also depends on the unit exterior normal vector $\mathbf{n}$ of $\partial\omega_\alpha$. In general, ${}^*\bar{\boldsymbol{\phi}}^\psi_\alpha$ could also depend on other differential geometric properties of the surface, say the mean and Gaussian curvatures, but Cauchy restricted the dependence to merely one on $\mathbf{n}$ (This is called the 'Cauchy assumption'). The flux and supply terms for mass, linear and angular momentum, energy and entropy are listed in columns 4 and 5 of Table 3.1.

The bar in the representation ${}^*\bar{\boldsymbol{\phi}}^\psi_\alpha(\mathbf{x},t,\mathbf{n})$ indicates that the flux density in (3.25) is referred to a unit mixture area. To assume the relation

$${}^*\bar{\boldsymbol{\phi}}^\psi_\alpha = \nu_\alpha \, {}^*\boldsymbol{\phi}^\psi_\alpha \ , \tag{3.26}$$

we must postulate that [10]

[A1] $\nu_\alpha(\mathbf{x},t) = \mu_\alpha(\mathbf{x},t)$

[10] With the symbol [**A N**], $\mathbf{N} = 1, 2, 3, \ldots$, we itemise *ad hoc* assumptions used in the sequel.

Table 3.1 Densities for the constituent balance relations

Balance	$\boldsymbol{\psi}_\alpha$	$\boldsymbol{\pi}_\alpha^\psi$	$\boldsymbol{\phi}_\alpha^\psi$	$\boldsymbol{\sigma}_\alpha^\psi$	γ_α^ψ
Mass	ρ_α	0	$\mathbf{0}$	0	$\rho_\alpha c_\alpha$
Momentum	$\rho_\alpha \mathbf{v}_\alpha$	$\mathbf{0}$	$\mathbf{T}_\alpha$	$\mathbf{b}_\alpha$	$\mathbf{m}_\alpha$
Moment of momentum	$\mathbf{x} \times \rho_\alpha \mathbf{v}_\alpha$	$\mathbf{0}$	$\mathbf{x} \times \mathbf{T}_\alpha$	$\mathbf{x} \times \mathbf{b}_\alpha$	$\mathbf{M}_\alpha$
Total energy	$\rho_\alpha \left(\varepsilon_\alpha + \frac{1}{2}\mathbf{v}_\alpha \cdot \mathbf{v}_\alpha\right)$	0	$\mathbf{T}_\alpha \mathbf{v}_\alpha - \mathbf{q}_\alpha$	$r_\alpha + \mathbf{b}_\alpha \cdot \mathbf{v}_\alpha$	e_α
Entropy	$\rho_\alpha \eta_\alpha$	$\pi_\alpha^{\rho\eta}$	$\boldsymbol{\phi}_\alpha^{\rho\eta}$	$\sigma_\alpha^{\rho\eta}$	$\gamma_\alpha^{\rho\eta}$

which states that *areal and volume fractions are the same*; this is in general an approximation.[11] Accepting [**A1**] we summarise the relations between partial and true densities,

$$\left\{\bar{\boldsymbol{\psi}}_\alpha,\ \bar{\boldsymbol{\pi}}_\alpha^\psi,\ {}^*\bar{\boldsymbol{\phi}}_\alpha^\psi,\ \bar{\boldsymbol{\sigma}}_\alpha^\psi,\ \bar{\gamma}_\alpha^\psi\right\} = \nu_\alpha \left\{\psi_\alpha,\ \boldsymbol{\pi}_\alpha^\psi,\ {}^*\boldsymbol{\phi}_\alpha^\psi,\ \boldsymbol{\sigma}_\alpha^\psi,\ \gamma_\alpha^\psi\right\} . \tag{3.27}$$

With the global quantities $\mathcal{G}_\alpha$, $\mathcal{P}_\alpha$, $\mathcal{S}_\alpha$ and $\mathcal{F}_\alpha$ expressed by the integrals (3.22)-(3.25), the global balance law for the constituent K_α as expressed in item (*i*) of the basic Postulate (see (3.18)) now takes the form

$$\frac{d}{dt}\int_{\omega_\alpha} \bar{\boldsymbol{\psi}}_\alpha dv = \int_{\omega_\alpha} \left(\bar{\boldsymbol{\pi}}_\alpha^\psi + \bar{\boldsymbol{\gamma}}_\alpha^\psi\right) dv + \int_{\omega_\alpha} \bar{\boldsymbol{\sigma}}_\alpha^\psi dv + \int_{\partial\omega_\alpha} {}^*\bar{\boldsymbol{\phi}}_\alpha^\psi \, da \, . \tag{3.28}$$

In view of the sum relations in item (*iii*) of the Postulate and (3.22)-(3.25), and because $\omega_\alpha = \omega$, $\partial\omega_\alpha = \partial\omega$ for all α, (3.28) also implies the mixture balance law in the form

$$\begin{aligned} \frac{d}{dt}\sum_{\alpha=1}^{n}\int_{\omega_\alpha} \bar{\boldsymbol{\psi}}_\alpha dv &= \sum_{\alpha=1}^{n}\int_{\omega_\alpha} \left(\bar{\boldsymbol{\pi}}_\alpha^\psi + \bar{\boldsymbol{\gamma}}_\alpha^\psi\right) dv \\ &+ \sum_{\alpha=1}^{n}\int_{\omega_\alpha} \bar{\boldsymbol{\sigma}}_\alpha^\psi dv + \sum_{\alpha=1}^{n}\int_{\partial\omega_\alpha} {}^*\bar{\boldsymbol{\phi}}_\alpha^\psi \, da \, . \end{aligned} \tag{3.29}$$

Up to this point in the derivation of (3.28) and (3.29) the only mathematical assumptions that were introduced beyond the basic Postulate were (*i*) the additivity assumption (3.22)-(3.25) and (*ii*) that the densities ${}^*\bar{\boldsymbol{\phi}}_\alpha^\psi$ depend on the differential geometric properties of the surface, on which ${}^*\bar{\boldsymbol{\phi}}_\alpha^\psi$

[11] Assumption [**A1**] has been differently introduced by MORLAND [91] who demonstrated the assumption to be very wrong. MORLAND's [92] recent work on anisotropic permeability in a structured matrix explicitly uses an area fraction depending on the surface normal. For granular materials with compact grains we believe [**A1**] to be acceptable.

is defined, via the unit normal vector $\mathbf{n}$ (CAUCHY assumption). This second assumption can be shown to imply that the flux quantity ${}^*\bar{\boldsymbol{\phi}}^{\psi}_{\alpha}$ is an affine function of $\mathbf{n}$, i. e., linear in $\mathbf{n}$ as follows:

$$ {}^*\bar{\boldsymbol{\phi}}^{\psi}_{\alpha}(\mathbf{x}, t, \mathbf{n}) = \bar{\boldsymbol{\phi}}^{\psi}_{\alpha}(\mathbf{x}, t)\mathbf{n} \ . \tag{3.30} $$

This equation is known as CAUCHY's *lemma*. Several remarks must be made in its connection. First, with this definition of $\bar{\boldsymbol{\phi}}^{\psi}_{\alpha}(\mathbf{x}, t)$ the flux quantity $\mathcal{F}_{\alpha}$ is now positive as an outflow from ω_{α} to its environment. Second, (3.30) is expressed in the partial densities because it emerges from the application of the general balance law of constituent K_{α} to an infinitesimal tetrahedron, see e. g., HUTTER & JÖHNK [62], GURTIN [49], CHADWICK [22] or any other book on continuum mechanics. However, in view of the definition (3.27) it also holds for the true density,

$$ {}^*\boldsymbol{\phi}^{\psi}_{\alpha}(\mathbf{x}, t, \mathbf{n}) = \boldsymbol{\phi}^{\psi}_{\alpha}(\mathbf{x}, t)\mathbf{n} \ . \tag{3.31} $$

Second, ${}^*\boldsymbol{\phi}^{\psi}_{\alpha}$ and $\boldsymbol{\phi}^{\psi}_{\alpha}$ are different quantities. Depending on the choice of the physical density $\boldsymbol{\psi}_{\alpha}$, $\boldsymbol{\phi}^{\psi}_{\alpha}$ constitutes a first order (for the energies), a second order (for the stresses), or a third order (for the moments of momentum) tensor (see fourth column in Table 3.1) and the multiplication on the right-hand side of (3.30) is a contraction. Third, in the literature it is often customary to define $\boldsymbol{\phi}^{\psi}_{\alpha}(\mathbf{x}, t)$ with a negative sign to assign to $\boldsymbol{\phi}^{\psi}_{\alpha}(-\mathbf{n})$ the meaning of an inflow, i. e., a gain of $\boldsymbol{\psi}_{\alpha}$ within the body. We have not done so here, to have positive quantities in the fourth column of Table 3.1. Finally we mention that the property (3.30) or (3.31) is mathematically convenient, since with it the flux term takes the form

$$ \mathcal{F}_{\alpha} = \int_{\partial\omega_{\alpha}} \bar{\boldsymbol{\phi}}^{\psi}_{\alpha}(\mathbf{x}, t)\mathbf{n}\, da \ , \tag{3.32} $$

which, with sufficient differentiability, can be transformed to a volume integral (divergence theorem, see below). This now brings the global balance law for constituent K_{α} into the form

$$ \frac{d}{dt}\int_{\omega_{\alpha}} \bar{\boldsymbol{\psi}}_{\alpha} dv = \int_{\omega_{\alpha}} \left(\bar{\boldsymbol{\pi}}^{\psi}_{\alpha} + \bar{\boldsymbol{\gamma}}^{\psi}_{\alpha}\right) dv + \int_{\omega_{\alpha}} \bar{\boldsymbol{\sigma}}^{\psi}_{\alpha} dv + \int_{\partial\omega_{\alpha}} \bar{\boldsymbol{\phi}}^{\psi}_{\alpha}\mathbf{n}\, da \ , \tag{3.33} $$

and by summing over all constituents the mixture balance law is obtained. With the help of Table 3.1 this global balance equation for a general physical field density $\boldsymbol{\psi}_{\alpha}$ can be specialized to the constituent mass, momentum, moment of momentum, energy and entropy densities.

For equation (3.33) to hold, it was assumed that the fields are integrable and summable. Other smoothness properties are not required for (3.33) to hold. If we request in the sequel nevertheless certain smoothness properties, special forms of the balance law (3.33) can be derived. In the continuum theories that are in focus in this work it will be assumed that the point fields $\bar{\boldsymbol{\psi}}_\alpha(\mathbf{x},t)$, $\bar{\boldsymbol{\pi}}^\psi_\alpha(\mathbf{x},t)$, $\bar{\gamma}^\psi_\alpha(\mathbf{x},t)$, $\bar{\boldsymbol{\sigma}}^\psi_\alpha(\mathbf{x},t)$ and $\bar{\boldsymbol{\phi}}^\psi_\alpha(\mathbf{x},t)$ and their true field counterparts are differentiable in any material part ω of a body *except* on material or non-material surfaces across which these fields may experience a finite jump[12]. In Fig. 3.4 such a material region ω is shown together with an orientable surface σ that traces the loci across which some of the fields (stated above) may experience a discontinuity. For this reason and for brevity we shall henceforth call such a surface a 'singular surface'. It divides the region ω into subregions ω^+ and ω^- which are bounded by $\partial\omega^+ \cup \sigma$ and $\partial\omega^- \cup \sigma$, respectively. Which subregions on the two sides of σ are denoted as ω^+ and ω^- is arbitrary, but once a denotation has been chosen, we agree that the unit vector, perpendicular to σ, $\mathbf{n}_\sigma$, points into the ω^+-subregion. As a non-material surface, σ possesses its own velocity, $\mathbf{s}$, different in general from the material velocities immediately on the positive and negative sides of σ; it is, however, clear that the only geometrically significant quantity is the normal speed $\mathbf{s} \cdot \mathbf{n}_\sigma$.

In the subregions $\omega^\pm$ with boundaries $\partial\omega^\pm \cup \sigma$, the divergence theorem (Gauss law) is separately applicable:

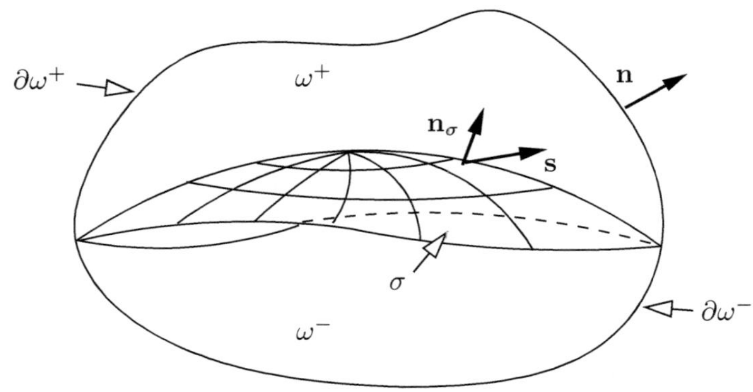

Fig. 3.4 Material volume ω that is divided in two subregions, ω^+ and ω^-, by a surface σ across which the physical fields may suffer a jump discontinuity. The positive and negative sides of σ are chosen arbitrarily, but when fixed, the unit vector $\mathbf{n}_\sigma$, perpendicular to σ points into ω^+. $\mathbf{s}$ is the velocity with which the singular surface moves. The exterior normal vector to ω is $\mathbf{n}$. So, on σ $\mathbf{n} = -\mathbf{n}_\sigma$ for ω^+ and $\mathbf{n} = \mathbf{n}_\sigma$ for ω^-.

[12] Certain weak, i. e., integrable singularities are also permissible. We restrict the attention here to finite jumps and leave the slightly more general case to the readers as an exercise.

$$\begin{aligned}\int_{\omega_\alpha^+} \left(\nabla \cdot \bar{\boldsymbol{\phi}}_\alpha^\psi\right)^+ dv &= \int_{\partial\omega_\alpha^+} \left(\bar{\boldsymbol{\phi}}_\alpha^\psi\right)^+ \mathbf{n}\, da - \int_\sigma \left(\bar{\boldsymbol{\phi}}_\alpha^\psi\right)^+ \mathbf{n}_\sigma\, da\,, \\ \int_{\omega_\alpha^-} \left(\nabla \cdot \bar{\boldsymbol{\phi}}_\alpha^\psi\right)^- dv &= \int_{\partial\omega_\alpha^-} \left(\bar{\boldsymbol{\phi}}_\alpha^\psi\right)^- \mathbf{n}\, da + \int_\sigma \left(\bar{\boldsymbol{\phi}}_\alpha^\psi\right)^- \mathbf{n}_\sigma\, da\,.\end{aligned} \tag{3.34}$$

Adding these equations, using the notation

$$[\![\bar{\boldsymbol{\phi}}_\alpha^\psi]\!] := \left(\bar{\boldsymbol{\phi}}_\alpha^\psi\right)^+ - \left(\bar{\boldsymbol{\phi}}_\alpha^\psi\right)^- \tag{3.35}$$

as the 'jump of $\bar{\boldsymbol{\phi}}_\alpha^\psi$ across σ', and rearranging, yields

$$\int_{\partial\omega_\alpha} \bar{\boldsymbol{\phi}}_\alpha^\psi(\mathbf{x},t)\mathbf{n} da = \int_{\omega_\alpha^+ \cup \omega_\alpha^-} \nabla \cdot \bar{\boldsymbol{\phi}}_\alpha^\psi(\mathbf{x},t) dv + \int_\sigma [\![\bar{\boldsymbol{\phi}}_\alpha^\psi(\mathbf{x},t)]\!]\mathbf{n}_\sigma da\,. \tag{3.36}$$

This transforms the flux term in the balance law.

In much the same way, the transport theorem is treated. We start with the decomposition

$$\frac{d}{dt}\int_{\omega_\alpha} \bar{\psi}_\alpha(\mathbf{x},t) dv = \frac{d}{dt}\int_{\omega_\alpha^+} \bar{\psi}_\alpha^+(\mathbf{x},t) dv + \frac{d}{dt}\int_{\omega_\alpha^-} \bar{\psi}_\alpha^-(\mathbf{x},t) dv\,, \tag{3.37}$$

and then may transform the two integrals on the right-hand side individually by applying the classical transport theorem for differentiable fields. In so doing it must be recognized that the normal flux of $\bar{\psi}_\alpha$ through the singular surface σ is $\mp\bar{\psi}_\alpha(\mathbf{v}_\alpha - \mathbf{s}) \cdot \mathbf{n}_\sigma$ on the $\pm$ sides of σ, respectively. With these remarks in mind, we obtain

$$\begin{aligned}\frac{d}{dt}\int_{\omega_\alpha^+} \bar{\psi}_\alpha^+ dv &= \int_{\omega_\alpha^+} \frac{\partial \bar{\psi}_\alpha^+}{\partial t} dv + \int_{\partial\omega_\alpha^+} \left(\bar{\psi}_\alpha \mathbf{v}_\alpha\right)^+ \mathbf{n}\, da - \int_\sigma \left(\bar{\psi}_\alpha \mathbf{s}\right)^+ \mathbf{n}_\sigma\, da \\ &= \int_{\omega_\alpha^+} \frac{\partial \bar{\psi}_\alpha^+}{\partial t} dv + \underbrace{\int_{\partial\omega_\alpha^+ \cup \sigma} \left(\bar{\psi}_\alpha \mathbf{v}_\alpha\right)^+ \mathbf{n}\, da}_{\int_{\omega_\alpha^+} \nabla\cdot\left(\bar{\psi}_\alpha \mathbf{v}_\alpha\right)^+ dv} - \int_\sigma \left(\bar{\psi}_\alpha(\mathbf{s} - \mathbf{v}_\alpha)\right)^+ \mathbf{n}_\sigma\, da\,.\end{aligned} \tag{3.38}$$

Here, we have used the fact that the exterior unit normal vector to ω_α^+ on σ is $-\mathbf{n}_\sigma$. Analogously, we obtain

$$\begin{aligned}\frac{d}{dt}\int_{\omega_\alpha^-} \bar{\psi}_\alpha^- dv = \int_{\omega_\alpha^-} \frac{\partial \bar{\psi}_\alpha^-}{\partial t} dv &+ \int_{\omega_\alpha^-} \nabla \cdot \left(\bar{\psi}_\alpha \mathbf{v}_\alpha\right)^- dv \\ &+ \int_\sigma \left(\bar{\psi}_\alpha(\mathbf{s} - \mathbf{v}_\alpha)\right)^- \mathbf{n}_\sigma\, da\,.\end{aligned} \tag{3.39}$$

Adding the last two equations yields the *generalised transport theorem* as follows:

$$\frac{d}{dt}\int_{\omega_\alpha} \bar{\psi}_\alpha(\mathbf{x},t)dv = \int_{\omega_\alpha} \Big\{\frac{\partial\bar{\psi}_\alpha(\mathbf{x},t)}{\partial t} + \nabla\cdot\left(\bar{\psi}_\alpha\mathbf{v}_\alpha\right)(\mathbf{x},t)\Big\}dv + \int_\sigma [\![\bar{\psi}_\alpha\left(\mathbf{v}_\alpha-\mathbf{s}\right)]\!](\mathbf{x},t)\mathbf{n}_\sigma da\ . \tag{3.40}$$

In the expressions (3.36) and (3.40) the products $[\![\mathbf{a}]\!]\mathbf{n}_\sigma$ are contractions, i. e., the scalar $[\![\mathbf{a}]\!]\cdot\mathbf{n}_\sigma$ if $\mathbf{a}$ is a vector, and the vector $[\![\mathbf{a}]\!]\mathbf{n}_\sigma$ if $\mathbf{a}$ is a second rank tensor, etc.

The transformations (3.36) and (3.40) allow the balance law for constituent K_α, (3.33) to be written as[13]

$$\int_{\omega_\alpha^+\cup\,\omega_\alpha^-} \Big\{\frac{\partial\bar{\psi}_\alpha}{\partial t} + \nabla\cdot\left(\bar{\psi}_\alpha\mathbf{v}_\alpha\right) - \nabla\cdot\bar{\boldsymbol{\phi}}_\alpha^\psi - \bar{\boldsymbol{\pi}}_\alpha^\psi - \bar{\boldsymbol{\gamma}}_\alpha^\psi - \bar{\boldsymbol{\sigma}}_\alpha^\psi\Big\}dv + \int_\sigma [\![\bar{\psi}_\alpha\left(\mathbf{v}_\alpha-\mathbf{s}\right) - \bar{\boldsymbol{\phi}}_\alpha^\psi]\!]\mathbf{n}_\sigma da = \mathbf{0}\ , \tag{3.41}$$

in which the multiplication $[\![\cdot]\!]\mathbf{n}_\sigma$ is a contraction.[14] The mathematical structure of expression (3.41) is a sum of a volume integral of which the integrand function consists of differentiable functions plus an integral of a jump quantity over the singular surface σ. It is a direct consequence of the additivity assumption that (3.41) holds for any material part of the body, no matter how large or small and irrespective whether this part is crossed by a singular surface. Consequently, if we take as the body part an local material volume element with no singular surface, (3.41) reduces to the statement

$$\partial\bar{\psi}_\alpha - \nabla\cdot\left(\bar{\boldsymbol{\phi}}_\alpha^\psi - \bar{\psi}_\alpha\otimes\mathbf{v}_\alpha\right) - \bar{\boldsymbol{\pi}}_\alpha^\psi - \bar{\boldsymbol{\sigma}}_\alpha^\psi - \bar{\boldsymbol{\gamma}}_\alpha^\psi = \mathbf{0}\ , \tag{3.42}$$

in which the notation

$$\partial(\cdot) := \frac{\partial(\cdot)}{\partial t} \tag{3.43}$$

---

[13] When splitting the volume integrals of the production densities into integrals over the regions ω_α^+ and ω_α^- one must, in general, also allow for the existence of an additional production term of the variable $\bar{\psi}_\alpha$ on the singular surface. Analogously, the flux of the variable at the cutting line of the surfaces $\partial\omega_\alpha^+$ and $\partial\omega_\alpha^-$ with the singular surface $\sigma(t)$ can give rise to an additional flux term that may represent surface stresses acting on singular surfaces. Such extensions of the jump conditions shall not be considered in this book.

[14] In this expression no surface terms arise because they are taken to be negligible. ALTS and HUTTER (1988) [1], [2], [3], (1989) [4], among others, derive a formulation of this complexity for single constituent continua as does GRAUEL [45] for mixtures. Moreover, MORLAND and SELLERS [93] show how the 'surface production terms' arise naturally from the limits of the continuous balances at either side.

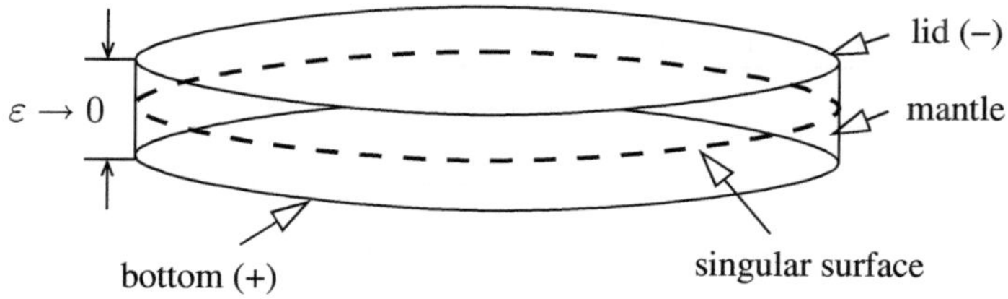

Fig. 3.5 Material pillbox with lid and bottom surfaces on either side of the singular surface and a mantle surface of height ε .

has been used. Equation (3.42) is called the *local balance law*, and it holds in any neighbourhood of a point where the fields $\bar{\psi}_\alpha$, $\bar{\phi}^\psi_\alpha$, $\bar{\pi}^\psi_\alpha$, $\bar{\sigma}^\psi_\alpha$ and $\bar{\gamma}^\psi_\alpha$ are differentiable.

In a similar way, if we choose for ω_α a 'pillbox' with lid and bottom on the positive and negative sides of the singular surface, but infinitely close to it, see Fig. 3.5, then the volume integral in (3.41) over $\omega^+_\alpha \cup \omega^-_\alpha$ vanishes as $\varepsilon \to 0$ since the integrand is differentiable and therefore bounded. So, in this limit only, the surface integral survives in (3.41). Thus, by making the pillbox diameter as small as we please, (3.41) reduces to

$$[\![\bar{\psi}_\alpha (\mathbf{v}_\alpha - \mathbf{s}) - \bar{\boldsymbol{\phi}}^\psi_\alpha]\!]\mathbf{n}_\sigma = \mathbf{0} \ . \tag{3.44}$$

Summation of (3.42) and (3.44) over all constituents yields the corresponding local balance law and jump condition for the mixture as a whole:

$$\sum_{\alpha=1}^{n} \left\{ \partial \bar{\psi}_\alpha - \bar{\boldsymbol{\pi}}^\psi_\alpha - \nabla \cdot \left(\bar{\boldsymbol{\phi}}^\psi_\alpha - \bar{\psi}_\alpha \otimes \mathbf{v}_\alpha \right) - \bar{\boldsymbol{\sigma}}^\psi_\alpha - \bar{\gamma}^\psi_\alpha \right\} = \mathbf{0} \ , \tag{3.45}$$

$$\sum_{\alpha=1}^{n} \left\{ [\![\bar{\psi}_\alpha (\mathbf{v}_\alpha - \mathbf{s}) - \bar{\boldsymbol{\phi}}^\psi_\alpha]\!]\mathbf{n}_\sigma \right\} = \mathbf{0} \ . \tag{3.46}$$

With the general balance equations and jump conditions (3.42)-(3.46) at hand we can deduce the physical balance equations and jump conditions on using Table 3.1. In addition to the independent variables, $\nu_\alpha(\mathbf{x},t)$, $\rho_\alpha(\mathbf{x},t)$, $\mathbf{v}_\alpha(\mathbf{x},t)$ and $\mathbf{x} \times \bar{\rho}_\alpha \mathbf{v}_\alpha(\mathbf{x},t)$ we have the *constituent* CAUCHY *stress tensors* $\mathbf{T}_\alpha(\mathbf{x},t)$, *specific internal energies* $\varepsilon_\alpha(\mathbf{x},t)$, *heat flux vectors* $\mathbf{q}_\alpha(\mathbf{x},t)$, *specific entropies* $\eta_\alpha(\mathbf{x},t)$, *entropy fluxes* $\boldsymbol{\phi}^{\rho\eta}_\alpha(\mathbf{x},t)$ and *intrinsic entropy production rate densities* $\pi^{\rho\eta}_\alpha(\mathbf{x},t)$ which represent the dependent quantities $(\alpha = 1,\ldots,n)$. Characteristic for mixture theories are the dependent *interaction supply rate densities* for *mass* $c_\alpha(\mathbf{x},t)$, *momentum* $\mathbf{m}_\alpha(\mathbf{x},t)$, *moment of momentum* $\mathbf{M}_\alpha(\mathbf{x},t)$, *energy* $e_\alpha(\mathbf{x},t)$ and *entropy* $\gamma^{\rho\eta}_\alpha(\mathbf{x},t)$. The *constituent external supply rate density* for *momentum* $\mathbf{b}_\alpha(\mathbf{x},t)$, *internal energy* $r_\alpha(\mathbf{x},t)$

Table 3.2 Densities for the mixture balance relations

Balance	ψ	$\boldsymbol{\pi}^{\psi}$	$\boldsymbol{\phi}^{\psi}$	$\boldsymbol{\sigma}^{\psi}$
Mass	ρ	0	$\mathbf{0}$	0
Momentum	$\rho\mathbf{v}$	$\mathbf{0}$	$\mathbf{T}$	$\mathbf{b}$
Moment of momentum	$\mathbf{x} \times \rho\mathbf{v}$	$\mathbf{0}$	$\mathbf{x} \times \mathbf{T}$	$\mathbf{x} \times \mathbf{b}$
Total energy	$\rho\left(\varepsilon + \frac{1}{2}\mathbf{v} \cdot \mathbf{v}\right)$	0	$\mathbf{T}\mathbf{v} - \mathbf{q}$	$r + \mathbf{b} \cdot \mathbf{v}$
Entropy	$\rho\eta$	$\pi^{\rho\eta}$	$\boldsymbol{\phi}^{\rho\eta}$	$\sigma^{\rho\eta}$

and *entropy* $\sigma_{\alpha}^{\rho\eta}(\mathbf{x}, t)$ are thought to be prescribed and thus determined by the environment of the mixture.[15]

One consequence of the fourth supposition of TRUESDELL is that the mixture as a whole obeys the local balance laws as if it were a single material, i. e. one may write down the usual general local balance relations and jump conditions for the mixture as a whole. In the spatial representation they have the form

$$\partial\psi = \boldsymbol{\pi}^{\psi} + \nabla \cdot \left(\boldsymbol{\phi}^{\psi} - \psi \otimes \mathbf{v}\right) + \boldsymbol{\sigma}^{\psi} \,, \tag{3.47}$$

within the regions ω^{+} and ω^{-} and

$$[\![\psi\,(\mathbf{v} - \mathbf{s}) - \boldsymbol{\phi}^{\psi}]\!]\mathbf{n}_{\sigma} = \mathbf{0} \tag{3.48}$$

on the singular surface σ. Here the velocity $\mathbf{v}$ must still be identified. The mixture fields have the same interpretation as their equivalents in (3.42), but as we think of the mixture as a single material no interaction supply rate densities can arise. The specific mixture balance equations can be read off from Table 3.2. TRUESDELL required in his second supposition that all properties of the mixture be mathematical consequences of properties of the constituents. Thus, and in view of item *(iii)* of the basic Postulate, the sum of all constituent balance equations (3.42) equals the mixture balance equation (3.47). If, in addition, the sum of the constituent interaction supply rate densities vanishes, i. e.

$$\sum_{\alpha=1}^{n} \bar{\gamma}_{\alpha}^{\psi} = 0 \,, \tag{3.49}$$

the following important relations between the constituent and the mixture quantities can be identified

[15] It is physically characteristic that supply rate densities of mass vanish identically. Such terms are not possible in Galileian mechanics. Despite of this, some authors have introduced mass supply terms, which may have arbitrary values. We take the position that this is unphysical.

$$
\begin{aligned}
\boldsymbol{\psi} &= \sum \bar{\boldsymbol{\psi}}_\alpha \;, \\
\boldsymbol{\phi}^\psi - \boldsymbol{\psi} \otimes \mathbf{v} &= \sum \left[\bar{\boldsymbol{\phi}}^\psi_\alpha - \bar{\boldsymbol{\psi}}_\alpha \otimes \mathbf{v}_\alpha \right] \;, \\
\boldsymbol{\pi}^\psi &= \sum \bar{\boldsymbol{\pi}}^\psi_\alpha \;, \\
\boldsymbol{\sigma}^\psi &= \sum \bar{\boldsymbol{\sigma}}^\psi_\alpha \;.
\end{aligned}
\tag{3.50}
$$

In (3.50) and the remainder of this work $\sum$ is an abbreviation for $\sum_{\alpha=1}^{n}$. If we further let $\boldsymbol{\varphi}_\alpha$ and $\boldsymbol{\varphi}$ be constituent and mixture specific thermodynamic fields, respectively and define them according to

$$
\boldsymbol{\psi}_\alpha =: \rho_\alpha \boldsymbol{\varphi}_\alpha \qquad \text{and} \qquad \boldsymbol{\psi} =: \rho \boldsymbol{\varphi} \;, \tag{3.51}
$$

relation $(3.50)_1$ takes the form

$$
\boldsymbol{\varphi} = \sum \bar{\xi}_\alpha \boldsymbol{\varphi}_\alpha \;, \tag{3.52}
$$

where

$$
\bar{\xi}_\alpha := \rho^{-1} \, \bar{\rho}_\alpha \tag{3.53}
$$

represents the *mass fraction* of constituent K_α in the mixture. Often $\bar{\xi}_\alpha$ is simply called *concentration*. We recall that ρ_α is the true density of constituent K_α and the density of the mixture as a whole is $\rho = \sum \bar{\rho}_\alpha$, see Table 3.2 and (3.50_1). This identification of ρ is denoted the mass density sum relation.

With these definitions at hand, we now turn to the physical quantities of Tables 3.1 and 3.2 and determine the relations between the constituent and mixture physical densities. To achieve this, we must choose the correct pairs of $(\boldsymbol{\varphi},\ \boldsymbol{\varphi}_\alpha)$ or $(\boldsymbol{\psi},\ \boldsymbol{\psi}_\alpha)$ and substitute them into (3.52) or $(3.50)_1$. The choice $(\boldsymbol{\varphi},\ \boldsymbol{\varphi}_\alpha) = (1, 1)$ or $(\boldsymbol{\psi},\ \boldsymbol{\psi}_\alpha) = (\rho,\ \rho_\alpha)$ results in

$$
1 = \sum \bar{\xi}_\alpha \qquad \text{or} \qquad \rho = \sum \bar{\rho}_\alpha \;. \tag{3.54}
$$

Equation $(3.54)_1$ is a natural constraint on the constituent mass fractions that follows from $(3.54)_2$. Another relation is obtained by identifying $\boldsymbol{\psi}$ and $\boldsymbol{\psi}_\alpha$ with mixture momentum $\rho \mathbf{v}$ and constituent momentum $\rho_\alpha \mathbf{v}_\alpha$, respectively. $\boldsymbol{\varphi}$ and $\boldsymbol{\varphi}_\alpha$ are chosen accordingly. The result

$$
\rho \mathbf{v} = \sum \bar{\rho}_\alpha \mathbf{v}_\alpha \tag{3.55}
$$

can be regarded as definition of the mixture velocity[16] $\mathbf{v}$, which is commonly called *barycentric velocity*. For mixtures it is also advantageous to define the

16 The term 'mixture velocity' is not precise and should be avoided since apart from the above mass weighted mixture velocity there is also a volume weighted mixture

constituent diffusion velocity

$$\mathbf{u}_\alpha := \mathbf{v}_\alpha - \mathbf{v} \ , \tag{3.56}$$

which, obviously, satisfies

$$\sum \bar{\rho}_\alpha \mathbf{u}_\alpha = 0 \ . \tag{3.57}$$

If we now take the gradient of (3.55), then

$$\rho \left(\nabla \mathbf{v} \right) = \sum \left\{ \bar{\rho}_\alpha \left(\nabla \mathbf{v}_\alpha \right) + \mathbf{u}_\alpha \otimes \left(\nabla \bar{\rho}_\alpha \right) \right\} \tag{3.58}$$

is obtained via the product rule of differentiation and the commutability of the operators ∇ and $\sum$. Also $(3.54)_2$ and (3.56) were used. As is possible for the constituent velocity gradient,

$$\nabla \mathbf{v}_\alpha = \mathbf{D}_\alpha + \mathbf{W}_\alpha \ , \tag{3.59}$$

the *mixture velocity gradient*, $\nabla \mathbf{v}$, can also be decomposed into a symmetric, $\mathbf{D}$, and skew-symmetric part, $\mathbf{W}$, corresponding to the *mixture stretching* and *mixture vorticity tensors*, respectively. This decomposition reads

$$\begin{aligned} &\mathbf{L} := \nabla \mathbf{v} = \mathbf{D} + \mathbf{W} \ , \\ &\mathbf{D} := \mathrm{sym} \left(\nabla \mathbf{v} \right) \ , \quad \mathbf{W} := \mathrm{skw} \left(\nabla \mathbf{v} \right) \ . \end{aligned} \tag{3.60}$$

Application of the operators sym$(\cdot)$ and skw$(\cdot)$ to relation (3.58) yields the sum relations

$$\rho \mathbf{D} = \sum \left\{ \bar{\rho}_\alpha \mathbf{D}_\alpha + \mathrm{sym} \left(\mathbf{u}_\alpha \otimes \nabla \bar{\rho}_\alpha \right) \right\} \tag{3.61}$$

and

$$\rho \mathbf{W} = \sum \left\{ \bar{\rho}_\alpha \mathbf{W}_\alpha + \mathrm{skw} \left(\mathbf{u}_\alpha \otimes \nabla \bar{\rho}_\alpha \right) \right\} \ . \tag{3.62}$$

Turning now to the energy-balance equation, by choosing

$$\left(\psi, \ \psi_\alpha \right) = \Big(\rho \left(\varepsilon + \tfrac{1}{2} \mathbf{v} \cdot \mathbf{v} \right), \ \rho_\alpha \left(\varepsilon_\alpha + \tfrac{1}{2} \mathbf{v}_\alpha \cdot \mathbf{v}_\alpha \right) \Big) \tag{3.63}$$

and substituting these into relation $(3.50)_1$, the mixture specific internal energy can be written as

$$\begin{aligned} &\varepsilon = \sum \bar{\xi}_\alpha \varepsilon_\alpha + \tfrac{1}{2} \sum \bar{\xi}_\alpha \mathbf{u}_\alpha \cdot \mathbf{u}_\alpha \ = \varepsilon_{\mathrm{I}} + \varepsilon_{\mathrm{D}}, \\ &\varepsilon_{\mathrm{I}} := \sum \bar{\xi}_\alpha \varepsilon_\alpha, \quad \varepsilon_{\mathrm{D}} := \tfrac{1}{2} \sum \bar{\xi}_\alpha \mathbf{u}_\alpha \cdot \mathbf{u}_\alpha \ . \end{aligned} \tag{3.64}$$

---

velocity which is quite popular (see CHEN & TAI) [27]). If ν_α denote the volume fractions, this volume weighted mixture velocity, $\mathbf{v}_{\mathrm{vol}}$, is given by $\mathbf{v}_{\mathrm{vol}} := \sum \nu_\alpha \mathbf{v}_\alpha$.

For the deduction of $(3.64)_1$, equations $(3.54)_1$ and (3.57) are needed. Definitions $(3.64)_2$ and $(3.64)_3$ specify the *'inner' mixture specific internal energy* and the *diffusive* contribution of ε, respectively.

We decompose the mixture fluxes $(3.50)_2$ in a similar way and define the *'inner'* and *diffusive mixture fluxes* as follows

$$\begin{aligned} \boldsymbol{\phi}^{\psi} &= \sum \left\{ \bar{\boldsymbol{\phi}}^{\psi}_{\alpha} - \bar{\boldsymbol{\psi}}_{\alpha} \otimes \mathbf{u}_{\alpha} \right\} = \boldsymbol{\phi}^{\psi}_{\mathrm{I}} + \boldsymbol{\phi}^{\psi}_{\mathrm{D}}, \\ \boldsymbol{\phi}^{\psi}_{\mathrm{I}} &:= \sum \bar{\boldsymbol{\phi}}^{\psi}_{\alpha}, \quad \boldsymbol{\phi}^{\psi}_{\mathrm{D}} := \sum -\bar{\boldsymbol{\psi}}_{\alpha} \otimes \mathbf{u}_{\alpha} \,. \end{aligned} \tag{3.65}$$

Here, again, definition (3.56) was used.

With the help of Table 3.1, Table 3.2 and the correct choices of the pairs $(\boldsymbol{\phi}^{\psi}, \boldsymbol{\phi}^{\psi}_{\alpha})$ and $(\boldsymbol{\psi}, \boldsymbol{\psi}_{\alpha})$, the general flux in (3.65) is specialized to the CAUCHY stress tensors (momentum fluxes) and to the heat flux vectors. The results are

$$\mathbf{T} = \sum \bar{\mathbf{T}}_{\alpha} - \sum \bar{\rho}_{\alpha} \mathbf{u}_{\alpha} \otimes \mathbf{u}_{\alpha} = \mathbf{T}_{\mathrm{I}} + \mathbf{T}_{\mathrm{D}}, \tag{3.66}$$

and

$$\mathbf{q} = \sum \bar{\mathbf{q}}_{\alpha} - \sum \bar{\mathbf{T}}_{\alpha} \mathbf{u}_{\alpha} + \sum \bar{\rho}_{\alpha} \left\{ \varepsilon_{\alpha} + \tfrac{1}{2} (\mathbf{u}_{\alpha} \cdot \mathbf{u}_{\alpha}) \right\} \mathbf{u}_{\alpha} = \mathbf{q}_{\mathrm{I}} + \mathbf{q}_{\mathrm{D}} \,, \tag{3.67}$$

where

$$\mathbf{T}_{\mathrm{I}} := \sum \bar{\mathbf{T}}_{\alpha} \,, \qquad \mathbf{T}_{\mathrm{D}} = -\sum \bar{\rho}_{\alpha} \mathbf{u}_{\alpha} \otimes \mathbf{u}_{\alpha} \tag{3.68}$$

and

$$\mathbf{q}_{\mathrm{I}} := \sum \bar{\mathbf{q}}_{\alpha} \,, \qquad \mathbf{q}_{\mathrm{D}} = -\sum \bar{\mathbf{T}}_{\alpha} \mathbf{u}_{\alpha} + \sum \bar{\rho}_{\alpha} \left\{ \varepsilon_{\alpha} + \tfrac{1}{2} (\mathbf{u}_{\alpha} \cdot \mathbf{u}_{\alpha}) \right\} \mathbf{u}_{\alpha} \,. \tag{3.69}$$

Here, the pairs

$$\left(\boldsymbol{\phi}^{\psi}, \boldsymbol{\phi}^{\psi}_{\alpha}\right) = (\mathbf{T}, \mathbf{T}_{\alpha}), \qquad \boldsymbol{\psi}_{\alpha} = \rho_{\alpha} \mathbf{v}_{\alpha} \tag{3.70}$$

for (3.66) and

$$\left(\boldsymbol{\phi}^{\psi}, \boldsymbol{\phi}^{\psi}_{\alpha}\right) = (\mathbf{T}\mathbf{v} - \mathbf{q}, \mathbf{T}_{\alpha} \mathbf{v}_{\alpha} - \mathbf{q}_{\alpha}), \qquad \boldsymbol{\psi}_{\alpha} = \rho_{\alpha} \left(\varepsilon_{\alpha} + \tfrac{1}{2} \mathbf{v}_{\alpha} \cdot \mathbf{v}_{\alpha} \right) \tag{3.71}$$

were substituted into (3.65) and relations (3.56), (3.57) and (3.66) were used.

Let us summarise the findings and make some remarks.

- We take the position, that priorities of theoretical formulations for continuous assemblages of a finite number of physical constituents, be these classical mixtures or multi-phase systems, cannot be objectively decided about prior to complete thermodynamic derivations of the formulations and comparison of two seemingly diverging theories.

- Consequently, we start with global balance laws as the backbones of classical continuum physics: balances of mass, linear and angular momentum, energy and entropy for each constituent, and take the view that the production terms for any physical quantity of the constituents may assume non-trivial values, but – with the exception of the entropy production – sum up to a zero value for the mixture as a whole.
- When writing down evolution equations for internal (hidden) variables, we regard these variables to describe exclusively certain *material behaviour* at the sub-RVE scale which cannot be affected by external source terms. In other words, a balance law for an internal variable has necessarily vanishing external supply rate density.
- On the basis that mass, momenta, energy and entropy are additive physical quantities (ψ) for which true (ψ_α) and partial ($\bar{\psi}_\alpha$) constituent densities can be introduced, the above fields obey local forms of balance laws of the form

$$\partial\bar{\boldsymbol{\psi}}_\alpha - \nabla\cdot\left(\bar{\boldsymbol{\phi}}^\psi_\alpha - \bar{\boldsymbol{\psi}}_\alpha\otimes\mathbf{v}_\alpha\right) - \bar{\boldsymbol{\pi}}^\psi_\alpha - \bar{\boldsymbol{\sigma}}^\psi_\alpha - \bar{\boldsymbol{\gamma}}^\psi_\alpha = \mathbf{0}\ , \tag{3.72}$$

at points where all variables are continuously differentiable, and

$$[\![\bar{\boldsymbol{\psi}}_\alpha\left(\mathbf{v}_\alpha - \mathbf{s}\right) - \bar{\boldsymbol{\phi}}^\psi_\alpha]\!]\mathbf{n}_\sigma = \mathbf{0} \tag{3.73}$$

on surfaces, across which variables suffer at most finite jump discontinuities.

- Identifying the variables arising in the above equations with the physical quantities stated in Table 3.1, explicit forms of the physical balance laws can be easily written down, namely

mass :

$$\bar{\rho}_\alpha c_\alpha = \partial\bar{\rho}_\alpha + \nabla\cdot\left(\bar{\rho}_\alpha\mathbf{v}_\alpha\right)\ , \tag{3.74}$$

$$[\![\mathcal{M}_\alpha]\!] := [\![\bar{\rho}_\alpha\left(\mathbf{v}_\alpha - \mathbf{s}\right)]\!]\cdot\mathbf{n}_\sigma = 0\ , \tag{3.75}$$

linear momentum :

$$\bar{\mathbf{m}}_\alpha = \partial\left(\bar{\rho}_\alpha\mathbf{v}_\alpha\right) - \nabla\cdot\left(\bar{\mathbf{T}}_\alpha - \bar{\rho}_\alpha\mathbf{v}_\alpha\otimes\mathbf{v}_\alpha\right) - \bar{\mathbf{b}}_\alpha\ , \tag{3.76}$$

$$\mathcal{M}_\alpha\ [\![\mathbf{v}_\alpha]\!] - [\![\bar{\mathbf{T}}_\alpha\mathbf{n}_\sigma]\!] = \mathbf{0}\ , \tag{3.77}$$

angular momentum :

$$\bar{\mathbf{M}}_\alpha = \mathbf{x}\times\bar{\mathbf{m}}_\alpha + 2\,\mathrm{skw}\left(\bar{\mathbf{T}}_\alpha\right)\ , \tag{3.78}$$

$$\text{energy}: \quad \bar{e}_\alpha = \partial\left\{\bar{\rho}_\alpha\left(\varepsilon_\alpha + \tfrac{1}{2}\mathbf{v}_\alpha\cdot\mathbf{v}_\alpha\right)\right\} - \nabla\cdot\left\{\bar{\mathbf{T}}_\alpha\mathbf{v}_\alpha - \bar{\mathbf{q}}_\alpha - \bar{\rho}_\alpha\left(\varepsilon_\alpha + \tfrac{1}{2}\mathbf{v}_\alpha\cdot\mathbf{v}_\alpha\right)\mathbf{v}_\alpha\right\} - \left(\bar{r}_\alpha + \bar{\mathbf{b}}_\alpha\cdot\mathbf{v}_\alpha\right), \quad (3.79)$$

$$[\![\varepsilon_\alpha + \tfrac{1}{2}\mathbf{v}_\alpha\cdot\mathbf{v}_\alpha]\!]\,\mathcal{M}_\alpha - [\![\left(\bar{\mathbf{T}}_\alpha\mathbf{v}_\alpha - \bar{\mathbf{q}}_\alpha\right)\cdot\mathbf{n}_\sigma]\!] = 0\,, \quad (3.80)$$

$$\text{entropy}: \quad \bar{\gamma}_\alpha^{\rho\eta} = \partial\left(\bar{\rho}_\alpha\eta_\alpha\right) - \bar{\pi}_\alpha^{\rho\eta} - \nabla\cdot\left(\bar{\boldsymbol{\phi}}_\alpha^{\rho\eta} - \bar{\rho}_\alpha\eta_\alpha\mathbf{v}_\alpha\right) - \bar{\sigma}_\alpha^{\rho\eta}, \quad (3.81)$$

$$[\![\bar{\rho}_\alpha\eta_\alpha\left(\mathbf{v}_\alpha - \mathbf{s}\right) - \bar{\boldsymbol{\phi}}_\alpha^{\rho\eta}]\!]\cdot\mathbf{n}_\sigma = 0\,, \quad (3.82)$$

where the jump condition for the constituent angular momentum is redundant because it is already satisfied by the jump condition of linear momentum.

- Given sum relations (3.50), the general balance law for the mixture as a whole takes the forms

$$\partial\boldsymbol{\psi} = \boldsymbol{\pi}^\psi + \nabla\cdot\left(\boldsymbol{\phi}^\psi - \boldsymbol{\psi}\otimes\mathbf{v}\right) + \boldsymbol{\sigma}^\psi\,, \quad (3.83)$$

and

$$[\![\boldsymbol{\psi}\left(\mathbf{v} - \mathbf{s}\right) - \bar{\phi}^\psi]\!]\mathbf{n}_\sigma = \mathbf{0} \quad (3.84)$$

at points where fields are differentiable and on singular surfaces, respectively. These laws hold for all fields stated in Table 3.2, provided that the sum of all constituent production rate densities except that for the entropy vanish identically. This requirement is the expression that the body as a whole does not recognise that it is composed of constituents, and it automatically generates conservation laws.

- If mass density weighted variables are introduced according to (3.51) then mixture fields are related to the constituent fields as given in (3.54), (3.55), (3.64) and (3.66)-(3.69), respectively. This defines mixture fields in terms of *density weighted constituent quantities*. We shall henceforth restrict considerations to this case, but mention that volume weighted mixture models are also fashionable, see CHEN & TAI [27].

Chapter 4
Constitutive Assumptions

Abstract This chapter explains the complexity of the mixture theory that will be used in the remainder of the book. We focus attention on mixtures whose constituents exhibit a single temperature only and whose internal processes do not involve exchanges of interaction energies among the constituents. So, balance laws for the constituent masses and momenta are needed, but it suffices to deal only with the conservation law of energy for the mixture as a whole. This excludes changes of aggregation between constituents but still allows mass changes due to (restricted) chemical reactions, fragmentation and abrasion. We also assume that the internal exchanges of angular momenta are due to the moments of the exchanges of linear momenta only. This leads to symmetric partial stress tensors.

The constitutive postulates are motivated by taking single material bodies as a basis. We argue that, owing to the functional dependence of the partial densities on the true densities and volume fractions, the latter must be given by their own evolution equations, and that both true densities and volume fractions and their gradients ought to enter as independent constitutive variables. Mechanically, the constituent velocities, deformation gradients (via the left CAUCHY-GREEN deformation tensors for isotropy), stretching and vorticity tensors are considered important for the description of the deformation and motion of the fluid and solid constituents as are thermodynamically the empirical temperature, its time rate of change and spatial gradient.

Classical hypo-plasticity is introduced as an evolution equation for the CAUCHY stress tensor, by equating an objective time derivative of $\mathbf{T}$ to a symmetric tensor-valued production term, which is an additive decomposition of linear and non-linear terms, which, in turn, are functions of the stress itself and the stretching. Since such an evolution equation is difficult to handle in a thermodynamic analysis, we propose to introduce for the description of the frictional processes an objective evolution equation for a symmetric stress-like tensor variable that may also serve as an independent constitutive variable. In Chapter 8 it will be shown to yield hypo-plasticity.

Further restrictive assumptions are the constancy of the true constituent density for a 'density-preserving' component and the saturation condition, according to which all constituents fill the entire geometric space of the mixture.

L. Schneider, K. Hutter, *Solid-Fluid Mixtures of Frictional Materials in Geophysical and Geotechnical Context*, Advances in Geophysical and Environmental Mechanics and Mathematics, DOI 10.1007/978-3-642-02968-4_4,

4.1 Selection of Balance Equations

So far we have dealt with a general mixture of n constituents that does not account for an internal structure. We even allowed for energy interactions between the constituents, so that they could exhibit different temperatures.

We shall now restrict the ensuing analysis to the special case that all constituents have the same temperature and that no constituent species in the mixture suffers a change in its aggregation state. Thus, we also suppose that no phase changes in any mixture component take place. In the geophysical context, there are a great number of such situations. Among these mention might be made of earthquake and typhoon or hurricane induced landslides of dry or water saturated soil movements, mud flows, particle laden transport in slurries, sediment transport in fluvial hydraulics and many others. However, pyroclastic flows such as lahars and debris flows from a volcanic eruption and an avalanche of hot pyroclastic material mixed with ice from a glacier may not fit into this category. This is so, because in pyroclastic turbulent gravity driven boundary layer flows, the air that is entrained and the eroded soil material both have different temperatures that must be adjusted to the temperature of the erupted volcanic material. Similarly, the ice from a mountain glacier that is overrun by the hot lahars will be melted and evaporated. Our limitation to the equal temperature case is a matter of reduction of theoretical complexity. Formally, the limiting assumption reads:

> [**A2**] The ensuing analysis restricts attention to mixtures of which all constituents have the same common temperature and no constituent changes its aggregation state.

This assumption is equivalent to the disregard of energy-interaction between the constituents[1]. Therefore, only the mixture reduced (internal) energy balance

$$0 = \partial\left(\rho\varepsilon\right) + \nabla\cdot\left(\mathbf{q} + \rho\varepsilon\mathbf{v}\right) - \mathbf{T}\cdot\left(\nabla\mathbf{v}\right) - r\ , \tag{4.1}$$

which is obtained from (3.47) and Table 3.2, must be considered. The restriction [**A2**] still leaves enough room for processes in which mass exchanges between the constituents can take place. Chemical reactions at a common temperature are possible and for solid constituents fragmentation or abrasion may qualify as such processes when particle size separation is significant. It is important that such processes are not associated with energy transfer from one constituent to another.

---

[1] More precisely we ignore volumetric phase changes that are associated with a 'Clausius-Clapeyron type equation'.

The present mixture model accounts for the constituent balance equations of mass, momentum and moment of momentum, i. e.

$$\bar{\rho}_\alpha c_\alpha = \partial \bar{\rho}_\alpha + \nabla \cdot (\bar{\rho}_\alpha \mathbf{v}_\alpha) \ , \tag{4.2}$$

$$\bar{\mathbf{m}}_\alpha = \partial (\bar{\rho}_\alpha \mathbf{v}_\alpha) - \nabla \cdot \left(\bar{\mathbf{T}}_\alpha - \bar{\rho}_\alpha \mathbf{v}_\alpha \otimes \mathbf{v}_\alpha\right) - \bar{\mathbf{b}}_\alpha \ , \tag{4.3}$$

$$\bar{\mathbf{M}}_\alpha = \mathbf{x} \times \bar{\mathbf{m}}_\alpha + 2\,\mathrm{skw}\left(\bar{\mathbf{T}}_\alpha\right) \ , \tag{4.4}$$

with $(\alpha = 1, \ldots, n)$. If we would e. g. consider the mixture momentum balance equations instead of those for the constituents, the constituent velocities could not be obtained independently, but only via $\mathbf{v}$. This shall not be the case in the approach pursued in this model.

To reduce the constituent momentum balance equations to the form

$$\begin{aligned} \bar{\mathbf{m}}^i_\alpha &= \bar{\rho}_\alpha \left(\partial \mathbf{v}_\alpha + (\nabla \mathbf{v}_\alpha)\,\mathbf{v}_\alpha\right) - \nabla \cdot \bar{\mathbf{T}}_\alpha - \bar{\mathbf{b}}_\alpha \\ &= \bar{\rho}_\alpha \, \acute{\mathbf{v}}_\alpha - \nabla \cdot \bar{\mathbf{T}}_\alpha - \bar{\mathbf{b}}_\alpha \ , \end{aligned} \tag{4.5}$$

we substitute (4.2) into (4.3) and use the definition of a new constituent momentum interaction supply rate density[2]

$$\mathbf{m}^i_\alpha := \mathbf{m}_\alpha - \rho_\alpha c_\alpha \mathbf{v}_\alpha \ . \tag{4.6}$$

The specialization of the sum relation (3.49) to the supply rate densities $\rho_\alpha c_\alpha$, $\mathbf{m}_\alpha$ and $\mathbf{M}_\alpha$ yields

$$\sum \bar{\rho}_\alpha c_\alpha = 0 \ , \quad \sum \bar{\mathbf{m}}_\alpha = \mathbf{0} \ , \quad \sum \bar{\mathbf{M}}_\alpha = \mathbf{0} \ . \tag{4.7}$$

From the algebraic equation (4.4) it does not in general follow that $\bar{\mathbf{T}}_\alpha$ is symmetric. This is only the case when

[A3] $$\bar{\mathbf{M}}_\alpha = \mathbf{x} \times \bar{\mathbf{m}}_\alpha$$

is prescribed. Assumption [**A3**] corresponds to the supposition that exchange of angular momentum is only due to exchange of linear momentum. This is a rather strong assumption, because it neglects the exchange of angular momentum that happens through rotation of the grains. This assumption is neither in agreement with the consideration of the constituent vorticity tensors as independent variables, because these tensors are also linked to the rotation of the grains (see Section 4.5). Unfortunately, there is no alternative

[2] We mention that c_α and $\mathbf{m}^i_\alpha$ (and not $\mathbf{m}_\alpha$) are objective scalar and vector valued variables.

to [**A3**] if we want $\bar{\mathbf{T}}_\alpha$ to be symmetric. The mixture stress tensor $\mathbf{T}$ is, however, automatically of symmetric form as it is assumed that the 'single' body balance equations are applicable to the mixture as a whole, and these are a priori assumed to be non-polar.

In view of relations (4.1) to (4.3) it is observed that there are only $(4n+1)$ equations to determine the $(14n+5)$ unknowns $(\bar{\rho}_1,\dots,\bar{\rho}_n,\ \mathbf{v}_1,\dots,\mathbf{v}_n,\ \bar{\mathbf{T}}_1,\dots,\bar{\mathbf{T}}_n,\ \bar{c}_1,\dots,\bar{c}_n,\ \bar{\mathbf{m}}_1,\dots,\bar{\mathbf{m}}_n,\ \mathbf{q},\ \varepsilon,\ \theta)$. Thus, the system of the balance laws has yet to be complemented by additional relations, first, to make the resulting system at least in principle solvable and second, to model the mixture behaviour. This system of balance laws and the additional relations are then called the field equations.

We aim to solve these equations for the *independent variables*

$$(\bar{\rho}_1,\dots,\bar{\rho}_n,\ \mathbf{v}_1,\dots,\mathbf{v}_n,\ \theta)\ , \tag{4.8}$$

where θ represents an *empirical temperature* which, later, will be connected to the *absolute temperature*. Any solution of the field equations is referred to as a *thermo-mechanical process*. Consequently, phenomenological relations will have to be found for the so-called *constitutive quantities* $\left(\bar{\mathbf{T}}_1,\dots,\bar{\mathbf{T}}_n,\ \bar{c}_1,\dots,\bar{c}_n,\ \bar{\mathbf{m}}_1,\dots,\bar{\mathbf{m}}_n,\ \mathbf{q},\ \varepsilon\right)$ to link them to the independent variables or derivatives of them. These variables, i. e. the independent variables and certain derivatives of them are denoted *constitutive variables*. In the present chapter we aim to find an appropriate set of constitutive variables to correctly model debris flows.

4.2 Constitutive Laws of Single-Material Bodies

We introduced TRUEDELL's suppositions (see Section 3.1) in order to employ the usual principles and 'tools' known from constitutive modelling of bodies composed of only one material. For the constitutive theory of these, the most general law reads

$$\mathsf{C}(X,t) = \underset{\substack{(Y,\ X)\in\mathcal{B} \\ 0\leqslant\ s\ <\infty}}{\hat{\mathsf{C}}}\Big(\rho(Y,t-s)\,,\ \boldsymbol{\chi}(Y,t-s)\,,\ \theta(Y,t-s)\,;\ X\Big)\ , \tag{4.9}$$

where C is a constitutive quantity.[3] The explicit dependence on the material point X expresses the inhomogeneity of the material. In (4.9) the *functional* $\hat{\mathsf{C}}$ is assumed to depend also on the mass density, ρ, the motion, $\boldsymbol{\chi}$, and the temperature, θ, which themselves are dependent on all elements Y of the

[3] It is known in continuum mechanics of single constituent bodies that $\rho(Y,t-s)$ does not have to be included as a variable if $\boldsymbol{\chi}(Y,t-s)$ is included. However, we keep both here because in mixture applications an interrelation between $\rho(\cdot)$ and $\boldsymbol{\chi}(\cdot)$ is not evident, and does not exist in general.

material body $\mathcal{B}$ (non-local effects) and its total thermo-mechanical history (hereditary effects). The latter contribution is formulated with the help of the scalar s that is subtracted from the time t.

Obviously, (4.9) is far too general to gain any solution from substituting it into the appropriate balance equations. Therefore, the following assumptions and general rules are often applied to the general constitutive equation (4.9):

1. The assumption of *homogeneity* leads to the loss of an explicit dependency on X.
2. The assumption of *local action* requires that the value of a certain constitutive quantity at a point $\mathbf{X}$ depends only on the behaviour of the independent variables in an immediate neighbourhood of the considered point.
3. The assumption of *fading memory* demands that long past events have weaker influences on the behaviour of a material than recent ones. (cf. TRUESDELL [121])
4. The principle of *material symmetry* allows for a determination of the structure of the constitutive equations and for the reduction of material parameters arising in these equations. A material is symmetric with respect to a rotation or a reflection if the same response to a thermo-mechanical process of a material is obtained from two reference configurations that differ by just the above rotation or reflection.
5. The principle of *material objectivity* states that constitutive equations must be frame indifferent under a change of observer frame. In other words, the constitutive equations are not allowed to change their form *when the observer is rigidly rotated or translated relative to a singled-out observer*, i. e. constitutive equations are unaffected by an observer transformation of the form[4]
$$\mathbf{x}^* = \mathbf{Q}\,(t)\,\mathbf{x} + \mathbf{c}\,(t)\,, \qquad \mathbf{Q}\mathbf{Q}^{\mathrm{T}} = \mathbf{I}\,. \tag{4.11}$$
6. The *entropy principle* embodies the main 'tool' to restrict the constitutive laws. The discussion of this important physical principle is postponed to Chapter 5.

In the realm of Taylor series expansions the second and third suppositions require a finite length of the series which are written down for the independent variables $(\rho, \chi, \theta)(\mathbf{x},\ t)$ in the arguments of (4.9). If only the first two terms of the Taylor expansions are kept for χ and θ and only the first for ρ, the material is called *simple*. For this case (4.9) reduces to

---

[4] One consequence of this principle is that scalar-, vector- or tensor-valued quantities $(a, \mathbf{a}, \mathbf{A})$ that are not objective, i. e. that do not satisfy

$$a^* = a, \qquad \mathbf{a}^* = \mathbf{Q}\mathbf{a}, \qquad \mathbf{A}^* = \mathbf{Q}\mathbf{A}\mathbf{Q}^{\mathrm{T}} \tag{4.10}$$

for an arbitrary time dependent orthogonal tensor $\mathbf{Q}$ are not allowed to appear in the constitutive laws. The starred quantities describe the quantities in the rotated system, those without a star in the unrotated one.

$$\mathsf{C} = \hat{\mathsf{C}}\,(\theta,\ \nabla\theta,\ \rho,\ \mathbf{B})\ . \tag{4.12}$$

In this relation, a dependence on the left CAUCHY-GREEN deformation tensor $\mathbf{B}$ instead of on the deformation gradient $\mathbf{F}$, see (3.10), excludes anisotropic behaviour, and the resulting simpler relations meet our needs.

For a detailed discussion of these issues the reader is referred e. g. to TRUESDELL & NOLL [121], MÜLLER [97], HUTTER & JÖHNK [62], GREVE [46] or HAUPT [51]. The application of the stated rules can be found there and thus shall not be discussed in detail here. However, the example

$$\mathsf{C} = \hat{\mathsf{C}}\left(\theta,\ \dot{\theta},\ \nabla\theta,\ \rho,\ \nabla\rho,\ \mathbf{v},\ \mathbf{B},\ \mathbf{D},\ \mathbf{W}\right)\ , \tag{4.13}$$

where only assumptions 1 to 3 and the material isotropy are used, will later become important. It accounts not only for a dependence on the temperature but also on its gradient and its material time derivative. As is shown by MÜLLER in 1971 [95], $\dot{\theta}$ is necessary for the linearised governing equation of θ to be hyperbolic. Even though this issue becomes significant for time scales much smaller than those we are looking at in debris flows, we will nevertheless account for it in the thermodynamic analysis. Furthermore, in (4.13) we consider a constitutive dependence on the left CAUCHY-GREEN tensor $\mathbf{B}$ and the stretching tensor $\mathbf{D}$ which model elastic and viscous effects of the material, respectively. Due to objectivity reasons the contributions of the velocity, $\mathbf{v}$, and vorticity, $\mathbf{W}$, are excluded in single-material constitutive theories. Moreover, as in (4.12), simultaneous incorporation of ρ and $\mathbf{B}$ as independent variables in a single constituent body is not necessary since $\rho = \rho_0(\det\mathbf{B})^{-1/2}$, but $\nabla\rho$ would account for a second order deformation gradient, viz.,

$$\nabla\rho = -\frac{\rho_0}{2}(\det\mathbf{B})^{-1/2}\mathbf{B}^{-1}(\nabla\mathbf{B}) = -\frac{\rho}{2}\mathbf{B}^{-1}(\nabla\mathbf{B})\,, \tag{4.14}$$

which, in a single constituent non-polar theory, can be ruled out on grounds of the Second Law of Thermodynamics.[5] For mixtures, such dependences cannot be ruled out, however. We will postpone the application of the principle of objectivity to Section 4.6, where it is introduced in the context of mixture theory.

4.3 Constitutive Laws in the Context of Mixture Theory

Let us now assume that for a mixture at least the assumptions of homogeneity, local action, fading memory and isotropy are valid for each constituent

[5] One could argue that $\dot{\rho}$ should also be included as an independent variable in (4.13), but that would, via the mass balance, correspond to the inclusion of $\operatorname{tr}\mathbf{D}$, which is already contained in (4.13).

material body $\mathcal{B}_\alpha$ (cf. TRUESDELL [119]). Then the constitutive law[6]

$$\begin{aligned}
\mathsf{C} = \ & \hat{\mathsf{C}}\left(\theta,\ \dot{\theta},\ \nabla\theta,\ \vec{\rho},\ \vec{\nabla\rho},\ \vec{\mathbf{v}},\ \vec{\mathbf{B}},\ \vec{\mathbf{D}},\ \vec{\mathbf{W}}\right) ,\\
& \vec{\rho} := \rho_1,\ldots,\rho_n , \quad \vec{\nabla\rho} := \nabla\rho_1,\ldots,\nabla\rho_n ,\\
& \vec{\mathbf{v}} := \mathbf{v}_1,\ldots,\mathbf{v}_n , \quad \vec{\mathbf{B}} := \mathbf{B}_1,\ldots,\mathbf{B}_n ,\\
& \vec{\mathbf{D}} := \mathbf{D}_1,\ldots,\mathbf{D}_n , \quad \vec{\mathbf{W}} := \mathbf{W}_1,\ldots,\mathbf{W}_n
\end{aligned} \tag{4.15}$$

constitutes the formal translation of relation (4.13) to mixtures. As ascertained in [**A2**] all constituents exhibit the same temperature θ and also the same temperature derivatives $\nabla\theta$ and $\dot{\theta}$, where the dot marks the material time derivative with respect to the barycentric velocity, $\mathbf{v}$. In the above constitutive law we accounted for the compressibility of the constituents and their different velocities through the dependences on $\vec{\mathbf{v}}$ and $\vec{\rho}$. The incorporation of $\mathbf{B}_\alpha$ and $\mathbf{D}_\alpha$ $(\alpha = 1,\ldots,n)$ allows for the description of elastic and viscous effects in all constituents. We also consider the constituent vorticity tensors $\mathbf{W}_\alpha$ $(\alpha = 1,\ldots,n)$ as independent constitutive variables, because they are needed for the description of hypo-plasticity introduced later in this chapter. By introducing $\mathbf{D}_\alpha$ and $\mathbf{W}_\alpha$ as arguments of (4.15) instead of $\mathbf{L}_\alpha$ $(\alpha = 1,\ldots,n)$ we gain additional degrees of freedom, because the stretching and vorticity tensors can be incorporated independently into the constitutive laws. On the other hand, we have to admit that $\mathbf{L}_\alpha$ $(\alpha = 1,\ldots,n)$ are the original constitutive variables and by considering all $\mathbf{D}_\alpha$ and $\mathbf{W}_\alpha$ independently we are losing information about their relation.[7]

In (4.15) we also introduced both the true mass densities ρ_α and the left CAUCHY-GREEN tensors $\mathbf{B}_\alpha$ $(\alpha = 1,\ldots,n)$ as independent variables. In single-material theories only one of $(\rho,\ \mathbf{B})$ is considered, because owing to mass conservation, i. e.

$$\rho_0 = \rho\sqrt{\det \mathbf{B}} , \tag{4.16}$$

ρ and $\mathbf{B}$ are not independent. Here, ρ_0 is the (temporally) constant density of the material in the reference configuration. For mixtures the situation is slightly different. Relation (4.16) can not be translated one-to-one to the constituents of a mixture, because the mass balance for constituent K_α

$$\partial\bar{\rho}_\alpha + \nabla\cdot(\bar{\rho}_\alpha\mathbf{v}_\alpha) = \bar{\rho}_\alpha c_\alpha , \tag{4.17}$$

---

6 At this point a careful reader may object against the choice of $\dot{\theta}$ as independent constitutive variable, since 'a constitutive-variable time derivative should follow the constituent and not the 'mean' motion' (L.W. Morland, pers. comm.). However, θ is the temperature of all variables, and its evolution is described by the energy equation of the mixture, so that the time derivative following the barycentric velocity is justified as an independent constitutive variable.

7 This issue will become important for the 'isotropic' expansion performed in Section 7.3.

is no longer conservative. To see this (in)dependence more clearly, let us consider the global analogue to (4.17) which is given by

$$\frac{d}{dt}\int_{\omega_\alpha} \bar{\rho}_\alpha dv = \int_{\omega_\alpha} \bar{\rho}_\alpha c_\alpha dv \ , \tag{4.18}$$

or when 'pulling this back' to the reference configuration $\kappa_{\alpha 0}$ (see [14])

$$\frac{d}{dt}\int_{\Omega_{\alpha\,0}} \bar{\rho}_\alpha \left(\det\mathbf{F}_\alpha\right) dV = \int_{\Omega_{\alpha\,0}} \bar{\rho}_\alpha c_\alpha \left(\det\mathbf{F}_\alpha\right) dV \ , \tag{4.19}$$

which implies via localization

$$\left\{\left(\det\mathbf{F}_\alpha\right) \bar{\rho}_\alpha\right\}' = \left(\det\mathbf{F}_\alpha\right) \bar{\rho}_\alpha c_\alpha \ . \tag{4.20}$$

From this we deduce

$$\bar{\rho}_{\alpha 0} = \bar{\rho}_\alpha \det\mathbf{F}_\alpha \qquad \text{or} \qquad \bar{\rho}_{\alpha 0} = \bar{\rho}_\alpha \sqrt{\det\mathbf{B}_\alpha} \ , \quad \mathbf{B}_\alpha = \mathbf{F}_\alpha \mathbf{F}_\alpha^{\mathrm{T}} \tag{4.21}$$

only if we assume that no mass interaction supply rate densities are present, i. e. $c_\alpha = 0, \forall\, \alpha$. Here $\bar{\rho}_{\alpha 0}$ corresponds to ρ_0 in (4.16). If we do not want to make this assumption, we have to consider as independent constitutive variables both the true mass densities ρ_α and the left CAUCHY-GREEN tensors $\mathbf{B}_\alpha$ $(\alpha = 1, \ldots, n)$.

If we now take a closer look at (4.15), the question arises, why we introduced the true mass densities ρ_α to describe the compressibility, but not the partial mass densities $\bar{\rho}_\alpha$ $(\alpha = 1, \ldots, n)$. Also, it is not clear yet how the internal structure of debris flows, i. e. the distribution and evolution of the grain structure is modeled. To bring light into these issues we first discuss the constituent mass balance equation (4.2). Clearly, (4.2) determines the evolution of the constituent partial mass densities and not of the true mass densities. Thus, all $\bar{\rho}_\alpha$ can be regarded as *true independent variables*, and ρ_α $(\alpha = 1, \ldots, n)$ as nothing more than internal variables (cf. SVENDSEN & HUTTER [115]). If we recall (see relation (3.27))

$$\bar{\rho}_\alpha = \nu_\alpha \rho_\alpha \ , \tag{4.22}$$

then the roles of the partial mass densities and the true densities as true independent and internal variables, respectively, are no longer so clearly separated. Indeed, $\bar{\rho}_\alpha$ may change because the true mass density ρ_α changes, which is due to material compressibility or else because the volume filled by constituent K_α changes. So, both sub-processes, material compressibility and changes of the volume which constituent K_α acquired in the mixture, determines the partial mass density. It follows that both, ρ_α and ν_α ought to be regarded as independent variables. With this interpretation the volume fraction density ν_α also plays the role of describing in a minimum fashion the distribution of the constituent within the mixture, and thus is the most simple form by which the sub-RVE structure of the mixture can be characterised. As such its dynamics ought to be described by an evolution equation. SVENDSEN & HUTTER [115] assumed, in analogy with the constituent mass balances, the simple form

$$\text{[A4]} \qquad \bar{n}_\alpha = \partial\nu_\alpha + \nabla\cdot(\nu_\alpha \mathbf{v}_\alpha)\,, \qquad \alpha = 1,\ldots,n$$

for the evolution of all ν_α, where the constitutive quantity $\bar{n}_\alpha$ expresses the *volume (fraction) production rate density* of the corresponding constituent. In contrast to relation $(4.7)_1$, the sum of $\bar{n}_\alpha$ over all constituents is in general non-zero.[8]

If we now combine the constituent mass balances (4.2) and [**A4**] we obtain

$$\rho_\alpha\,(c_\alpha - n_\alpha) = \partial\rho_\alpha + (\nabla\rho_\alpha)\cdot\mathbf{v}_\alpha, \qquad \alpha = 1,\ldots,n \tag{4.23}$$

which can be interpreted as evolution relations for the constituent true mass densities $\rho_1,\ldots,\rho_n$. With [**A4**] and (4.2) or [**A4**] and (4.23) $2n$ evolution equations are available for the $3n$ variables $\bar{\rho}_\alpha$, ρ_α and ν_α ($\alpha = 1,\ldots,n$). Due to (4.22) we can choose $2n$ of the above variables to be members of the set of independent variables; [**A4**] then assures that any $2n$ of $\{\vec{\bar{\rho}},\, \vec{\rho},\, \vec{\nu}\}$ are true independent variables. Therefore, the constituent true mass densities ρ_α ($\alpha = 1,\ldots,n$) are permissible variables in (4.15), but as the above discussion also shows, one of the variables $\bar{\rho}_\alpha$ and ν_α should equally be an independent constitutive quantity. We choose ν_α ($\alpha = 1,\ldots,n$). The extended general constitutive law therefore reads

$$\begin{aligned} \mathsf{C} &= \hat{\mathsf{C}}\left(\theta,\ \dot{\theta},\ \nabla\theta,\ \vec{\rho},\ \vec{\nabla\rho},\ \vec{\nu},\ \vec{\nabla\nu},\ \vec{\mathbf{v}},\ \vec{\mathbf{B}},\ \vec{\mathbf{D}},\ \vec{\mathbf{W}}\right), \\ \vec{\nu} &= \nu_1,\ldots,\nu_n\ , \quad \vec{\nabla\nu} = \nabla\nu_1,\ldots,\nabla\nu_n\ , \end{aligned} \tag{4.24}$$

where, as a consequence of the assumption of local action, $\vec{\nabla\nu}$ was added.

In the constitutive law (4.24) the constituent volume fraction densities were introduced as independent constitutive quantities and associated evolution equations, [**A4**], were postulated to account for the sub-macroscopic, i. e. sub-RVE structure that is not distinguished by the macroscopic mixture quantities. We also alluded to the fact that the volume fraction and volume fraction gradient dependencies represent one, perhaps, minimal form to account for the description of this subscale structural behaviour. It is expected with the postulate of such new continuous, so-called internal field variables,

---

[8] This statement can be shown by summing [**A4**] over all constituents and using the saturation condition (see [**A7**] in Section 4.5).

that the observed subtleties, which cannot be described by the macroscopic mixture quantities, can be described when such internal variables are included. In the description of the deformation of soils and the motion of dry and wet debris, effects, such as particle size segregation, fluidisation, fragmentation and abrasion, are of sub-macroscopic scale. We chose above the volume fractions and their first gradients as the decisive variables. In more complex formulations, second gradients $\nabla^2 \nu_\alpha$ may enter as measures of higher order variability.

In the material sciences of granular bodies, there are also proposals with fewer or more variables and evolution equations different from [**A4**] to describe such phenomena. WILMAŃSKI [124] uses the porosity (that is the volume fraction of all fluid constituents together) as such a variable and changes the evolution equation in [**A4**] by adding on the right-hand side a flux term. Geotechnical engineers prefer to work with the void ratio, which is simply related to the porosity via

$$e_f = \frac{\nu_f}{1 - \nu_f}, \tag{4.25}$$

where f stands for the union of all fluid constituents. If the details of the flow of the fluids through the pore space are important, then the 'curvature tensor' of the volume fractions $\nabla^2 \nu_\alpha$ may have to be added as an additional independent variable that could enter the constitutive relation (4.24). However, one may also introduce additional internal variables, not related to the volume fractions. The tortuosity of the pore space may be such a variable. Irrespective of this choice, it is necessary to postulate new evolution equations for these internal variables, because their evolution is not described by the classical balance equations of mixture theory.

From this point of view we expect or hope that the contributions ν_α and $\nabla \nu_\alpha$ $(\alpha = 1, \ldots, n)$ in the constitutive law (4.24) and the balance equations for the volume fractions, [**A4**], have the potential of describing fluidisation and particle size segregation in debris flows. The hope, that with the help of the quantities $\nabla \nu_\alpha$ $(\alpha = 1, \ldots, n)$ we are able to characterise the resistance of geophysical materials (e. g. soil) to shear stress in equilibrium ('heap problem') is tenuous, because, first, shear stresses would be absent if the volume fractions were uniformly distributed, and second, in simulations the variables $\nabla \nu_\alpha$ $(\alpha = 1, \ldots, n)$ pose the difficult task of finding boundary conditions for the volume fractions (see FANG [39] and FANG, WANG & HUTTER [41, 42]). Although $\nabla \nu_1, \ldots, \nabla \nu_n$ are included in the set of constitutive variables, in the derivation of the reduced model we aim to avoid them in favour of a frictional contribution which is presented in the following section.

Before we turn our attention to that, let us address the choice of dependence of (4.24) on the stretching tensor $\mathbf{D}$. Most dense granular or particle laden flows in the geophysical context use constitutive relations which fit into subclasses of (4.24). There are, however, also stress-deformation relations in use which do not fit into the class (4.24). As an example, one may extend the dependence on $\mathbf{D}$ by higher order RIVLIN-ERICKSEN tensors

$$\mathbf{A}^{(r+1)} = \dot{\mathbf{A}}^{(r)} + \mathbf{L}\mathbf{A}^{(r)} + \mathbf{A}^{(r)}\mathbf{L}^{\mathrm{T}} \,, \qquad \mathbf{A}^{(1)} = 2\mathbf{D} \tag{4.26}$$

which are objective time derivatives of $\mathbf{D}$. Such a proposal has been made by NOREM et al. [100] who use an extended version of the CRIMINALE-FILBEY-ERICKSEN fluid as a basis for their avalanche model. This model employs the first and the second RIVLIN-ERICKSEN tensors. We shall not dwell upon such added complexity; the analysis is complex enough for constitutive models of the class (4.24).

4.4 A First Attempt to Incorporate Hypo-plasticity

In this subsection it will be explained how hypo-plasticity is theoretically treated in the above theoretical formulations. The treatment is different from how frictional effects are incorporated in the mixture theory that follows. We present it here primarily to familiarize the reader with the subject.

The elasto-visco-plastic behaviour of soils is nowadays described by two classes of frictional material behaviour. In one class, the transition from small elastic and reversible behaviour to rapid deformations is accomplished by an abrupt change of the deformation into a regime of flow. This transition takes place when the state of stress reaches a yield condition. In stress space this is characterised by a (closed) yield surface and the abrupt change of flow at yield is expressed by a second rule, the so-called flow rule, which describes the direction of flow in stress space. In the second class of material behaviour the existence of a sharp transition of the flow state at yield is rejected. Rather, the transition from slow creeping behaviour to rapid failure is still smooth, but accompanied by relatively steep gradients. We are here concerned with this second class of material response. For soils, the theory that describes these effects of localisations and apparent instabilities is of a mathematical structure which does not employ the concepts of a yield surface and a flow rule. The theory is of a mathematical structure described below and was founded by KOLYMBAS in 1978 [74] and coined by him hypo-plasticity. In the subsequent years, many improved and extended constitutive models based on the theory of hypo-plasticity (cf. KOLYMBAS [75]) have been developed for granular materials such as sand and gravel. VON WOLFFERSDORFF [126] summarises these models as of 1996. Later, the hypo-plastic theories were also applied and extended to fine grained soils and clays.[9] The mathematical structure of hypo-plasticity is expressed by the constitutive relation for the CAUCHY stress tensor. It has the form of an objective evolution equation of the stress tensor and reads

---

[9] Important steps of the development were done by KOLYMBAS [75], GUDEHUS [48], NIEMUNIS [99], HERLE & GUDEHUS [54], BAUER [8], WU [127], MASIN [84] and others. A detailed reference list is given in the Ph. D. dissertation of MASIN [84].

$$\dot{\mathbf{T}} - [\mathbf{W}, \mathbf{T}] = \boldsymbol{\Phi}, \tag{4.27}$$

where $\mathbf{W}$ represents the vorticity tensor, $[\cdot\,,\,\cdot]$ is the LIE bracket, see (2.10), and the left-hand side of (4.27) is the objective JAUMANN time derivative of $\mathbf{T}$. On the right-hand side, $\boldsymbol{\Phi}$ is regarded as a constitutive quantity that measures the time rate of change of $\mathbf{T}$.

A *hypo-plastic* material is commonly defined by (i), assuming the second rank tensor $\boldsymbol{\Phi}$ to be a function of $\mathbf{T}$ and $\mathbf{D}$, plus possibly other variables i. e.

$$\boldsymbol{\Phi} = \hat{\boldsymbol{\Phi}}(\mathbf{T},\ \mathbf{D},\ \cdot)\ , \tag{4.28}$$

and (ii), supposing that $\boldsymbol{\Phi}$ is continuously differentiable for all $\mathbf{D}$, except at $\mathbf{D} = \mathbf{0}$ (cf. e. g. WU & KOLYMBAS [129]). The point in (4.28) indicates other dependencies that have been suggested in the course of development of the theory of hypo-plasticity. Such a dependence is e. g. that on void ratio. These dependencies are here structurally not important, but will be incorporated later on. To further confine the form of $\boldsymbol{\Phi}$ the following restrictions are made. First, (4.27) is required to be rate independent. So, in a process accelerated by λ, $\dot{\mathbf{T}}$ and $\mathbf{W}$ change into $\lambda\dot{\mathbf{T}}$ and $\lambda\mathbf{W}$, while $\mathbf{T}$ remains unchanged. So, $\boldsymbol{\Phi}$ must be positively homogeneous of the first order in $\mathbf{D}$. In other words,

$$\boldsymbol{\Phi}(\mathbf{T},\ \lambda\mathbf{D}, \cdot) = \lambda\boldsymbol{\Phi}(\mathbf{T},\ \mathbf{D}, \cdot) \tag{4.29}$$

must be satisfied for all positive scalars λ. Second, $\boldsymbol{\Phi}$ is thought to be also positively homogeneous in $\mathbf{T}$, i. e.

$$\boldsymbol{\Phi}(\lambda\mathbf{T},\ \mathbf{D}, \cdot) = \lambda\boldsymbol{\Phi}(\mathbf{T},\ \mathbf{D}, \cdot) \tag{4.30}$$

for arbitrary positive scalars λ which also follows from (4.27). In a next step towards a concrete formulation of the constitutive equation WU & KOLYMBAS [129] decompose $\boldsymbol{\Phi}$ into two parts, i. e.

$$\boldsymbol{\Phi} = \boldsymbol{\mathcal{L}}(\mathbf{T},\ \mathbf{D}, \cdot) + \boldsymbol{\mathcal{N}}(\mathbf{T},\ \mathbf{D}, \cdot)\ , \tag{4.31}$$

where the first operator, $\boldsymbol{\mathcal{L}}$, is considered to be linear in $\mathbf{D}$ and is of class $C^1 \in [0,\ \infty)$. The second operator, $\boldsymbol{\mathcal{N}}$, accounts for the frictional behaviour of the material, and for hypo-plasticity, ought to be non-linear in $\mathbf{D}$. It is of class $C^1 \in (0,\ \infty)$ but not differentiable at $\mathbf{D} = \mathbf{0}$. The specification of this non-linearity is not arbitrary but has to satisfy the requirement of rate-independence. The choice

$$\boldsymbol{\Phi} = \mathrm{f}_1(\cdot)\big(\mathsf{L}(\mathbf{T})\mathbf{D} + \mathrm{f}_2(\cdot)\mathbf{N}(\mathbf{T})\,|\mathbf{D}|\,\big) \tag{4.32}$$

is characteristic for hypo-plasticity. It is in agreement with the above requirements and uses the norm

$$|\mathbf{D}| := \sqrt{\mathrm{tr}(\mathbf{D}^2)}\ . \tag{4.33}$$

We have also incorporated in the representation (4.32) two scalar functions, $f_1(\cdot)$ and $f_2(\cdot)$, which indicate additional dependencies on scalar valued variables, e. g. the void ratio in the most simple case. $f_1(\cdot)$ is said in the hypoplasticity community to describe *barotropy*[10] and $f_2(\cdot)$ the *pyknotropy* of the material. These notions are introduced here, to familiarize the reader with the jargon that is employed.

The equivalent form of (4.32),

$$\boldsymbol{\Phi} = f_1(\cdot)\Big\{\mathsf{L} + f_2(\cdot)\mathbf{N} \otimes \frac{\mathbf{D}}{|\mathbf{D}|}\Big\}\mathbf{D} \tag{4.34}$$

is convenient, because in the constitutive law for the CAUCHY stress tensor (to be derived), the derivative $\boldsymbol{\Phi}_{,\mathbf{D}}$ has to be performed. We also observe from (4.32) that $\boldsymbol{\Phi}_{,\mathbf{D}}$ is singular for $\mathbf{D} = \mathbf{0}$ and thus, for the 'simplest' case, namely stationary flow, the stress can not be determined. Regularisations are called for.

The main task of this theory is to find representations for L, $\mathbf{N}$ and $f_1(\cdot)$, $f_2(\cdot)$ that allow for the correct description of the behaviour of the respective material. The existing formulations are derived through simulations and trial and error techniques. In this process, parameter identifications are performed with results taken from carefully conducted triaxial experiments. In addition, (4.32) also suffers from the shortcoming that it is not clear how the exploitation of the entropy principle in the version of MÜLLER & LIU can be carried out (cf. TEUFEL [117]). This problem will be addressed in Chap. 8.

Trying to circumvent the latter drawback SVENDSEN et al.[11] [116] formulated a hypo-plastic theory for a single-material body that does not prescribe an evolution law for the CAUCHY stress tensor but for an objective, 'stress-like', symmetric-tensor-valued, spatial internal variable, $\mathbf{Z}$. This law has the form

$$\mathring{\mathbf{Z}} := \dot{\mathbf{Z}} - [\boldsymbol{\Omega}, \mathbf{Z}] = \tilde{\boldsymbol{\Phi}}, \qquad \boldsymbol{\Omega} = -\boldsymbol{\Omega}^{\mathrm{T}}\,, \tag{4.35}$$

where $\mathring{\mathbf{Z}}$ represents a general objective time derivative of $\mathbf{Z}$, $\boldsymbol{\Omega}$ being a corresponding spin tensor. We observe that due to the skew-symmetry of $\boldsymbol{\Omega}$, $\mathring{\mathbf{Z}}$ remains symmetric and e. g. for $\boldsymbol{\Omega} = \mathbf{W}$, the JAUMANN time derivative for $\mathbf{Z}$ is recovered. Again, $\tilde{\boldsymbol{\Phi}}$ is a constitutive quantity but should not be confused with the constitutive quantity $\boldsymbol{\Phi}$ in (4.27).

SVENDSEN et al. assumed further that the collective behaviour of the class of materials under consideration, and in particular that of granular materials, is elasto-visco-plastic. Thus, they incorporated the internal variable, $\mathbf{Z}$, into the set of constitutive variables. For an interpretation of $\mathbf{Z}$ the reader is referred to KIRCHNER [71],[72] and TEUFEL [117].

Motivated by SVENDSEN's idea we now postulate as follows:

[10] It has nothing to do with the notion of 'barotropy' in fluid mechanics.

[11] The paper is by SVENDSEN et al., but the idea of introducing (4.35) is SVENDSEN's alone.

[A5] We introduce *constituent, partial, objective, 'stress-like', symmetric-tensor-valued internal variables,* $\bar{\mathbf{Z}}_\alpha(\mathbf{x}, t)$ $(\alpha = 1, \ldots, n)$, into the set of constitutive variables. Their objective time rates, $\overset{\circ}{\bar{\mathbf{Z}}}_\alpha$ which include the material time derivatives $\acute{\bar{\mathbf{Z}}}_\alpha$ and corresponding constituent spin tensors, $\boldsymbol{\Omega}_\alpha$, are balanced by the symmetric, tensor-valued constitutive quantities $\bar{\boldsymbol{\Phi}}_\alpha$ $(\alpha = 1, \ldots, n)$, i. e.

$$\overset{\circ}{\bar{\mathbf{Z}}}_\alpha := \acute{\bar{\mathbf{Z}}}_\alpha - \left[\boldsymbol{\Omega}_\alpha, \bar{\mathbf{Z}}_\alpha\right] = \bar{\boldsymbol{\Phi}}_\alpha, \quad \boldsymbol{\Omega}_\alpha = -\boldsymbol{\Omega}_\alpha^{\mathrm{T}}, \qquad \alpha = 1, \ldots, n \,. \tag{4.36}$$

In principle, $\boldsymbol{\Omega}_\alpha$ can be any corresponding spin tensor, e. g. the mixture $\mathbf{W}$, or constituent $\mathbf{W}_\alpha$, vorticity tensor, respectively, or the spin tensor $\acute{\mathbf{R}}_\alpha \mathbf{R}_\alpha^{\mathrm{T}}$ constructed out of the rotational part, $\mathbf{R}_\alpha$, of the deformation gradient, $\mathbf{F}_\alpha$. For the time-being we leave the spin tensor, $\boldsymbol{\Omega}_\alpha$, unspecified but will choose suitable representations later.

Being aware of the inadequacy of the introduction of internal variables $\bar{\mathbf{Z}}_\alpha$ for all constituents, i. e. also for fluids, we will still keep them formally in the thermodynamic analysis. In Chapter 8 which is dealing with reduced models, those $\bar{\mathbf{Z}}_\alpha$ that are related to fluids will be abandoned. In this way, only the granular part of debris flows is described by such internal variables.

Adding all $\bar{\mathbf{Z}}_\alpha$ to the constitutive law (4.24), it turns into

$$\begin{gathered}\mathsf{C} = \hat{\mathsf{C}}\left(\theta,\ \dot{\theta},\ \nabla\theta,\ \vec{\rho},\ \nabla\vec{\rho},\ \vec{\nu},\ \nabla\vec{\nu},\ \vec{\mathbf{v}},\ \vec{\mathbf{B}},\ \vec{\mathbf{D}},\ \vec{\mathbf{W}},\ \vec{\bar{\mathbf{Z}}}\right), \\ \vec{\bar{\mathbf{Z}}} = \bar{\mathbf{Z}}_1, \ldots, \bar{\mathbf{Z}}_n \,.\end{gathered} \tag{4.37}$$

So far, we have not established a connection of $\bar{\boldsymbol{\Phi}}_\alpha$ $(\alpha = 1, \ldots, n)$ with hypo-plasticity and we will not do so until we reduce the model further to the special case of debris flows. Indeed [**A5**] and (4.37) classify a frictional material far more generally than does hypo-plasticity.

4.5 Further Assumptions

The constitutive assumptions will now be complemented by constraints on constituent mass densities and volume fractions. In the present mixture theory it is assumed that some or all of the constituent true mass densities are constant, i. e.

[A6] $\rho_\alpha = \text{const.}\,, \qquad \alpha = m+1,\ldots,n\,, \quad (0 \leqslant m \leqslant n)\,.$

BLUHM [13] doubts the interpretation that **[A6]** is a consequence of the resistance of each of the $(n-m)$ constituents to volume changes. He argues that the condition $\det \mathbf{F}_\alpha = 1$ $(\alpha = m+1,\ldots,n)$ which is analogous to **[A6]** but not equivalent, forms a condition for a macroscopic field and thus can not be correct. From his point of view resistance to volume changes is not linked to macroscopic but to microscopic fields. Therefore, only the condition $\det \mathbf{F}^\alpha_{\text{micro}} = 1$ $(\alpha = m+1,\ldots,n)$ for a real non-compatible deformation gradient, $\mathbf{F}^\alpha_{\text{micro}}$, should be used as a constraint for density-preserving constituents. This remark is the starting point of BLUHM's mixture theory. This procedure requires homogenisation rules to be applied for the upscaling from the micro scale to the macro scale description, which we try to avoid.

Even though BLUHM's arguments exhibit a convincing element, we will, for simplicity, assume that **[A6]** allows the interpretation that it expresses the resistance of constituents to density changes, but does not express an obvious notion of volume preserving. The two concepts are not indentical in general for mixtures. This *constituent density constraint* reduces the number of independent variables by $(n-m)$. If we substitute this constraint into the mass balances of constituents K_{m+1} to K_n, what emerges is the reduced form

$$\bar{c}_\alpha = \partial \nu_\alpha + \nabla\cdot(\nu_\alpha \mathbf{v}_\alpha)\,, \qquad \alpha = m+1,\ldots,n\,. \tag{4.38}$$

The right-hand side of (4.38) is in agreement with that of **[A4]** and consequently

$$\bar{c}_\alpha = \bar{n}_\alpha\,, \qquad \text{or} \qquad c_\alpha = n_\alpha\,, \qquad \alpha = m+1,\ldots,n \tag{4.39}$$

is required for all density-preserving constituents.

The *saturation constraint* in this mixture theory reads

[A7] $1 = \sum \nu_\alpha$

which states that there is no (material free) void space within the mixture.[12] [**A7**] implies that there are only $(n-1)$ independent volume fractions and thus only, say, $\nu_1, \dots, \nu_{n-1}$, arise as independent variables in the constitutive laws. On the other hand, we still have n independent volume fraction balance equations, but of course in the nth balance equation of [**A4**], ν_n is replaced according to [**A7**] by $\nu_n = 1 - \sum_{\alpha=1}^{n-1} \nu_\alpha$ (cf. BAUER [10]). Collecting the balance equations of mass and volume fraction that are left we obtain

$$\bar{\rho}_\alpha c_\alpha = \partial \bar{\rho}_\alpha + \nabla \cdot (\bar{\rho}_\alpha \mathbf{v}_\alpha), \qquad \alpha = 1, \dots, m, \tag{4.40}$$

$$\bar{n}_\alpha = \partial \nu_\alpha + \nabla \cdot (\nu_\alpha \mathbf{v}_\alpha), \qquad \alpha = 1, \dots, n-1, \tag{4.41}$$

and

$$\bar{n}_\alpha = -\partial \Big(\sum_{\beta=1}^{n-1} \nu_\beta \Big) + \nabla \cdot \mathbf{v}_\alpha - \nabla \cdot \Big(\sum_{\beta=1}^{n-1} \nu_\beta \mathbf{v}_\alpha \Big), \qquad \alpha = n\,, \tag{4.42}$$

for the independent variables ρ_1, ..., ρ_m and ν_1, ..., ν_{n-1}. Therefore, we have $(n+m)$ balance laws for $(m+n-1)$ unknowns. Furthermore, (4.39) is valid for the density-preserving constituents K_α $(\alpha = m+1, \dots, n)$.

Following the arguments of SVENDSEN & HUTTER [115] we state that the constituent density constraint, [**A6**], reduces the number of evolution equations and independent variables equally. [**A7**], on the other hand, has no impact on the number of evolution equations but lowers the number of independent variables by one. Therefore, [**A7**] can be understood as a 'true' constraint, inducing a new, unknown field, whereas for [**A6**] no such field is justified. This property is characteristic for the structure of balance equations as formulated by SVENDSEN & HUTTER. In the context of internal constraints, introduced by TRUESDELL & NOLL [121], the new constraint field is necessary to maintain the saturation [**A7**].

Summarising the rules in Section 4.2 and considering assumptions [**A2**] to [**A7**] we now choose the formulation of the final constitutive law and the definition of the set of independent constitutive variables, $\mathbb{S}$, that will be used in the thermodynamic analysis. They read

[12] This definition of the saturation condition requires clarification. In the mechanics literature, saturation always means that there is no empty space present in the mixture. That is the constituents are solids, liquids and gases. So, in a non-saturated or partly-saturated mixture, there is one 'constituent' present which is massless and to which no physical laws can be applied. It is simply empty space. In the geotechnical and geophysical literature 'saturated' means that all pore space is filled with liquid, whilst in partly or non-saturated soils, the empty void space is filled with a gas. In the mechanics community this case is also called a saturated mixture. In this work we adopt the definition of the mechanicists.

[**A8**]

$$\mathsf{C} = \hat{\mathsf{C}}\,(\mathbb{S}),$$

$$\mathbb{S} = \Big\{\theta,\ \dot{\theta},\ \nabla\theta,\ \vec{\rho},\ \nabla\vec{\rho},\ \vec{\nu},\ \nabla\vec{\nu},\ \vec{\mathbf{v}},\ \vec{\mathbf{B}},\ \vec{\mathbf{D}},\ \vec{\mathbf{W}},\ \vec{\bar{\mathbf{Z}}}\Big\},$$

$$\vec{\rho} := \rho_1,\ldots,\rho_m, \quad \vec{\nu} := \nu_1,\ldots,\nu_{n-1},$$
$$\vec{\mathbf{v}} := \mathbf{v}_1,\ldots,\mathbf{v}_n, \quad \vec{\mathbf{B}} := \mathbf{B}_1,\ldots,\mathbf{B}_n,$$
$$\vec{\mathbf{D}} := \mathbf{D}_1,\ldots,\mathbf{D}_n, \quad \vec{\mathbf{W}} := \mathbf{W}_1,\ldots,\mathbf{W}_n,$$
$$\vec{\bar{\mathbf{Z}}} := \bar{\mathbf{Z}}_1,\ldots,\bar{\mathbf{Z}}_n\,.$$

C denotes any dependent constitutive variable and $\hat{\mathsf{C}}\,(\mathbb{S})$ is the function describing it. That the independent set $\mathbb{S}$ is assumed to be the same for all dependent constitutive variables is the expression of the rule of equipresence, which states that all constitutive quantities initially share the same dependences. Possible variable reductions ought to be deduced by proof. In [**A8**] we already used the assumption that each constituent behaves as an isotropic material, i. e. an arbitrary rotation or a mirror reflection does not alter the material behaviour of a constituent (cf. HUTTER & JÖHNK [62]). Thus, the shape of the constitutive law, [**A8**], shall not be affected by these rotations or reflections.
By combining the material isotropy with the requirement that every constitutive law must be objective, it is observed that all constitutive quantities must be *isotropic functions* of their dependent variables, i. e.,

$$\begin{aligned}\mathbf{Q}_*\hat{\mathsf{C}} = \ \hat{\mathsf{C}}\Big(\theta,\ \dot{\theta},\ \mathbf{Q}(\nabla\theta),\ \vec{\rho},\ \mathbf{Q}(\nabla\vec{\rho}),\ \vec{\nu},\ \mathbf{Q}(\nabla\vec{\nu}),\ \mathbf{Q}(\vec{\mathbf{v}}),\\ \mathbf{Q}\vec{\mathbf{B}}\mathbf{Q}^{\mathrm{T}},\ \mathbf{Q}\vec{\mathbf{D}}\mathbf{Q}^{\mathrm{T}},\ \mathbf{Q}\vec{\mathbf{W}}\mathbf{Q}^{\mathrm{T}},\ \mathbf{Q}\vec{\bar{\mathbf{Z}}}\mathbf{Q}^{\mathrm{T}}\Big)\,, \qquad (4.43)\end{aligned}$$

for all orthogonal time-dependent tensors $\mathbf{Q}\,(t)$, where $\mathbf{Q}_*$ denotes the action

$$\left(\mathbf{Q}_*\hat{\mathsf{C}}\right)_{i_1 i_2 \ldots i_n} = \mathsf{Q}_{i_1 j_1}\,\mathsf{Q}_{i_2 j_2}\,\ldots\,\mathsf{Q}_{i_n j_n}\left(\hat{\mathsf{C}}\right)_{j_1 j_2 \ldots j_n} \qquad (4.44)$$

on the tensor valued quantity $\hat{\mathsf{C}}$.

4.6 Remarks on the Principle of Objectivity

In Section 4.2 we stated the principle of objectivity for a single-material body. In mixture theory the general constitutive law has to satisfy the equality

$$\hat{\mathsf{C}}(\,\cdots,\ \vec{\mathbf{v}},\ \cdots,\ \vec{\mathbf{W}},\ \cdots) = \hat{\mathsf{C}}(\,\cdots,\ \vec{\mathbf{v}} + \vec{\mathbf{a}},\ \cdots,\ \vec{\mathbf{W}} + \vec{\Omega}^*,\ \cdots)\,, \quad (4.45)$$

identically for all three-dimensional vectors $\vec{\mathbf{a}} = (\mathbf{a}, \cdots, \mathbf{a})$ and all skew-symmetric tensors $\vec{\boldsymbol{\Omega}}^* = (\boldsymbol{\Omega}^*, \cdots, \boldsymbol{\Omega}^*)$ in order to be in accordance with the principle of objectivity (cf. SVENDSEN & HUTTER [115]). The added terms $\vec{\mathbf{a}}$ and $\vec{\boldsymbol{\Omega}}^*$ in (4.45) are necessary to transform $\vec{\mathbf{v}}$ and $\vec{\mathbf{W}}$ into objective quantities. The remaining constitutive variables are already objective. As stated in SVENDSEN & HUTTER [115] a common choice for $\mathbf{a}$ is the negative mixture velocity $-\mathbf{v}$ and for $\boldsymbol{\Omega}^*$, the mixture vorticity tensor $\mathbf{W}$ may be chosen. For this choice and with the definition (3.56) of the diffusion velocities, $\mathbf{u}_\alpha$, and diffusion vorticities, $\mathbf{U}_\alpha$, defined by

$$\mathbf{u}_\alpha := \mathbf{v}_\alpha - \mathbf{v}, \qquad \mathbf{U}_\alpha := \mathbf{W}_\alpha - \mathbf{W} , \tag{4.46}$$

the principle of objectivity for mixtures can be formulated as

$$\hat{\mathsf{C}}(\ \cdots,\ \vec{\mathbf{v}},\ \cdots,\ \vec{\mathbf{W}},\ \cdots) = \hat{\mathsf{C}}(\ \cdots,\ \vec{\mathbf{u}},\ \cdots,\ \vec{\mathbf{U}},\ \cdots) . \tag{4.47}$$

If we take into account the identity[13]

$$\mathbf{u}_{\beta,\mathbf{v}_\alpha} \overset{(4.46)}{=} \mathbf{v}_{\beta,\mathbf{v}_\alpha} - \sum_{\gamma=1}^{n} \bar{\xi}_\gamma (\mathbf{v}_\gamma)_{,\mathbf{v}_\alpha} = \left(\delta_{\beta\alpha} - \bar{\xi}_\alpha\right) \mathbf{I} , \tag{4.48}$$

in which $\mathbf{I}$ is the (3×3) unit matrix and if we differentiate (4.47) with respect to $\mathbf{v}_\alpha$, then, the chain rule of differentiation yields

$$\mathsf{C}_{,\mathbf{v}_\alpha} = \sum_{\beta=1}^{n} \mathsf{C}_{,\mathbf{u}_\beta}\, \mathbf{u}_{\beta,\mathbf{v}_\alpha} \overset{(4.48)}{=} \mathsf{C}_{,\mathbf{u}_\alpha} - \bar{\xi}_\alpha \sum_{\beta=1}^{n} \mathsf{C}_{,\mathbf{u}_\beta} , \qquad \alpha = 1, \ldots, n . \tag{4.49}$$

Summing these relations over all constituents K_α we obtain the restriction

$$\sum \mathsf{C}_{,\mathbf{v}_\alpha} = \mathbf{0} . \tag{4.50}$$

In a similar fashion we derive

$$\sum \mathsf{C}_{,\mathbf{W}_\alpha} = \mathbf{0} , \tag{4.51}$$

which follows from

---

[13] Perhaps the derivation of (4.48) in Cartesian tensor notation is easier to follow:

$$(\mathbf{u}_{\beta,\mathbf{v}_\alpha})_{ij} = \frac{\partial(\mathbf{u}_\beta)_i}{\partial(\mathbf{v}_\alpha)_j} \overset{(4.46)}{=} \frac{\partial(\mathbf{v}_\beta)_i}{\partial(\mathbf{v}_\alpha)_j} - \sum_{\gamma=1}^{n} \bar{\xi}_\gamma \frac{\partial(\mathbf{v}_\gamma)_i}{\partial(\mathbf{v}_\alpha)_j}$$
$$= \delta_{\beta\alpha}\delta_{ij} - \sum_{\gamma=1}^{n} \bar{\xi}_\gamma \delta_{\gamma\alpha}\delta_{ij} = \left(\delta_{\beta\alpha} - \bar{\xi}_\alpha\right)\delta_{ij} .$$

This Cartesian notation may be used as an alternative for many derivations.

$$\mathsf{C}_{,\mathbf{W}_\alpha} = \mathsf{C}_{,\mathbf{U}_\alpha} - \bar{\xi}_\alpha \sum_{\beta=1}^{n} \mathsf{C}_{,\mathbf{U}_\beta}, \qquad \alpha = 1, \ldots, n \tag{4.52}$$

and

$$\begin{aligned} \mathbf{U}_{\beta,\mathbf{W}_\alpha} &\overset{(4.46)}{=} \mathbf{W}_{\beta,\mathbf{W}_\alpha} - \mathbf{W}_{,\mathbf{W}_\alpha} \\ &\overset{(3.62)}{=} \delta_{\beta\alpha}\boldsymbol{\mathcal{I}} - \sum_{\gamma=1}^{n} \bar{\xi}_\gamma (\mathbf{W}_\gamma)_{,\mathbf{W}_\alpha} = \left(\delta_{\beta\alpha} - \bar{\xi}_\alpha\right)\boldsymbol{\mathcal{I}}, \end{aligned} \tag{4.53}$$

where $\boldsymbol{\mathcal{I}}$ is the fourth order unit tensor.

In the thermodynamic analysis below, the principle of objectivity will not be applied until it is necessary for the progress of the calculations. Such a procedure is allowed, because $\hat{\mathsf{C}}$ can be chosen arbitrarily (cf. SVENDSEN & HUTTER [115]).

Chapter 5
Entropy Principle and Transformation of the Entropy Inequality

Abstract After a very brief introduction into the recent developments of modern rational thermodynamics with essentially two competing mathematical postulates for the exploitation of the Second Law of Thermodynamics we concentrate on the entropy principle of I. MÜLLER with its LAGRANGE multipliers technique of exploitation by LIU. We sketch LIU's proof of how the entropy inequality, augmented by the 'LAGRANGE multiplied balance laws' is reduced to the so-called LIU identities and the reduced entropy inequality. In this process the physical assumption [**A10**] that external source terms cannot affect the material behaviour is significant. Computations for the fluid-solid saturated mixture with an arbitrary number of constituents of which some may be compressible are rather involved and are partly deferred into Appendices. The chapter ends with inequality (5.33) from which the concrete thermodynamic analysis ensues.

A person not familiar with thermodynamics, who is confronted with debris flows, would presumably not guess that thermodynamics are relevant to the behaviour of these flows. That this is indeed the case becomes clear, if we assume that friction influences debris flows. Following the argument of PLANCK [112] who said:

> 'Alle mit Reibung behafteten Prozesse sind irreversibel'[1],

we immediately realise that thermodynamic considerations should be applied to modelling debris flows. The above quotation is one possible early statement of the entropy principle or Second Law of Thermodynamics. There are several mathematical formulations of this principle,[2] among which that of CLAUSIUS

---

[1] English: 'All processes influenced by friction are irreversible'

[2] The most popular formulations of the entropy principle in modern thermodynamics are that of CLAUSIUS & DUHEM, applied to materials according to the COLEMAN-NOLL approach [28] and that of MÜLLER & LIU (cf. MÜLLER [97, 96, 95], LIU [78, 79, 81] and HUTTER & JÖHNK [62]). Deviating approaches like that of CASEY [21] are also in use, yet less popular. These various different laws or axioms lead to the same inferences in

L. Schneider, K. Hutter, *Solid-Fluid Mixtures of Frictional Materials in Geophysical and Geotechnical Context*, Advances in Geophysical and Environmental Mechanics and Mathematics, DOI 10.1007/978-3-642-02968-4_5,

& DUHEM is probably best known. Unfortunately, their entropy inequality in connection with the approach of COLEMAN & NOLL to exploit it, is not feasible for mixtures (cf. MÜLLER [97], HUTTER & JÖHNK [62]). This is due to the fact that first, strong a priori assumptions on the entropy supply and entropy flux are made which are not plausible for mixtures and second, the balance laws of momentum and energy are thought to be always satisfied by appropriate choices of supply terms.[3] In his entropy principle MÜLLER was able to soften these assumptions by basically (i) regarding the entropy flux as a constitutive quantity, (ii) assuming that external supply terms which appear in the balance equations can not influence the material behaviour and (iii) using the technique of LAGRANGE multipliers to express the restricted optimisation problem which arises from the non-negativity of the entropy production rate density, the balance laws and constraints. In the context of mixtures, we will discuss in the following section the MÜLLER-LIU entropy principle.

The difficulties with the different forms of the Second Law of Thermodynamics lies at two different levels, (i) the fundamental statement in terms of an axiom and (ii) mathematical technicalties how inferences are drawn from the basic axiom. Different scientists generally agree on the fundamental statement, at least when it is formulated as an entropy principle; differences arise in the mathematical technicalties. Of these, the most serious ones are the choice of the entropy flux vector and the choice whether the physical space in which the material bodies 'live' allows these bodies to exist in isolation, not subject to exterior sources, or whether they are always affected by sources acting in their environment. The different approaches yield different implications, sometimes in conflict with common intuition or observations. So far experience has shown that in these circumstances MÜLLER formulation of the entropy principle is generating the most convincing results, for a detailed explanation of these issues see [58] [65] [98].

---

simple situations but imply subtle differences in more complex situations. Experience has shown that the results obtained with the MÜLLER & LIU form are the most general ones. Reviews addressed to the 'axiomatic' structure of the Second Law of Thermodynamics with emphasis on the physical implications are by HUTTER [58], HUTTER & WANG [65] and MUSCHIK et al. [98], among others.

[3] More importantly, in the classical COLEMAN-NOLL approach the evolution postulates for internal variables such as e. g. (4.36) in [**A5**] are complemented by a source term, so that $\bar{\boldsymbol{\Phi}}_\alpha = \bar{\boldsymbol{\Phi}}_\alpha^{\rm int} + \bar{\boldsymbol{\Phi}}_\alpha^{\rm ext}$, where $\bar{\boldsymbol{\Phi}}_\alpha^{\rm ext}$ may have any arbitrarily assigned value. It follows in this case, as we shall soon see, that the evolution equations for the internal variables do not exert any influence when exploiting the Second Law of Thermodynamics. This may be correct or false – it is a priori not known. Otherwise stated, the Second Law of Thermodynamics must be fulfilled subject to the conditions that all field equations (i. e. balance laws, internal variable evolution equations and all constitutive relations) are equally obeyed.

5.1 Mixture Entropy Principle According to Müller & Liu

In Section 3.3 we have already introduced the mixture and constituent specific entropies η and η_α ($\alpha = 1, \ldots, n$), respectively, which are assumed to be additive. The mixture specific entropy satisfies the balance relation

$$\pi^{\rho\eta} = \partial\left(\rho\eta\right) - \nabla \cdot \left(\boldsymbol{\phi}^{\rho\eta} - \rho\eta\mathbf{v}\right) - \sigma^{\rho\eta} \tag{5.1}$$

and the constituent specific entropies obey the non-conservative balance law

$$\bar{\pi}_\alpha^{\rho\eta} = \partial\left(\bar{\rho}_\alpha\eta_\alpha\right) - \nabla \cdot \left(\bar{\boldsymbol{\phi}}_\alpha^{\rho\eta} - \bar{\rho}_\alpha\eta_\alpha\mathbf{v}_\alpha\right) - \bar{\sigma}_\alpha^{\rho\eta} - \bar{\gamma}_\alpha^{\rho\eta}\ , \tag{5.2}$$

where (5.1) and (5.2) are obtained from Table 3.2 and 3.1 and equations (3.47) and (3.42), respectively. The constituent specific entropy η_α, its flux $\boldsymbol{\phi}_\alpha^{\rho\eta}$, its supply rate density $\sigma_\alpha^{\rho\eta}$ and its interaction rate density $\gamma_\alpha^{\rho\eta}$ are thought to be constitutive quantities following the rules introduced in Chapter 4, but its supply rate density $\sigma_\alpha^{\rho\eta}$ is of external origin and not of constitutive nature. The Second Law is stated as follows:

The *mixture entropy principle* demands that the mixture intrinsic entropy production rate density, $\pi^{\rho\eta}$, must be non-negative during any thermo-mechanical process which the mixture may undergo, i. e. the imbalance

$$\pi^{\rho\eta} \geqslant 0 \tag{5.3}$$

is required for all solutions ($\theta,\ \rho_1, \ldots, \rho_m,\ \nu_1, \ldots, \nu_{n-1},\ \mathbf{v}_1, \ldots, \mathbf{v}_n,\ \bar{\mathbf{Z}}_1, \ldots, \bar{\mathbf{Z}}_n$) of the field equations (cf. Truesdell [119]). This solution set is denoted a *thermo-mechanical admissible process.*

In other words, the above entropy principle asks for the simultaneous satisfaction of (5.3), the balance laws (4.1) to (4.4), the volume fraction evolution laws [**A4**], the evolution laws of the stress like symmetric tensor variables (4.36) in [**A5**], the density preserving constraints [**A6**], the saturation constraint [**A7**] and the constitutive laws [**A8**]. The substitution of (5.1) into (5.3) yields the inequality

$$\pi^{\rho\eta} = \partial\left(\rho\eta\right) - \nabla \cdot \left(\boldsymbol{\phi}^{\rho\eta} - \rho\eta\mathbf{v}\right) - \sigma^{\rho\eta} \geqslant 0\ . \tag{5.4}$$

For the mathematical formulation of this (restricted optimisation) problem let us, in thought, substitute the constitutive relations, [**A8**], into inequality (5.4). Along with this, the balance laws (4.1) to (4.4), [**A4**], and the evolution law (4.36) in [**A5**] must also be satisfied. If we use the chain rule for those constitutive quantities that are differentiated in this process, we obtain spatial and temporal differentiations of all constitutive variables, to which we will now assign the symbol Y_I $(I = 1, \ldots, N)$. These quantities are of higher order than the temporal and spatial derivatives entering the constitutive and evolution laws. It follows that the above problem can be written in the general form

$$A_{JK} Y_K + B_J = 0, \qquad J = 1, \ldots, P \,, \tag{5.5}$$

$$a_K Y_K + \beta \geqslant 0 \,, \tag{5.6}$$

in which EINSTEIN's summation convention applies to the index K and where the equations in (5.5) stand for the balance laws and constraint equations, and inequality (5.6) is a recast of inequality (5.4). Thus, $A_{JK} Y_K$ and $a_K Y_K$ are linear in Y_K, $(K = 1, \ldots, R \;;\; R \leqslant N)$. The coefficient matrix A_{JK}, the coefficient vectors B_J and a_K, and the coefficient scalar β then comprise all remaining terms which do not involve Y_K. LIU proved in his Lemma [78, Section 3] that the satisfaction of the entropy principle and thus, relations (5.5) and (5.6), are equivalent to anyone of the statements

(i) There exists a $\mathbf{\Lambda} = (\Lambda_{J=1}, \ldots, \Lambda_{J=P})$ with $\mathbf{\Lambda} \neq \mathbf{0}$ such that

$$a_K Y_K + \beta - \Lambda_J \left(A_{JK} Y_K + B_J \right) \geqslant 0 \qquad \forall \; Y_K \,. \tag{5.7}$$

(ii) There exists a $\mathbf{\Lambda} = (\Lambda_{J=1}, \ldots, \Lambda_{J=P})$ with $\mathbf{\Lambda} \neq \mathbf{0}$ such that

$$a_K - \Lambda_J A_{JK} = 0, \qquad K = 1, \ldots, R \,, \tag{5.8}$$

$$\beta - \Lambda_J B_J \geqslant 0 \,. \tag{5.9}$$

$\mathbf{\Lambda}$ does not depend on any Y_K, $K = 1, \ldots, R$.

We call Λ_J the LAGRANGE multiplier for the Jth equation; moreover, relations (5.7) and (5.8) are called *entropy inequality* and LIU *identities*, respectively. As will become clear in a moment, (5.9) represents the *residual entropy inequality*.

For the evaluation of (5.7) the mixture external entropy supply rate density $\sigma^{\rho\eta}$ and the external supply rate densities for momentum $\mathbf{b}_\alpha$ $(\alpha = 1, \ldots, n)$ and for internal energy r, arising in B_J and β, are to be specified. MÜLLER [97] achieved this by assuming that fields which are determined by the environment of the mixture are not allowed to play a role in restricting the constitutive relations. Consequently, as it is done in the next section, all external supply terms have to be eliminated from the entropy inequality, be they physically justified or not. This requirement simlifies how field equations with or without external supply terms are treated in the entropy principle.

For single-material theories the entropy principle of MÜLLER is completed by the assumption that between two continua special material singular surfaces, so-called *ideal walls*, exist across which *the temperature and the tangential velocity are continuous.* This assumption is necessary to make the temperature measurable by means of a thermometer. For multiphase mixtures exhibiting a single temperature WILMAŃSKI [123] showed that, in general, ideal walls do not exist. In [125] he found conditions for which the temperature is continuous across a material singular surface in a fluid-solid mixture. The first of these conditions allows no entropy production on the singular surface and the second requires either impermeability of the surface or continuity of the fluid GIBBS free energy (see Section 6.3) across the surface. *We will assume that our mixture model is such that conditions do exist for which ideal wall conditions prevail.* Thus, we will assume that

> [**A9**] Under sufficiently general conditions material singular surfaces do exist, across which the temperature and the tangential velocity are continuous. In addition, we assume that these conditions are sufficiently weak to not affect the debris-flow model presented here.

5.2 Mixture Entropy Inequality

Let us now demonstrate how the entropy principle is used to restrict the constitutive law, [**A8**]. To this end, we combine inequality (5.4) and the balance equations of mass, momenta, energy and volume fractions (see (4.1) to (4.3), [**A4**]), and the evolution laws (4.36) in [**A5**] to obtain the entropy inequality

$$
\begin{aligned}
\pi^{\rho\eta} \;=\; & \partial(\rho\eta) - \nabla\cdot(\boldsymbol{\phi}^{\rho\eta} - \rho\eta\mathbf{v}) - \sigma^{\rho\eta} \\
& - \sum \lambda_\alpha^\rho \left\{\partial\bar{\rho}_\alpha + \nabla\cdot(\bar{\rho}_\alpha\mathbf{v}_\alpha) - \bar{\rho}_\alpha c_\alpha\right\} \\
& - \sum \boldsymbol{\lambda}_\alpha^v \cdot \left\{\partial(\bar{\rho}_\alpha\mathbf{v}_\alpha) - \nabla\cdot\left(\bar{\mathbf{T}}_\alpha - \bar{\rho}_\alpha\mathbf{v}_\alpha\otimes\mathbf{v}_\alpha\right) - \bar{\mathbf{b}}_\alpha - \bar{\mathbf{m}}_\alpha\right\} \\
& - \lambda^\varepsilon \left\{\partial(\rho\varepsilon) + \nabla\cdot(\mathbf{q} + \rho\varepsilon\mathbf{v}) - \mathbf{T}\cdot(\nabla\mathbf{v}) - r\right\} \\
& - \sum \lambda_\alpha^\nu \left\{\partial\nu_\alpha + \nabla\cdot(\nu_\alpha\mathbf{v}_\alpha) - \bar{n}_\alpha\right\} \\
& - \sum \boldsymbol{\lambda}_\alpha^Z \cdot \left\{\overset{\triangle}{\bar{\mathbf{Z}}}_\alpha - [\boldsymbol{\Omega}_\alpha, \bar{\mathbf{Z}}_\alpha] - \bar{\boldsymbol{\Phi}}_\alpha\right\} \\
\geqslant 0\,, &
\end{aligned}
\tag{5.10}
$$

where the technique of LAGRANGE multipliers[4] was introduced. In this technique the balance laws and the evolution law for the internal variables $\bar{\mathbf{Z}}_\alpha$ $(\alpha = 1, \ldots, n)$ are regarded as constraints. They are pre-multiplied by their respective coefficients or so-called LAGRANGE multipliers and subtracted from inequality (5.4). This modification of the original entropy inequality (5.4) does not change its meaning, because only 'zeros' are subtracted. Conversely, by requesting the inequality (5.10) to hold for *arbitrary* fields makes the balance equations to hold identically. In doing so, we introduced the constituent LAGRANGE multipliers for mass λ^ρ_α, momentum $\boldsymbol{\lambda}^v_\alpha$, volume fraction λ^ν_α and frictional behaviour $\boldsymbol{\lambda}^Z_\alpha$ and the mixture LAGRANGE multiplier for energy λ^ε.

The constitutive quantities in (5.10) could, in general, also depend on the external supply terms, but, as stated in the last section, MÜLLER & LIU ruled out this situation. They specified the mixture entropy supply rate density, $\sigma^{\rho\eta}$, in such a way that no fields determined by the mixture environment appear in the entropy inequality (5.10), i. e.[5]

[**A10**] $$\sigma^{\rho\eta} = \sum \boldsymbol{\lambda}^{\mathbf{v}}_\alpha \cdot \bar{\mathbf{b}}_\alpha + \lambda^\varepsilon\, r\,.$$

This is perhaps the appropriate place to mention that other authors, following the COLEMAN-NOLL approach in the exploitation of the entropy principle (usually in the form of the CLAUSIUS-DUHEM inequality) also introduce external supply terms in the balance equations for mass, π^{ρ_α}, volume fraction, π^{ν_α}, and frictional variable, $\boldsymbol{\pi}^{\mathbf{Z}_\alpha}$, or some of these and assume that these terms can take any values desired. For this situation the analogue to [**A10**] would be

$$\sigma^{\rho\eta} = \sum \boldsymbol{\lambda}^{\mathbf{v}}_\alpha \cdot \bar{\mathbf{b}}_\alpha + \lambda^\varepsilon\, r + \sum \lambda^{\rho_\alpha}_\alpha \pi^{\rho_\alpha} + \sum \lambda^{\nu_\alpha}_\alpha \pi^{\nu_\alpha} + \sum \lambda^{\mathbf{Z}_\alpha} \cdot \boldsymbol{\pi}^{\mathbf{Z}_\alpha}\,. \quad (5.11)$$

It follows that the exploitation of the entropy principle according to MÜLLER & LIU is insensitive to the external source terms, be the latter physically justified or not.

In (5.10) no constituent density constraint ([**A6**]) and no saturation constraint ([**A7**]) have been used yet. To consider those, we recall Section 4.5, where we argue that not [**A6**], but [**A7**] requires a new constraint field. In

[4] In optimisation theory this technique is denoted as 'multiplier-rule of F. JOHN'. HAUSER & KIRCHNER [52] perform a precise mathematical classification of the technique of LAGRANGE multipliers. They show that the technique goes back to work by FARKAS [44] in 1894 and MINKOWSKI [88] in 1896.

[5] For historical correctness it should be mentioned that MÜLLER in his original papers [96, 95] only considered source free conditions. It was later that LIU [79] looked at open thermodynamic systems and required [**A10**].

Section 4.5 we have already eliminated the nth volume fraction from the set of constitutive variables and thus, there is no reason to consider the saturation condition and a corresponding LAGRANGE multiplier explicitly in the entropy inequality. We even observe that, if λ_n^ν is introduced, the above constraint field arises naturally from the nth volume fraction balance equation. It will be shown later that its LAGRANGE multiplier, λ_n^ν, is independent of the constitutive variables and therefore inherits the rôle of a constraint field for the saturation.[6] The above form of the entropy inequality is due to the special structure of the SVENDSEN & HUTTER-balance equations.

The density preserving assumption [**A6**], i. e. $\rho_\alpha = \text{const.}$ $(\alpha = m+1,\dots,n)$, and the symmetry of the CAUCHY stress tensors, $\mathbf{T}_\alpha = \mathbf{T}_\alpha^{\text{T}}$ $(\alpha = 1,\dots,n)$, are regarded as side conditions on the mass densities and the skew-symmetric parts of all $\mathbf{T}_\alpha$. These conditions are simply inserted into (5.10) without requiring new constraint fields.

Now, we substitute the saturation constraint (see (4.42)) and relation [**A10**] into (5.10) and obtain

$$
\begin{aligned}
\pi^{\rho\eta} \;=\;& \partial\left(\rho\eta\right) - \nabla\cdot\left(\boldsymbol{\phi}^{\rho\eta} - \rho\eta\mathbf{v}\right)\\
&-\sum \lambda_\alpha^\rho \left\{\partial\bar{\rho}_\alpha + \nabla\cdot\left(\bar{\rho}_\alpha\mathbf{v}_\alpha\right) - \bar{\rho}_\alpha c_\alpha\right\}\\
&-\sum \boldsymbol{\lambda}_\alpha^v\cdot\left\{\partial\left(\bar{\rho}_\alpha\mathbf{v}_\alpha\right) - \nabla\cdot\left(\bar{\mathbf{T}}_\alpha - \bar{\rho}_\alpha\mathbf{v}_\alpha\otimes\mathbf{v}_\alpha\right) - \bar{\mathbf{m}}_\alpha\right\}\\
&-\lambda^\varepsilon\left\{\partial\left(\rho\varepsilon\right) + \nabla\cdot\left(\mathbf{q} + \rho\varepsilon\mathbf{v}\right) - \mathbf{T}\cdot\left(\nabla\mathbf{v}\right)\right\}\\
&-\sum_{\alpha=1}^{n-1}\lambda_\alpha^\nu\left\{\partial\nu_\alpha + \nabla\cdot\left(\nu_\alpha\mathbf{v}_\alpha\right) - \bar{n}_\alpha\right\}\\
&-\lambda_n^\nu\left\{-\partial\Big(\sum_{\beta=1}^{n-1}\nu_\beta\Big) + \nabla\cdot\mathbf{v}_n - \nabla\cdot\Big(\sum_{\beta=1}^{n-1}\nu_\beta\mathbf{v}_n\Big) - \bar{n}_n\right\}\\
&-\sum \boldsymbol{\lambda}_\alpha^Z\cdot\left\{\overset{\triangle}{\mathbf{Z}}_\alpha - \left[\boldsymbol{\Omega}_\alpha, \bar{\mathbf{Z}}_\alpha\right] - \bar{\boldsymbol{\Phi}}_\alpha\right\}\\
\geqslant\;& 0\ ,
\end{aligned}
\tag{5.12}
$$

Before substituting the constitutive laws, [**A8**], into (5.12), this inequality has to be manipulated in order, (i), to introduce the constituent density constraint, [**A6**], (ii), to make the linear dependencies of (5.12) on the constitutive variables and their derivatives explicitly apparent, (iii), to allow for

[6] BAUER [10] presents a detailed discussion on the role of λ_n^ν in a saturated mixture. In addition, by mathematical reasoning he answers the question on how the saturation condition must be incorporated into the system of mixture balance equations such that the resulting system is solvable in a way that is unique in the sense of CAUCHY & KOVALEVSKAYA [77].

the identification of the constraint field for the saturation which is denoted *saturation field*, s, and (iv), to allow for the introduction of the scalar-valued and the vector-valued one-forms (cf. SVENDSEN & HUTTER [115])[7]

$$\mathcal{P} = \sum_{I=1}^{K} \mathcal{P}_{x_I} \mathrm{d}x_I := \mathrm{d}\left(\rho\eta\right) - \lambda^{\varepsilon}\mathrm{d}\left(\rho\varepsilon\right) \tag{5.13}$$

and

$$\boldsymbol{\mathcal{F}} = \sum_{I=1}^{K} \boldsymbol{\mathcal{F}}_{x_I} \mathrm{d}x_I := \mathrm{d}\boldsymbol{\phi}^{\rho\eta} + \lambda^{\varepsilon}\left(\mathrm{d}\mathbf{q}\right) - \sum\left(\mathrm{d}\bar{\mathbf{T}}_{\alpha}\right)\boldsymbol{\lambda}_{\alpha}^{v} \tag{5.14}$$

into the entropy inequality. Here, the operator $\mathrm{d}(\cdot)$ denotes the exterior derivative (see Section 2.2), and the coefficients $\mathcal{P}_{x_I}$ and $\boldsymbol{\mathcal{F}}_{x_I}$ represent the following abbreviations

$$\mathcal{P}_{x_I} = \left(\rho\eta\right)_{,x_I} - \lambda^{\varepsilon}\left(\rho\varepsilon\right)_{,x_I} , \tag{5.15}$$

$$\boldsymbol{\mathcal{F}}_{x_I} = \left(\boldsymbol{\phi}^{\rho\eta}\right)_{,x_I} + \lambda^{\varepsilon}\left(\boldsymbol{q}\right)_{,x_I} - \sum \boldsymbol{\lambda}_{\alpha}^{v}\left(\bar{\mathbf{T}}_{\alpha}\right)_{,x_I} . \tag{5.16}$$

The quantity $(\cdot)_{,x_I}$ marks a partial derivative with respect to x_I which also includes t, but $\mathcal{P}_{x_I}$ and $\boldsymbol{\mathcal{F}}_{x_I}$ are, in general, not differentials of a certain potential function. Moreover, the contractions in the last summation terms in (5.14) and (5.16), $\left(\mathrm{d}\bar{\mathbf{T}}_{\alpha}\right)\boldsymbol{\lambda}_{\alpha}^{v}$ and $\boldsymbol{\lambda}_{\alpha}^{v}\left(\bar{\mathbf{T}}_{\alpha}\right)_{,x_I}$ are understood in the sense that

$$\mathbf{a} * d\mathbf{A} := d\left(\mathbf{A}^{\mathrm{T}}\mathbf{a}\right) , \tag{5.17}$$

where $\mathbf{A}$, $\mathbf{a}$ and $*$ are a second rank tensor, a *constant* vector and the contraction operation.

The result of these manipulations reads

[7] The notation of Chapter 2 is used.

$$\pi^{\rho\eta} = \overbrace{\partial(\rho\eta) - \lambda^{\varepsilon}\partial(\rho\varepsilon)}^{\mathcal{P}_t} + \overbrace{\{\nabla(\rho\eta) - \lambda^{\varepsilon}\nabla(\rho\varepsilon)\}}^{\mathcal{P}_{\mathbf{x}}} \cdot \mathbf{v}$$

$$- \overbrace{\left\{\nabla \cdot \boldsymbol{\phi}^{\rho\eta} + \lambda^{\varepsilon}\nabla \cdot \boldsymbol{q} - \sum \left(\boldsymbol{\lambda}^{v}_{\alpha} \cdot (\nabla \cdot \bar{\mathbf{T}}_{\alpha})\right)\right\}}^{\left(\mathcal{F}_{\mathbf{x}}\right)_{ii}}$$

$$- \sum_{\alpha=1}^{m} \left\{ \bar{l}^{\rho}_{\alpha} \, (\partial\rho_{\alpha}) \; + \; \nu_{\alpha} \left(\, l^{\rho}_{\alpha}\mathbf{v}_{\alpha} - \boldsymbol{\Gamma}\mathbf{u}_{\alpha}\right) \cdot (\nabla\rho_{\alpha})\right\}$$

$$- \sum_{\alpha=1}^{n-1} \left\{\left(\, l^{\nu}_{\alpha} + s\right)(\partial\nu_{\alpha}) + \left(\, l^{\nu}_{\alpha}\mathbf{v}_{\alpha} - \rho_{\alpha}\boldsymbol{\Gamma}\mathbf{u}_{\alpha} + \mathbf{s}\right) \cdot (\nabla\nu_{\alpha})\right\}$$

$$- \sum_{\alpha=1}^{n} \left\{\bar{\rho}_{\alpha} \, \boldsymbol{\lambda}^{v}_{\alpha} \cdot (\partial\mathbf{v}_{\alpha}) + \left\{\nu_{\alpha} \, l^{\nu}_{\alpha} \, \mathbf{I} - \bar{\rho}_{\alpha} \left(\boldsymbol{\Gamma} - \boldsymbol{\lambda}^{v}_{\alpha} \otimes \mathbf{v}_{\alpha}\right) \right\} \cdot (\nabla\mathbf{v}_{\alpha})\right\}$$

$$- \sum_{\alpha=1}^{n} \left\{\boldsymbol{\lambda}^{Z}_{\alpha} \cdot (\partial\bar{\mathbf{Z}}_{\alpha}) + \left(\boldsymbol{\lambda}^{Z}_{\alpha} \otimes \mathbf{v}_{\alpha}\right) \cdot (\nabla\bar{\mathbf{Z}}_{\alpha})\right\}$$

$$- \sum_{\alpha=1}^{n} \left[\bar{\mathbf{Z}}_{\alpha} \, , \, \boldsymbol{\lambda}^{Z}_{\alpha}\right] \cdot \boldsymbol{\Omega}_{\alpha} + \sum_{\alpha=1}^{n} \bar{\boldsymbol{\Phi}}_{\alpha} \cdot \boldsymbol{\lambda}^{Z}_{\alpha}$$

$$+ \sum_{\alpha=1}^{n} \left\{\boldsymbol{\lambda}^{v}_{\alpha} \cdot \bar{\mathbf{m}}^{i}_{\alpha} + l^{\rho}_{\alpha}\bar{\rho}_{\alpha}c_{\alpha} + \lambda^{\nu}_{\alpha}\bar{n}_{\alpha}\right\}$$

$$\geqslant 0 \ , \tag{5.18}$$

where $\bar{\mathbf{m}}^{i}_{\alpha}$ is given in (4.6) and the inserted quantities l^{ρ}_{α}, l^{ν}_{α}, $\boldsymbol{\Gamma}$, s, $\mathbf{s}$ are abbreviations defined as

$$l^{\rho}_{\alpha} := \lambda^{\rho}_{\alpha} + \boldsymbol{\lambda}^{v}_{\alpha} \cdot \mathbf{v}_{\alpha} \ , \tag{5.19}$$

$$l^{\nu}_{\alpha} := \rho_{\alpha} \, l^{\rho}_{\alpha} + \lambda^{\nu}_{\alpha} \ , \tag{5.20}$$

$$\boldsymbol{\Gamma} := \lambda^{\varepsilon}\rho^{-1}\mathbf{T} + (\eta - \lambda^{\varepsilon}\varepsilon)\,\mathbf{I} = \boldsymbol{\Gamma}^{\mathrm{T}} \ , \tag{5.21}$$

$$s := -l^{\nu}_{n} \ , \tag{5.22}$$

$$\mathbf{s} := s\mathbf{v}_{n} + \rho_{n}\boldsymbol{\Gamma}\mathbf{u}_{n} \ . \tag{5.23}$$

The derivation of inequality (5.18) can be found in Appendix B.1.[8]

---

[8] Some computations in this chapter are rather involved. Detailed and long computations are deferred to Appendix B. The reader may omit these in a first reading and simply accept the results.

The only assumption which has not yet been considered is the constitutive law, [**A8**]. In the last subsection we manipulated the entropy inequality in such a way, that we can now easily substitute the one-forms $\mathcal{P}$ and $\boldsymbol{\mathcal{F}}$ into the first two lines of (5.18). We observe that, except for the two one-forms, no differentiation has to be performed for any other constitutive quantity. From the definitions (5.13) and (5.14) and the constitutive law, [**A8**], we obtain

$$\begin{aligned}\mathcal{P} = \mathcal{P}_\theta \,\mathrm{d}\theta + \mathcal{P}_{\dot\theta}\,\mathrm{d}\dot\theta + \mathcal{P}_{\nabla\theta}\cdot \mathrm{d}(\nabla\theta) + \sum_{\alpha=1}^{m} \mathcal{P}_{\rho_\alpha}\mathrm{d}\rho_\alpha + \sum_{\alpha=1}^{m} \mathcal{P}_{\nabla\rho_\alpha}\cdot \mathrm{d}(\nabla\rho_\alpha)\\ + \sum_{\alpha=1}^{n-1} \mathcal{P}_{\nu_\alpha}\mathrm{d}\nu_\alpha + \sum_{\alpha=1}^{n-1} \mathcal{P}_{\nabla\nu_\alpha}\cdot \mathrm{d}(\nabla\nu_\alpha) + \sum \mathcal{P}_{\mathbf{v}_\alpha}\cdot \mathrm{d}\mathbf{v}_\alpha\\ + \sum \mathcal{P}_{\mathbf{B}_\alpha}\cdot \mathrm{d}\mathbf{B}_\alpha + \sum \mathcal{P}_{\mathbf{D}_\alpha}\cdot \mathrm{d}\mathbf{D}_\alpha + \sum \mathcal{P}_{\mathbf{W}_\alpha}\cdot \mathrm{d}\mathbf{W}_\alpha + \sum \mathcal{P}_{\bar{\mathbf{Z}}_\alpha}\cdot \mathrm{d}\bar{\mathbf{Z}}_\alpha\end{aligned} \tag{5.24}$$

and

$$\begin{aligned}\boldsymbol{\mathcal{F}} = \boldsymbol{\mathcal{F}}_\theta \,\mathrm{d}\theta + \boldsymbol{\mathcal{F}}_{\dot\theta}\,\mathrm{d}\dot\theta + \boldsymbol{\mathcal{F}}_{\nabla\theta}\mathrm{d}(\nabla\theta) + \sum_{\alpha=1}^{m} \boldsymbol{\mathcal{F}}_{\rho_\alpha}\mathrm{d}\rho_\alpha + \sum_{\alpha=1}^{m} \boldsymbol{\mathcal{F}}_{\nabla\rho_\alpha}\mathrm{d}(\nabla\rho_\alpha)\\ + \sum_{\alpha=1}^{n-1} \boldsymbol{\mathcal{F}}_{\nu_\alpha}\mathrm{d}\nu_\alpha + \sum_{\alpha=1}^{n-1} \boldsymbol{\mathcal{F}}_{\nabla\nu_\alpha}\mathrm{d}(\nabla\nu_\alpha) + \sum \boldsymbol{\mathcal{F}}_{\mathbf{v}_\alpha}\mathrm{d}\mathbf{v}_\alpha\\ + \sum \boldsymbol{\mathcal{F}}_{\mathbf{B}_\alpha}\mathrm{d}\mathbf{B}_\alpha + \sum \boldsymbol{\mathcal{F}}_{\mathbf{D}_\alpha}\mathrm{d}\mathbf{D}_\alpha + \sum \boldsymbol{\mathcal{F}}_{\mathbf{W}_\alpha}\mathrm{d}\mathbf{W}_\alpha + \sum \boldsymbol{\mathcal{F}}_{\bar{\mathbf{Z}}_\alpha}\mathrm{d}\bar{\mathbf{Z}}_\alpha\,.\end{aligned} \tag{5.25}$$

In (5.25), the summed elements in the third line have to be interpreted as $\mathcal{F}_{iB^\alpha_{jk}}\mathrm{d}B^\alpha_{jk}$, etc. The exterior derivatives $\mathrm{d}(\cdot)$ arising in $\mathcal{P}$ and $\boldsymbol{\mathcal{F}}$ have to be adapted to the temporal and spatial derivatives of inequality (5.18). These are given by

$$\mathcal{P}_t = \mathcal{P}_\theta\ \partial\theta + \mathcal{P}_{\dot\theta}\ \partial\dot\theta + \mathcal{P}_{\nabla\theta}\cdot\partial(\nabla\theta) + \sum_{\alpha=1}^{m}\mathcal{P}_{\rho_\alpha}\partial\rho_\alpha + \cdots + \sum\mathcal{P}_{\bar{\mathbf{Z}}_\alpha}\cdot\partial\bar{\mathbf{Z}}_\alpha\ ,$$

$$\begin{aligned}\mathcal{P}_{\mathbf{x}} = {} & \mathcal{P}_\theta\ \nabla\theta + \mathcal{P}_{\dot\theta}\ \nabla\dot\theta + \mathcal{P}_{\nabla\theta}\nabla(\nabla\theta) + \sum_{\alpha=1}^{m}\mathcal{P}_{\rho_\alpha}\nabla\rho_\alpha + \sum_{\alpha=1}^{m}\mathcal{P}_{\nabla\rho_\alpha}\nabla(\nabla\rho_\alpha)\\ & + \sum_{\alpha=1}^{n-1}\mathcal{P}_{\nu_\alpha}(\nabla\nu_\alpha) + \sum_{\alpha=1}^{n-1}\mathcal{P}_{\nabla\nu_\alpha}\nabla(\nabla\nu_\alpha) + \sum\mathcal{P}_{\mathbf{v}_\alpha}\mathbf{D}_\alpha + \sum\mathcal{P}_{\mathbf{v}_\alpha}\mathbf{W}_\alpha\\ & + \sum\mathcal{P}_{\mathbf{B}_\alpha}\nabla\mathbf{B}_\alpha + \sum\mathcal{P}_{\mathbf{D}_\alpha}\nabla\mathbf{D}_\alpha + \sum\mathcal{P}_{\mathbf{W}_\alpha}\nabla\mathbf{W}_\alpha + \sum\mathcal{P}_{\bar{\mathbf{Z}}_\alpha}\nabla\bar{\mathbf{Z}}_\alpha\ ,\end{aligned}$$

$$\begin{aligned}(\boldsymbol{\mathcal{F}}_{\mathbf{x}})_{ii} = {} & \boldsymbol{\mathcal{F}}_\theta\ \cdot\nabla\theta + \boldsymbol{\mathcal{F}}_{\dot\theta}\ \cdot\nabla\dot\theta + (\boldsymbol{\mathcal{F}}_{\nabla\theta})^{\mathrm{T}}\cdot\nabla(\nabla\theta) + \sum_{\alpha=1}^{m}\boldsymbol{\mathcal{F}}_{\rho_\alpha}\cdot\nabla\rho_\alpha\\ & + \sum_{\alpha=1}^{m}(\boldsymbol{\mathcal{F}}_{\nabla\rho_\alpha})^{\mathrm{T}}\cdot\nabla(\nabla\rho_\alpha) + \sum_{\alpha=1}^{n-1}\boldsymbol{\mathcal{F}}_{\nu_\alpha}\cdot(\nabla\nu_\alpha) + \sum_{\alpha=1}^{n-1}(\boldsymbol{\mathcal{F}}_{\nabla\nu_\alpha})^{\mathrm{T}}\cdot\nabla(\nabla\nu_\alpha)\\ & + \sum(\boldsymbol{\mathcal{F}}_{\mathbf{v}_\alpha})^{\mathrm{T}}\cdot\mathbf{D}_\alpha + \sum(\boldsymbol{\mathcal{F}}_{\mathbf{v}_\alpha})^{\mathrm{T}}\cdot\mathbf{W}_\alpha + \sum(\boldsymbol{\mathcal{F}}_{\mathbf{B}_\alpha})^{\mathrm{T}}\cdot\nabla\mathbf{B}_\alpha\\ & + \sum(\boldsymbol{\mathcal{F}}_{\mathbf{D}_\alpha})^{\mathrm{T}}\cdot\nabla\mathbf{D}_\alpha - \sum(\boldsymbol{\mathcal{F}}_{\mathbf{W}_\alpha})^{\mathrm{T}}\cdot\nabla\mathbf{W}_\alpha + \sum(\boldsymbol{\mathcal{F}}_{\bar{\mathbf{Z}}_\alpha})^{\mathrm{T}}\cdot\nabla\bar{\mathbf{Z}}_\alpha\ ,\end{aligned} \tag{5.26}$$

where the transpose of a third order tensor $\mathbf{A}$ is defined here as

$$\mathbf{c}\cdot\mathbf{A}^{\mathrm{T}}(\mathbf{a}\otimes\mathbf{b}) = \mathbf{A}\mathbf{c}\cdot(\mathbf{a}\otimes\mathbf{b})\,, \qquad \forall\ \mathbf{a},\,\mathbf{b},\,\mathbf{c}\in\mathcal{W}\ , \tag{5.27}$$

which, for any $\mathbf{A}:\mathcal{V}\to\mathcal{L}(\mathcal{V})$, interprets the transpose of $\mathbf{A}$ as a mapping $\mathbf{A}^{\mathrm{T}}:\mathcal{L}(\mathcal{V})\to\mathcal{V}$. Trivially then $\mathrm{A}^{\mathrm{T}}_{ijk} = \mathrm{A}_{jki}$.

If we substitute (5.26) into (5.18), then partial time and partial space derivatives of the temperature, θ, arise which can be combined to yield the material derivative, $\dot\theta$. In fact, what is needed is the combination on the left-hand side of (5.28) below. In Appendix B.2 it is shown that this identity can be put into the following form:

$$
\begin{aligned}
&\Big\{\mathcal{P}_\theta\ \partial\theta + \mathcal{P}_{\dot\theta}\ \partial\dot\theta + \mathcal{P}_{\nabla\theta}\cdot\partial(\nabla\theta)\Big\} + \big\{\mathcal{P}_\theta\ \nabla\theta + \mathcal{P}_{\dot\theta}\ \nabla\dot\theta + \mathcal{P}_{\nabla\theta}\nabla(\nabla\theta)\big\}\cdot\mathbf{v}\\
&\quad - \Big\{\boldsymbol{\mathcal{F}}_\theta\cdot\nabla\theta + \boldsymbol{\mathcal{F}}_{\dot\theta}\cdot\nabla\dot\theta + \boldsymbol{\mathcal{F}}_{\nabla\theta}\cdot\nabla(\nabla\theta)\Big\}\\
&= \mathcal{P}_\theta\ \dot\theta + \mathcal{P}_{\dot\theta}\ \ddot\theta - \boldsymbol{\mathcal{F}}_\theta\cdot(\nabla\theta) - \boldsymbol{\mathcal{F}}_{\nabla\theta}\cdot\nabla(\nabla\theta) + \big\{\mathcal{P}_{\nabla\theta} - \boldsymbol{\mathcal{F}}_{\dot\theta}\big\}\cdot(\nabla\dot\theta)\\
&\quad -\rho^{-1}\sum\bar\rho_\alpha\,(\mathcal{P}_{\nabla\theta}\otimes\nabla\theta)\cdot(\mathbf{D}_\alpha - \mathbf{W}_\alpha)\\
&\quad -\rho^{-1}\sum_{\alpha=1}^{n-1}\{\rho_\alpha\,(\mathcal{P}_{\nabla\theta}\otimes\nabla\theta)\,\mathbf{u}_\alpha - \rho_n\,(\mathcal{P}_{\nabla\theta}\otimes\nabla\theta)\,\mathbf{u}_n\}\cdot\nabla\nu_\alpha\\
&\quad -\rho^{-1}\sum_{\alpha=1}^{m}\{\nu_\alpha\,(\mathcal{P}_{\nabla\theta}\otimes\nabla\theta)\ \mathbf{u}_\alpha\}\cdot(\nabla\rho_\alpha)\ .
\end{aligned}
\tag{5.28}
$$

We further point out that $\mathbf{v}\cdot(\mathcal{P}_{\mathbf{x}_I}\nabla\mathbf{x}_I)$, which arises in the first line of (5.18), can be written as

$$
(\mathcal{P}_{\mathbf{x}_I}\nabla\mathbf{x}_I)\cdot\mathbf{v} = \Big((\mathcal{P}_{\mathbf{x_I}})_{\{\cdots\}}\ (\mathbf{x_I})_{\{\cdots\},i}\Big)v_i = (\mathcal{P}_{\mathbf{x}_I}\otimes\mathbf{v})\ \ \cdot\nabla\mathbf{x}_I\ , \tag{5.29}
$$

where $\mathbf{x}_I$ stands for all vector and tensor-valued constitutive variables as indicated in the first expression of (5.24). If we now substitute these relations into (5.18), we obtain after lengthy manipulations and resorting of terms

$$
\begin{aligned}
\pi^{\rho\eta} = \ & \mathcal{P}_\theta(\dot\theta) + \mathcal{P}_{\dot\theta}(\ddot\theta) - \boldsymbol{\mathcal{F}}_\theta\cdot(\nabla\theta) - \boldsymbol{\mathcal{F}}_{\nabla\theta}\cdot\nabla(\nabla\theta) + \big\{\mathcal{P}_{\nabla\theta} - \boldsymbol{\mathcal{F}}_{\dot\theta}\big\}\cdot(\nabla\dot\theta)\\
&+\sum_{\alpha=1}^{m}\{\mathcal{P}_{\rho_\alpha} - \bar l^\rho_\alpha\}\ \ (\ \partial\rho_\alpha) + \sum_{\alpha=1}^{m}\mathcal{P}_{\nabla\rho_\alpha}\cdot(\ \partial\nabla\rho_\alpha)\\
&+\sum_{\alpha=1}^{n-1}\{\mathcal{P}_{\nu_\alpha} - l^\nu_\alpha - s\}\ \ (\partial\nu_\alpha) + \sum_{\alpha=1}^{n-1}\mathcal{P}_{\nabla\nu_\alpha}\cdot(\ \partial\nabla\nu_\alpha)\\
&+\sum_{\alpha=1}^{n}\{\mathcal{P}_{\mathbf{v}_\alpha} - \bar\rho_\alpha\boldsymbol{\lambda}^v_\alpha\}\cdot(\partial\mathbf{v}_\alpha) + \sum_{\alpha=1}^{n}\mathcal{P}_{\mathbf{D}_\alpha}\cdot(\partial\mathbf{D}_\alpha)\\
&+\sum_{\alpha=1}^{n}\mathcal{P}_{\mathbf{W}_\alpha}\cdot(\partial\mathbf{W}_\alpha) + \sum_{\alpha=1}^{n}\left(\mathcal{P}_{\bar{\mathbf{Z}}_\alpha} - \boldsymbol{\lambda}^Z_\alpha\right)\cdot\partial\bar{\mathbf{Z}}_\alpha
\end{aligned}
$$

(cont.)

$$
\begin{aligned}
&+\sum_{\alpha=1}^{n}\left\{\mathcal{P}_{\mathbf{B}_\alpha}\cdot(\partial\mathbf{B}_\alpha)+(\mathcal{P}_{\mathbf{B}_\alpha}\otimes\mathbf{v})\cdot(\nabla\mathbf{B}_\alpha)-(\boldsymbol{\mathcal{F}}_{\mathbf{B}_\alpha})^{\mathrm{T}}\cdot(\nabla\mathbf{B}_\alpha)\right\}\\
&+\sum_{\alpha=1}^{m}\Big\{\{\mathcal{P}_{\rho_\alpha}\mathbf{v}-\boldsymbol{\mathcal{F}}_{\rho_\alpha}\}-\nu_\alpha\,(\,l_\alpha^{\rho}\mathbf{v}_\alpha-\boldsymbol{\Gamma}\;\mathbf{u}_\alpha)\\
&\qquad\qquad\qquad -\rho^{-1}\nu_\alpha\,(\mathcal{P}_{\nabla\theta}\otimes\nabla\theta)\,\mathbf{u}_\alpha\Big\}\cdot(\nabla\rho_\alpha)\\
&+\sum_{\alpha=1}^{m}\left\{\mathcal{P}_{\nabla\rho_\alpha}\otimes\mathbf{v}-(\boldsymbol{\mathcal{F}}_{\nabla\rho_\alpha})^{\mathrm{T}}\right\}\cdot\nabla(\nabla\rho_\alpha)\\
&+\sum_{\alpha=1}^{n-1}\Big\{\{\mathcal{P}_{\nu_\alpha}\mathbf{v}-\boldsymbol{\mathcal{F}}_{\nu_\alpha}\}-\;l_\alpha^{\nu}\mathbf{v}_\alpha+\rho_\alpha\boldsymbol{\Gamma}\;\mathbf{u}_\alpha-\mathbf{s}\\
&\qquad +\rho^{-1}\rho_\alpha\,(\mathcal{P}_{\nabla\theta}\otimes\nabla\theta)\,\mathbf{u}_\alpha+\rho^{-1}\rho_n\,(\mathcal{P}_{\nabla\theta}\otimes\nabla\theta)\,\mathbf{u}_n\Big\}\cdot(\nabla\nu_\alpha)\\
&+\sum_{\alpha=1}^{n-1}\left\{\mathcal{P}_{\nabla\nu_\alpha}\otimes\mathbf{v}-(\boldsymbol{\mathcal{F}}_{\nabla\nu_\alpha})^{\mathrm{T}}\right\}\cdot\nabla(\nabla\nu_\alpha)\\
&+\sum_{\alpha=1}^{n}\Big\{\{\mathcal{P}_{\mathbf{v}_\alpha}\otimes\mathbf{v}-(\boldsymbol{\mathcal{F}}_{\mathbf{v}_\alpha})^{\mathrm{T}}\}-\nu_\alpha\;l_\alpha^{\nu}\mathbf{I}+\bar{\rho}_\alpha\,(\boldsymbol{\Gamma}-\boldsymbol{\lambda}_\alpha^{v}\otimes\mathbf{v}_\alpha)\\
&\qquad\qquad\qquad -\rho^{-1}\bar{\rho}_\alpha\,(\mathcal{P}_{\nabla\theta}\otimes\nabla\theta)\Big\}\cdot(\mathbf{D}_\alpha)\\
&+\sum_{\alpha=1}^{n}\Big\{\{\mathcal{P}_{\mathbf{v}_\alpha}\otimes\mathbf{v}-(\boldsymbol{\mathcal{F}}_{\mathbf{v}_\alpha})^{\mathrm{T}}\}-\nu_\alpha\;l_\alpha^{\nu}\mathbf{I}+\bar{\rho}_\alpha\,(\boldsymbol{\Gamma}-\boldsymbol{\lambda}_\alpha^{v}\otimes\mathbf{v}_\alpha)\\
&\qquad\qquad\qquad +\rho^{-1}\bar{\rho}_\alpha\,(\mathcal{P}_{\nabla\theta}\otimes\nabla\theta)\Big\}\cdot(\mathbf{W}_\alpha)\\
&+\sum_{\alpha=1}^{n}\left\{\mathcal{P}_{\mathbf{D}_\alpha}\otimes\mathbf{v}-(\boldsymbol{\mathcal{F}}_{\mathbf{D}_\alpha})^{\mathrm{T}}\right\}\cdot(\nabla\mathbf{D}_\alpha)\\
&+\sum_{\alpha=1}^{n}\left\{\mathcal{P}_{\mathbf{W}_\alpha}\otimes\mathbf{v}-(\boldsymbol{\mathcal{F}}_{\mathbf{W}_\alpha})^{\mathrm{T}}\right\}\cdot(\nabla\mathbf{W}_\alpha)\\
&+\sum_{\alpha=1}^{n}\left\{\{\mathcal{P}_{\bar{\mathbf{Z}}_\alpha}\otimes\mathbf{v}-(\boldsymbol{\mathcal{F}}_{\bar{\mathbf{Z}}_\alpha})^{\mathrm{T}}\}-\left(\boldsymbol{\lambda}_\alpha^{Z}\otimes\mathbf{v}_\alpha\right)\right\}\cdot(\nabla\bar{\mathbf{Z}}_\alpha)\\
&-\sum_{\alpha=1}^{n}\left[\bar{\mathbf{Z}}_\alpha\;,\;\boldsymbol{\lambda}_\alpha^{Z}\right]\cdot\boldsymbol{\Omega}_\alpha+\sum_{\alpha=1}^{n}\bar{\boldsymbol{\Phi}}_\alpha\cdot\boldsymbol{\lambda}_\alpha^{Z}\\
&+\sum_{\alpha=1}^{n}\left\{\boldsymbol{\lambda}_\alpha^{v}\cdot\bar{\mathbf{m}}_\alpha^{i}+l_\alpha^{\rho}\bar{\rho}_\alpha c_\alpha+\lambda_\alpha^{\nu}\bar{n}_\alpha\right\}\\
&\geqslant 0\;, \qquad\qquad (5.30)
\end{aligned}
$$

where $\mathbf{L}_\alpha = \nabla \mathbf{v}_\alpha$ has been decomposed into its symmetric and skew-symmetric parts. This is necessary, because in [**A8**] we have equally separated $\mathbf{L}_\alpha$ into $\mathbf{D}_\alpha$ and $\mathbf{W}_\alpha$, respectively. These terms also arise in the sixth line of inequality (5.30). Indeed we show in Appendix B.2 that

$$\begin{aligned} &\mathcal{P}_{\mathbf{B}_\alpha} \cdot (\partial \mathbf{B}_\alpha) + (\mathcal{P}_{\mathbf{B}_\alpha} \otimes \mathbf{v}) \cdot (\nabla \mathbf{B}_\alpha) \\ &\quad = \langle \mathbf{B}_\alpha, \mathcal{P}_{\mathbf{B}_\alpha} \rangle \cdot \mathbf{D}_\alpha - [\mathbf{B}_\alpha, \mathcal{P}_{\mathbf{B}_\alpha}] \cdot \mathbf{W}_\alpha - (\mathcal{P}_{\mathbf{B}_\alpha} \otimes \mathbf{u}_\alpha) \cdot (\nabla \mathbf{B}_\alpha) \ . \end{aligned} \tag{5.31}$$

Moreover, it is also convenient to introduce the abbreviations

$$\boldsymbol{\mathcal{Q}}_{x_I} := \mathcal{P}_{x_I} \otimes \mathbf{v} - (\boldsymbol{\mathcal{F}}_{x_I})^{\mathrm{T}} \tag{5.32}$$

and arrange the terms in such a way, that derivatives of those constitutive variables which are not members of the set $\mathbb{S}$ [9] appear at the beginning of the inequality.

In so doing the entropy inequality takes the final form

$$\begin{aligned} \pi^{\rho\eta} = {} & \mathcal{P}_{\dot\theta}(\ddot\theta) - \boldsymbol{\mathcal{F}}_{\nabla\theta} \cdot \nabla(\nabla\theta) + \left\{\mathcal{P}_{\nabla\theta} - \boldsymbol{\mathcal{F}}_{\dot\theta}\right\} \cdot (\nabla\dot\theta) \\ & + \sum_{\alpha=1}^{m} \{\mathcal{P}_{\rho_\alpha} - \bar{l}^{\rho}_{\alpha}\} \ (\partial\rho_\alpha) + \sum_{\alpha=1}^{m} \mathcal{P}_{\nabla\rho_\alpha} \cdot (\ \partial\nabla\rho_\alpha) \\ & + \sum_{\alpha=1}^{n-1} \{\mathcal{P}_{\nu_\alpha} - l^{\nu}_{\alpha} - s\} \ (\partial\nu_\alpha) + \sum_{\alpha=1}^{n-1} \mathcal{P}_{\nabla\nu_\alpha} \cdot (\ \partial\nabla\nu_\alpha) \\ & + \sum_{\alpha=1}^{n} \{\mathcal{P}_{\mathbf{v}_\alpha} - \bar\rho_\alpha \boldsymbol{\lambda}^{v}_{\alpha}\} \cdot (\partial\mathbf{v}_\alpha) + \sum_{\alpha=1}^{n} \mathcal{P}_{\mathbf{D}_\alpha} \cdot (\partial\mathbf{D}_\alpha) + \sum_{\alpha=1}^{n} \mathcal{P}_{\mathbf{W}_\alpha} \cdot (\partial\mathbf{W}_\alpha) \\ & + \sum_{\alpha=1}^{n} \left(\mathcal{P}_{\bar{\mathbf{Z}}_\alpha} - \boldsymbol{\lambda}^{Z}_{\alpha}\right) \cdot \partial\bar{\mathbf{Z}}_\alpha \\ & + \sum_{\alpha=1}^{m} \boldsymbol{\mathcal{Q}}_{\nabla\rho_\alpha} \cdot \nabla(\nabla\rho_\alpha) + \sum_{\alpha=1}^{n-1} \boldsymbol{\mathcal{Q}}_{\nabla\nu_\alpha} \cdot \nabla(\nabla\nu_\alpha) \\ & + \sum_{\alpha=1}^{n} \boldsymbol{\mathcal{Q}}_{\mathbf{D}_\alpha} \cdot (\nabla\mathbf{D}_\alpha) + \sum_{\alpha=1}^{n} \boldsymbol{\mathcal{Q}}_{\mathbf{W}_\alpha} \cdot (\nabla\mathbf{W}_\alpha) \end{aligned}$$

(cont.)

[9] In Section 5.1 we denoted these variables as Y_K $(K = 1, \ldots, R)$.

$$+\sum_{\alpha=1}^{n}\left\{\boldsymbol{\mathcal{Q}}_{\bar{\mathbf{Z}}_\alpha}-\left(\boldsymbol{\lambda}_\alpha^Z\otimes\mathbf{v}_\alpha\right)\right\}\cdot(\nabla\bar{\mathbf{Z}}_\alpha)$$

$$-\sum_{\alpha=1}^{n}\left\{(\mathcal{P}_{\mathbf{B}_\alpha}\otimes\mathbf{u}_\alpha)+(\boldsymbol{\mathcal{F}}_{\mathbf{B}_\alpha})^{\mathrm{T}}\right\}\cdot(\nabla\mathbf{B}_\alpha)$$

$$+\mathcal{P}_\theta(\dot{\theta})-\boldsymbol{\mathcal{F}}_\theta\cdot(\nabla\theta)$$

$$+\sum_{\alpha=1}^{m}\left\{\boldsymbol{\mathcal{Q}}_{\rho_\alpha}-\nu_\alpha\,(\,l_\alpha^\rho\mathbf{v}_\alpha-\boldsymbol{\Gamma}\,\mathbf{u}_\alpha)-\rho^{-1}\nu_\alpha\,(\mathcal{P}_{\nabla\theta}\otimes\nabla\theta)\,\mathbf{u}_\alpha\right\}\cdot(\nabla\rho_\alpha)$$

$$+\sum_{\alpha=1}^{n-1}\Big\{\boldsymbol{\mathcal{Q}}_{\nu_\alpha}-\,l_\alpha^\nu\mathbf{v}_\alpha+\rho_\alpha\boldsymbol{\Gamma}\,\mathbf{u}_\alpha-\mathbf{s}$$

$$-\rho^{-1}\rho_\alpha\,(\mathcal{P}_{\nabla\theta}\otimes\nabla\theta)\,\mathbf{u}_\alpha+\rho^{-1}\rho_n\,(\mathcal{P}_{\nabla\theta}\otimes\nabla\theta)\,\mathbf{u}_n\Big\}\cdot(\nabla\nu_\alpha)$$

$$+\sum_{\alpha=1}^{n}\Big\{\boldsymbol{\mathcal{Q}}_{\mathbf{v}_\alpha}-\nu_\alpha\,l_\alpha^\nu\mathbf{I}+\bar{\rho}_\alpha\,(\boldsymbol{\Gamma}-\boldsymbol{\lambda}_\alpha^v\otimes\mathbf{v}_\alpha)$$

$$+\langle\mathbf{B}_\alpha\;,\;\mathcal{P}_{\mathbf{B}_\alpha}\rangle-\rho^{-1}\bar{\rho}_\alpha\,(\mathcal{P}_{\nabla\theta}\otimes\nabla\theta)\Big\}\cdot(\mathbf{D}_\alpha)$$

$$+\sum_{\alpha=1}^{n}\Big\{\boldsymbol{\mathcal{Q}}_{\mathbf{v}_\alpha}-\nu_\alpha\,l_\alpha^\nu\mathbf{I}+\bar{\rho}_\alpha\,(\boldsymbol{\Gamma}-\boldsymbol{\lambda}_\alpha^v\otimes\mathbf{v}_\alpha)$$

$$-[\mathbf{B}_\alpha\;,\;\mathcal{P}_{\mathbf{B}_\alpha}]+\rho^{-1}\bar{\rho}_\alpha\,(\mathcal{P}_{\nabla\theta}\otimes\nabla\theta)\Big\}\cdot(\mathbf{W}_\alpha)$$

$$-\sum_{\alpha=1}^{n}\left[\bar{\mathbf{Z}}_\alpha\;,\;\boldsymbol{\lambda}_\alpha^Z\right]\cdot\boldsymbol{\Omega}_\alpha+\sum_{\alpha=1}^{n}\bar{\boldsymbol{\Phi}}_\alpha\cdot\boldsymbol{\lambda}_\alpha^Z$$

$$+\sum_{\alpha=1}^{n}\left\{\boldsymbol{\lambda}_\alpha^v\cdot\bar{\mathbf{m}}_\alpha^i+l_\alpha^\rho\bar{\rho}_\alpha c_\alpha+\lambda_\alpha^\nu\bar{n}_\alpha\right\}$$

$$\geqslant 0\;. \tag{5.33}$$

This formidably looking inequality is now in the form from which all subsequent inferences are drawn. This will be done in the following two chapters.

Chapter 6
Thermodynamic Analysis I Liu Identities, One-Forms and Integrability Conditions

Abstract The extended entropy inequality, derived and stated at the end of Chap. 5 is used to derive the LIU identities and the reduced entropy inequality in Sect. 6.1. Further reductions of the former are only possible if the implications of the symmetry group of the material are accounted for. This is done here for isotropy of the mixture, but also requires a number of ad-hoc assumptions. The most significant ones of these suppose (i) that the LAGRANGE multiplier of the mixture energy equation is a universal function of the (empirical) temperature and its time rate of change and (ii) that the LAGRANGE multiplier of the constituent momentum equation is proportional to the negative constituent diffusion velocity with the LAGRANGE multiplier of the energy as proportionality factor. On the basis of a number of Lemmas proved by LIU and an additional theorem motivated by him and a few technical assumptions restricting the functional dependence of two-forms arising in the LIU identities, we are then able to define the constituent thermodynamic pressures, the constituent configuration pressures and constituent free enthalpies as quantities that are derivable from a HELMHOLTZ-like free energy (whose number of independent variables is drastically reduced) and the saturation pressure. Moreover, all LAGRANGE multipliers can be expressed in terms of these variables; more specifically, they all have the LAGRANGE multiplier of the energy as a common factor. Some of the above mentioned ad-hoc assumptions are physically motivated, others are mathematically enforced, but all clearly state the conditions for which the validity of the theory ensues.

6.1 Liu Identities and Residual Entropy Inequality

We recall LIU's Lemma stated in relations (5.5) to (5.9). Inequality (5.33) corresponds to (5.7), the LIU identities and the residual entropy inequality follow from (5.8) and (5.9). To deduce these, note that this inequality has been ordered such that the first nine lines are linear in $\mathbf{Y} = \{\ddot{\theta},\ \nabla(\nabla\theta),\ \nabla\dot{\theta},\ \dots\ \nabla\bar{\mathbf{Z}}_\alpha,\ \nabla\mathbf{B}_\alpha\}$ whereas the remaining terms do not contain elements of $\mathbf{Y}$. If we now follow LIU's Lemma (see Section 5.1), a first

L. Schneider, K. Hutter, *Solid-Fluid Mixtures of Frictional Materials in Geophysical and Geotechnical Context*, Advances in Geophysical and Environmental Mechanics and Mathematics, DOI 10.1007/978-3-642-02968-4_6,

set of the LIU identities, i. e.

$$
\begin{aligned}
&\mathcal{P}_{\dot\theta} = 0, && && \mathcal{P}_{\rho_\alpha} = \bar l^{\rho}_{\alpha}, && \alpha = 1,\dots,m,\\
&\mathcal{P}_{\nabla\rho_\alpha} = \mathbf{0}, && \alpha = 1,\dots,m, && \mathcal{P}_{\nu_\alpha} = l^{\nu}_{\alpha} + s, && \alpha = 1,\dots,n-1,\\
&\mathcal{P}_{\nabla\nu_\alpha} = \mathbf{0}, && \alpha = 1,\dots,n-1, && \mathcal{P}_{\mathbf{v}_\alpha} = \bar\rho_\alpha \boldsymbol{\lambda}^{v}_{\alpha}, && \alpha = 1,\dots,n,\\
&\mathcal{P}_{\mathbf{D}_\alpha} = \mathbf{0}, && \alpha = 1,\dots,n, && \mathcal{P}_{\bar{\mathbf{Z}}_\alpha} = \boldsymbol{\lambda}^{Z}_{\alpha}, && \alpha = 1,\dots,n,\\
&\mathcal{P}_{\mathbf{W}_\alpha} = \mathbf{0}, && \alpha = 1,\dots,n\ , && &&
\end{aligned}
\tag{6.1}
$$

can be read off from (5.33). Using these relations, the remaining LIU identities for the vector-valued one-form $\boldsymbol{\mathcal{F}}$ reduce to

$$
\begin{aligned}
&\boldsymbol{\mathcal{F}}_{\nabla\theta}\cdot\nabla(\nabla\theta) = 0, && \forall\ \nabla(\nabla\theta),\\
&\{\mathcal{P}_{\nabla\theta} - \boldsymbol{\mathcal{F}}_{\dot\theta}\}\cdot\nabla\dot\theta = 0, && \forall\ \nabla\dot\theta,\\
&\boldsymbol{\mathcal{Q}}_{\nabla\rho_\alpha}\cdot\nabla(\nabla\rho_\alpha) \overset{(6.1)_3}{=} -\boldsymbol{\mathcal{F}}_{\nabla\rho_\alpha}\cdot\nabla(\nabla\rho_\alpha) = 0, && \forall\ \nabla(\nabla\rho_\alpha),\\
& && \alpha = 1,\dots,m,\\
&\boldsymbol{\mathcal{Q}}_{\nabla\nu_\alpha}\cdot\nabla(\nabla\nu_\alpha) \overset{(6.1)_5}{=} -\boldsymbol{\mathcal{F}}_{\nabla\nu_\alpha}\cdot\nabla(\nabla\nu_\alpha) = 0, && \forall\ \nabla(\nabla\nu_\alpha),\\
& && \alpha = 1,\dots,n-1,\\
&\boldsymbol{\mathcal{Q}}_{\mathbf{D}_\alpha}\cdot(\nabla\mathbf{D}_\alpha) \overset{(6.1)_7}{=} -(\boldsymbol{\mathcal{F}}_{\mathbf{D}_\alpha})^{\mathrm{T}}\cdot\nabla\mathbf{D}_\alpha = 0, && \forall\ \nabla\mathbf{D}_\alpha,\\
& && \alpha = 1,\dots,n,\\
&\boldsymbol{\mathcal{Q}}_{\mathbf{W}_\alpha}\cdot(\nabla\mathbf{W}_\alpha) \overset{(6.1)_9}{=} -(\boldsymbol{\mathcal{F}}_{\mathbf{W}_\alpha})^{\mathrm{T}}\cdot\nabla\mathbf{W}_\alpha = 0, && \forall\ \nabla\mathbf{W}_\alpha,\\
& && \alpha = 1,\dots,n,\\
&\Big\{\boldsymbol{\mathcal{Q}}_{\bar{\mathbf{Z}}_\alpha} - \Big(\boldsymbol{\lambda}^{Z}_{\alpha}\otimes\mathbf{v}_\alpha\Big)\Big\}\cdot\nabla\bar{\mathbf{Z}}_\alpha\\
&\quad\overset{(6.1)_8}{=} \{-(\boldsymbol{\mathcal{F}}_{\bar{\mathbf{Z}}_\alpha})^{\mathrm{T}} - (\mathcal{P}_{\bar{\mathbf{Z}}_\alpha}\otimes\mathbf{u}_\alpha)\}\cdot\nabla\bar{\mathbf{Z}}_\alpha = 0, && \forall\ \nabla\bar{\mathbf{Z}}_\alpha,\\
& && \alpha = 1,\dots,n,\\
&\{\mathcal{P}_{\mathbf{B}_\alpha}\otimes\mathbf{u}_\alpha + (\boldsymbol{\mathcal{F}}_{\mathbf{B}_\alpha})^{\mathrm{T}}\}\cdot\nabla\mathbf{B}_\alpha = 0, && \forall\ \nabla\mathbf{B}_\alpha,\\
& && \alpha = 1,\dots,n\ .
\end{aligned}
\tag{6.2}
$$

Relations $(6.2)_{1,3,4}$ are equivalent to the restrictions that

$$\begin{aligned}&\mathcal{F}_{\nabla\theta},\ \mathcal{F}_{\nabla\rho_\alpha},\ \mathcal{F}_{\nabla\nu_\beta} \qquad \text{are all } \textit{skew-symmetric}\ ,\\ &\qquad \text{for } \alpha=1,\dots,m \text{ and } \beta=1,\dots,n-1\ ,\end{aligned} \tag{6.3}$$

where, the symmetry of the expressions $\nabla(\nabla\theta)$, $\nabla(\nabla\rho_\alpha)$ and $\nabla(\nabla\nu_\alpha)$ is used.

In $(6.2)_5$ we can use the symmetry of $\mathbf{D}_\alpha$ in $\mathcal{F}_{\mathbf{D}_\alpha}$ and $\nabla\mathbf{D}_\alpha$ to show that $\mathcal{F}_{\mathbf{D}_\alpha}$ has both properties, symmetry and skew symmetry. This can only be satisfied if $\mathcal{F}_{\mathbf{D}_\alpha}$ vanishes.[1] Similarly, we can take the skew symmetry of $\mathbf{W}_\alpha$ in $\mathcal{F}_{\mathbf{W}_\alpha}$ and $\nabla\mathbf{W}_\alpha$ into account to conclude that $\mathcal{F}_{\mathbf{W}_\alpha}$ must vanish. With the third order zero tensor $\mathbf{0}^3$ we write

$$\mathcal{F}_{\mathbf{D}_\alpha}=\mathbf{0}^3,\quad \mathcal{F}_{\mathbf{W}_\alpha}=\mathbf{0}^3,\qquad \alpha=1,\dots,n\ . \tag{6.4}$$

Furthermore, $(6.2)_2$ simply implies

$$\mathcal{F}_{\dot\theta}=\mathcal{P}_{\nabla\theta} \tag{6.5}$$

and the symmetry of $\bar{\mathbf{Z}}_\alpha$ and $\mathbf{B}_\alpha$ require the brackets in $(6.2)_{7,8}$ to be skew-symmetric. Thus, we infer that

$$\begin{aligned}&\left\{(\mathcal{F}_{\bar{\mathbf{Z}}_\alpha})^{\mathrm{T}}+(\boldsymbol{\lambda}^Z_\alpha\otimes\mathbf{u}_\alpha)\right\},\quad \left\{(\mathcal{F}_{\mathbf{B}_\alpha})^{\mathrm{T}}+(\mathcal{P}_{\mathbf{B}_\alpha}\otimes\mathbf{u}_\alpha)\right\}\\ &\qquad\text{are both skew-symmetric for } \alpha=1,\dots,n\ .\end{aligned} \tag{6.6}$$

Via the symmetries of $\bar{\mathbf{Z}}_\alpha$ and $\mathbf{B}_\alpha$ in $(\mathcal{F}_{\bar{\mathbf{Z}}_\alpha})^{\mathrm{T}}$ and $(\mathcal{F}_{\mathbf{B}_\alpha})^{\mathrm{T}}$, respectively, it is shown in Appendix B.3 that the following relations hold

$$\begin{aligned}\mathcal{F}_{\bar{\mathbf{Z}}_\alpha}&=-\left(\mathbf{u}_\alpha\otimes\mathcal{P}_{\bar{\mathbf{Z}}_\alpha}\right)\overset{(6.1)_8}{=}-\left(\mathbf{u}_\alpha\otimes\boldsymbol{\lambda}^Z_\alpha\right), &&\alpha=1,\dots,n,\\ \mathcal{F}_{\mathbf{B}_\alpha}&=-\left(\mathbf{u}_\alpha\otimes\mathcal{P}_{\mathbf{B}_\alpha}\right), &&\alpha=1,\dots,n\ .\end{aligned} \tag{6.7}$$

It cannot be denied that the analysis from (5.12) to (5.33) is cumbersome, prone to errors and demanding patience. However, the reward, expressed by the Liu identities (6.1) to (6.7), that is gained from it, is far reaching and justifies the effort. This should be kept in mind if such cumbersome computations have to be conducted.

One immediate consequence of the statements in the left column of (6.1) is that λ^ε is a function of the independent constitutive variables and no more. Indeed, from (5.24) and the definition

---

[1] The proof is as follows: in index-notation $(6.2)_5$ reads

$$\frac{\partial\mathcal{F}_i}{\partial D^\alpha_{jk}}\frac{\partial D^\alpha_{jk}}{\partial x_i}=0$$

which implies that $\partial\mathcal{F}_i/\partial D^\alpha_{jk}$ must be skewsymmetric in j,k, since D^α_{jk} is symmetric in these variables. However, D^α_{jk} is symmetric in j,k, so $\partial\mathcal{F}_i/\partial D^\alpha_{jk}$ is also symmetric in j,k.

$$\mathcal{P}_{,x_I} := (\rho\eta)_{,x_I} - \lambda^\varepsilon(\rho\varepsilon)_{,x_I} \,, \quad x_I \in \mathbb{S} \,, \tag{6.8}$$

it follows e. g. that

$$\lambda^\varepsilon = \frac{(\rho\eta)_{,x_I}}{(\rho\varepsilon)_{,x_I}} \,, \quad \text{for } x_I \in \left\{\dot{\theta},\, \vec{\nabla}\rho,\, \vec{\nabla}\nu,\, \vec{\mathbf{D}},\, \vec{\mathbf{W}}\right\} \,, \tag{6.9}$$

subject to the condition that for at least those x_I for which neither the numerator nor the denominator vanish, λ^ε is defined by the right-hand side of (6.9). It follows, since the right-hand side of (6.9) is a function of the constitutive variables, the left-hand side is such a function as well. This is motivation for us to assume that λ^ε is a function of constitutive class also when the right-hand side of (6.9) is not determined.[2]

Having shown this, it then also follows from the right column of (6.1) that λ^ρ_α, λ^ν_α and $\boldsymbol{\lambda}^Z_\alpha$ $(\alpha = 1, \ldots, n)$ are functions of the same class.

To write down the residual inequality, which comprises the terms below line 9 of (5.33), it is advantageous to introduce the abbreviations

$$\boldsymbol{\Gamma}^* := \boldsymbol{\Gamma} - \rho^{-1}\,\mathrm{sym}\,(\mathcal{P}_{\nabla\theta} \otimes \nabla\theta) = \boldsymbol{\Gamma}^{*\mathrm{T}} \tag{6.10}$$

$$\left(= \lambda^\varepsilon\rho^{-1}\mathbf{T} + (\eta - \lambda^\varepsilon\varepsilon)\mathbf{I} - \rho^{-1}\,\mathrm{sym}\,(\mathcal{P}_{\nabla\theta} \otimes \nabla\theta) \right) \,,$$

$$\mathbf{s}^* := s\mathbf{v}_n + \rho_n\boldsymbol{\Gamma}^*\mathbf{u}_n \,. \tag{6.11}$$

With them, the residual entropy inequality takes the form

[2] It will be shown in the course of the developments that numerator and denominator of the right-hand side of equation (6.9) vanish for $x_I \in \{\vec{\nabla}\rho,\, \vec{\mathbf{D}},\, \vec{\mathbf{W}}\}$ but differ from zero for $x_I = \dot{\theta}$ if $\dot{\theta}$ is an independent constitutive variable. If $\dot{\theta}$ should not be an independent constitutive variable, then $\lambda^\varepsilon = (\rho\eta)_{,\theta} / \rho\varepsilon_{,\theta}$ of which again the numerator and denominator on the right-hand side do *not* vanish. So, the claim that λ^ε is a function of the constitutive class is safe.

$$
\begin{aligned}
\pi^{\rho\eta} = \quad & \mathcal{P}_\theta\,(\dot\theta) - \boldsymbol{\mathcal{F}}_\theta \cdot (\nabla\theta) \\
& + \sum_{\alpha=1}^{m} \{\nu_\alpha\,(\boldsymbol{\Gamma}^* - l^\rho_\alpha \mathbf{I})\,\mathbf{u}_\alpha - \boldsymbol{\mathcal{F}}_{\rho_\alpha} \\
& \qquad\qquad - \rho^{-1}\nu_\alpha\,\mathrm{skw}\,(\mathcal{P}_{\nabla\theta} \otimes \nabla\theta)\,\mathbf{u}_\alpha\} \cdot (\nabla\rho_\alpha) \\
& + \sum_{\alpha=1}^{n-1} \Big\{ (\rho_\alpha \boldsymbol{\Gamma}^* - l^\nu_\alpha \mathbf{I})\,\mathbf{u}_\alpha - \boldsymbol{\mathcal{F}}_{\nu_\alpha} + s\mathbf{v} - \mathbf{s}^* \\
& \qquad\qquad - \mathrm{skw}\,(\mathcal{P}_{\nabla\theta} \otimes \nabla\theta)\,(\xi_\alpha \mathbf{u}_\alpha - \xi_n \mathbf{u}_n) \Big\} \cdot (\nabla\nu_\alpha) \\
& + \sum_{\alpha=1}^{n} \Big\{ \nu_\alpha\,(\rho_\alpha \boldsymbol{\Gamma}^* - l^\nu_\alpha \mathbf{I}) - \bar\rho_\alpha\,\mathrm{sym}\,(\boldsymbol{\lambda}^v_\alpha \otimes \mathbf{u}_\alpha) \\
& \qquad\qquad - (\boldsymbol{\mathcal{F}}_{\mathbf{v}_\alpha})^{\mathrm{T}} + \langle \mathbf{B}_\alpha\,,\ \mathcal{P}_{\mathbf{B}_\alpha} \rangle \Big\} \cdot (\mathbf{D}_\alpha)
\end{aligned}
$$

(cont.)

$$
\left.
\begin{aligned}
& + \sum_{\alpha=1}^{n} \Big\{ - \bar\rho_\alpha\,\mathrm{skw}\,(\boldsymbol{\lambda}^v_\alpha \otimes \mathbf{u}_\alpha) - \mathrm{skw}\,\big((\boldsymbol{\mathcal{F}}_{\mathbf{v}_\alpha})^{\mathrm{T}}\big) - [\mathbf{B}_\alpha, \mathcal{P}_{\mathbf{B}_\alpha}] \\
& \qquad\qquad + \rho^{-1}\bar\rho_\alpha\,\mathrm{skw}\,(\mathcal{P}_{\nabla\theta} \otimes \nabla\theta) \Big\} \cdot (\mathbf{W}_\alpha) \\
& - \sum_{\alpha=1}^{n} \Big[\bar{\mathbf{Z}}_\alpha, \boldsymbol{\lambda}^Z_\alpha \Big] \cdot \boldsymbol{\Omega}_\alpha
\end{aligned}
\right\} (*)
$$

$$
\begin{aligned}
& + \sum_{\alpha=1}^{n} \boldsymbol{\lambda}^Z_\alpha \cdot \bar{\boldsymbol{\Phi}}_\alpha + \sum_{\alpha=1}^{n} \{ \boldsymbol{\lambda}^v_\alpha \cdot \bar{\mathbf{m}}^i_\alpha + l^\rho_\alpha \bar\rho_\alpha c_\alpha + \lambda^\nu_\alpha \bar n_\alpha \} \\
\geqslant 0\,, &
\end{aligned}
\tag{6.12}
$$

where the obvious identities

$$
\begin{aligned}
\mathrm{skw}\,(\mathcal{P}_{\nabla\theta} \otimes \nabla\theta) \cdot \mathbf{D}_\alpha &= 0\,, \\
\mathrm{skw}\,(\boldsymbol{\lambda}^v_\alpha \otimes \mathbf{u}_\alpha) \cdot \mathbf{D}_\alpha &= 0\,, \\
(\bar\rho_\alpha \boldsymbol{\Gamma}^* - l^\nu_\alpha \mathbf{I}) \cdot \mathbf{W}_\alpha &= 0\,, \\
\mathrm{sym}\,(\boldsymbol{\lambda}^v_\alpha \otimes \mathbf{u}_\alpha) \cdot \mathbf{W}_\alpha &= 0\,, \\
(\boldsymbol{\mathcal{F}}_{\mathbf{v}_\alpha})^{\mathrm{T}} \cdot \mathbf{W}_\alpha &= \mathrm{skw}\,\big((\boldsymbol{\mathcal{F}}_{\mathbf{v}_\alpha})^{\mathrm{T}}\big) \cdot \mathbf{W}_\alpha
\end{aligned}
\tag{6.13}
$$

have been used. Those lines in (6.12) which are marked with an asterisk will be set to zero in the later assumptions [**A13a**] or [**A13b**]. However, before we address further exploitations of this inequality, let us analyse in greater depth the inferences that can be drawn from the LIU identities.

6.2 Exploiting the Isotropy of the Vector-valued One-form

Owing to the assumption of material isotropy of the debris flow mixture and due to the principle of objectivity, we concluded that equation (4.43) applies to all constitutive quantities. Thus, the vector-valued one-form (5.14), viz.,

$$\boldsymbol{\mathcal{F}} := \mathrm{d}\boldsymbol{\phi}^{\rho\eta} + \lambda^{\varepsilon}\ (\mathrm{d}\mathbf{q}) - \sum (\mathrm{d}\bar{\mathbf{T}}_{\alpha})\boldsymbol{\lambda}^{v}_{\alpha}\ , \tag{6.14}$$

consists of elements which are of constitutive class and must satisfy the relation

$$\mathbf{Q}\hat{\boldsymbol{\mathcal{F}}} = \hat{\boldsymbol{\mathcal{F}}}\Big(\theta,\ \dot{\theta},\ \mathbf{Q}(\nabla\theta),\ \vec{\rho},\ \mathbf{Q}(\vec{\nabla}\rho),\ \vec{\nu},\ \mathbf{Q}(\vec{\nabla}\nu),\ \mathbf{Q}(\vec{\mathbf{v}}),\\ \mathbf{Q}\vec{\mathbf{B}}\mathbf{Q}^{\mathrm{T}},\ \mathbf{Q}\vec{\mathbf{D}}\mathbf{Q}^{\mathrm{T}},\ \mathbf{Q}\vec{\mathbf{W}}\mathbf{Q}^{\mathrm{T}},\ \mathbf{Q}\vec{\vec{\mathbf{Z}}}\mathbf{Q}^{\mathrm{T}}\Big) \tag{6.15}$$

for all orthogonal tensors $\mathbf{Q}(t)$. For the ensuing thermodynamic analysis it is advantageous to define the *extra entropy flux*

$$\mathbf{k} := \boldsymbol{\phi}^{\rho\eta} + \lambda^{\varepsilon}\mathbf{q} - \sum \bar{\mathbf{T}}_{\alpha}\boldsymbol{\lambda}^{v}_{\alpha}\ , \tag{6.16}$$

which is also of constitutive class. The determination of the connection between $\mathbf{k}$ and $\boldsymbol{\mathcal{F}}$ is facilitated if one further assumption is made, namely

[**A11**] $\boldsymbol{\lambda}^{v}_{\alpha} = -\lambda^{\varepsilon}\mathbf{u}_{\alpha}\,;$

its introduction is not absolutely necessary, but its adequacy will be discussed in Section 6.3, cf. also SVENDSEN & HUTTER [115] and KIRCHNER [71]. From the application of the exterior derivative to $\mathbf{k}$ and the product rule of differentiation, we then obtain via (6.14)

$$\mathrm{d}\mathbf{k} = \boldsymbol{\mathcal{F}} + \mathrm{d}\lambda^{\varepsilon}\Big\{\mathbf{q} + \sum \bar{\mathbf{T}}_{\alpha}\mathbf{u}_{\alpha}\Big\} + \lambda^{\varepsilon}\sum \bar{\mathbf{T}}_{\alpha}(\mathrm{d}\mathbf{u}_{\alpha})\ . \tag{6.17}$$

In this section we aim to exploit the isotropy of $\boldsymbol{\mathcal{F}}$ in order to find further restrictions on $\boldsymbol{\mathcal{F}}_{x_i}$, $x_i \in \mathbb{S}$, and also on $\mathbf{k}$. For convenience we recall the LIU identities for $\boldsymbol{\mathcal{F}}$, given in (6.3)-(6.7),

$$\begin{array}{llll} \boldsymbol{\mathcal{F}}_{\nabla\theta} & \text{is skew-symmetric}, & \boldsymbol{\mathcal{F}}_{\mathbf{W}_\gamma} = \mathbf{0}^3, & \\ \boldsymbol{\mathcal{F}}_{\nabla\rho_\alpha} & \text{is skew-symmetric}, & \boldsymbol{\mathcal{F}}_{\mathbf{D}_\gamma} = \mathbf{0}^3, & \\ \boldsymbol{\mathcal{F}}_{\nabla\nu_\beta} & \text{is skew-symmetric}, & \boldsymbol{\mathcal{F}}_{\bar{\mathbf{Z}}_\gamma} = -\left(\mathbf{u}_\gamma \otimes \mathcal{P}_{\bar{\mathbf{Z}}_\gamma}\right), & \\ \boldsymbol{\mathcal{F}}_{\dot{\theta}} = \mathcal{P}_{\nabla\theta}, & & \boldsymbol{\mathcal{F}}_{\mathbf{B}_\gamma} = -\left(\mathbf{u}_\gamma \otimes \mathcal{P}_{\mathbf{B}_\gamma}\right) & \end{array} \tag{6.18}$$

for $\alpha = 1,\dots,m,\quad \beta = 1,\dots,n-1,\quad \gamma = 1,\dots,n$. The ensuing analysis is concerned with the identification of the conditions, according to which one may claim that

$$\boldsymbol{\mathcal{F}}_{\nabla\theta} = \mathbf{0}, \quad \boldsymbol{\mathcal{F}}_{\nabla\rho_\alpha} = \mathbf{0}, \quad \boldsymbol{\mathcal{F}}_{\nabla\nu_\beta} = \mathbf{0} \tag{6.19}$$

for $\alpha = 1,\dots,m$, $\beta = 1,\dots,n-1$. We have failed to give an unconditional proof that would generally hold, but we can show, under which conditions (6.19) hold true. These conditions are stated as assumptions [**A12**], [**A13 a,b**] and [**A14**] and are statements on certain constitutive dependencies and/or interrelations, which, of course, must be obeyed if (6.19) are used (as we will do). The restricted proof of (6.19) follows in the next 11 pages, involves mathematical theorems and ends with (6.55). In a first reading, the proof may be skipped and only the assumptions [**A12**] - [**A14**] be carefully looked at. We now outline the proof.

In the context of single-material bodies, Liu [80] proved useful theorems in which restrictions on $\boldsymbol{\mathcal{F}}_{x_I}$, $x_I \in \mathbb{S}$, and $\mathbf{k}$ were found via the isotropy of $\boldsymbol{\mathcal{F}}$. Unfortunately, in our approach, we cannot directly rely on Liu's theorems, because (i) in the one-form, $\boldsymbol{\mathcal{F}}$, and in the extra entropy flux, $\mathbf{k}$, the additional flux terms $\sum(\mathrm{d}\bar{\mathbf{T}}_\alpha)\boldsymbol{\lambda}^v_\alpha$ and $\sum\bar{\mathbf{T}}_\alpha\boldsymbol{\lambda}^v_\alpha$ arise which are absent in the single-material analysis of Liu [80], (ii) there is no restriction on $\boldsymbol{\mathcal{F}}_{\mathbf{v}_\alpha}$ which is necessary for Liu's derivations and (iii) the restrictions on $\boldsymbol{\mathcal{F}}_{\dot{\theta}}$, $\boldsymbol{\mathcal{F}}_{\bar{\mathbf{Z}}_\alpha}$ and $\boldsymbol{\mathcal{F}}_{\mathbf{B}_\alpha}$ imply additional complications.

To circumvent these problems, let us, first, invoke the principle of objectivity to find an additional restriction on $\boldsymbol{\mathcal{F}}_{\mathbf{v}_\alpha}$ $(\alpha = 1,\dots,n)$. Even though we recognised that $\boldsymbol{\mathcal{F}}_{\mathbf{v}_\alpha}$ is in general not the gradient of a potential and thus, $\sum\boldsymbol{\mathcal{F}}_{\mathbf{v}_\alpha}$ does not vanish directly by using objectivity reasons, we are, however, able to use the result (4.50), deduced for any constitutive quantity from the principle of objectivity for $\sum\boldsymbol{\phi}^{\rho\eta}_{,\mathbf{v}_\alpha}$ and $\sum\mathbf{q}_{,\mathbf{v}_\alpha}$ which are parts of $\sum\boldsymbol{\mathcal{F}}_{\mathbf{v}_\alpha}$ (see (5.14)), i. e.,

$$\begin{aligned} \sum_{\alpha=1}^{n}\boldsymbol{\mathcal{F}}_{\mathbf{v}_\alpha} &\overset{(4.50)}{=} \underbrace{\sum_{\alpha=1}^{n}\boldsymbol{\phi}^{\rho\eta}_{,\mathbf{v}_\alpha}}_{\mathbf{0}} + \lambda^\varepsilon\underbrace{\sum_{\alpha=1}^{n}\mathbf{q}_{,\mathbf{v}_\alpha}}_{\mathbf{0}} - \sum_{\alpha=1}^{n}\sum_{\beta=1}^{n}\left(\boldsymbol{\lambda}^v_\beta\bar{\mathbf{T}}_{\beta,\mathbf{v}_\alpha}\right) \\ &= -\sum_{\beta=1}^{n}\sum_{\alpha=1}^{n}\left(\boldsymbol{\lambda}^v_\beta\bar{\mathbf{T}}_\beta\right)_{,\mathbf{v}_\alpha} + \sum_{\alpha=1}^{n}\sum_{\beta=1}^{n}\boldsymbol{\lambda}^v_{\beta,\mathbf{v}_\alpha}\bar{\mathbf{T}}_\beta\,. \end{aligned} \tag{6.20}$$

We can apply the principle of objectivity to $\sum_{\alpha=1}^{n}\left(\boldsymbol{\lambda}_{\beta}^{v}\bar{\mathbf{T}}_{\beta}\right)_{,\mathbf{v}_{\alpha}}$ and $\sum_{\alpha=1}^{n}\lambda^{\varepsilon}_{,\mathbf{v}_{\alpha}}$, and claim them to vanish, because $\boldsymbol{\lambda}_{\beta}^{v}$ and λ^{ε} have been proven to be constitutive quantities. Therefore,

$$\sum_{\alpha=1}^{n}\boldsymbol{\mathcal{F}}_{\mathbf{v}_{\alpha}}=\sum_{\alpha=1}^{n}\sum_{\beta=1}^{n}\boldsymbol{\lambda}_{\beta,\mathbf{v}_{\alpha}}^{v}\bar{\mathbf{T}}_{\beta}\ . \tag{6.21}$$

If we again apply [**A11**] to the right-hand side of (6.21) and use the identity

$$\mathbf{u}_{\beta,\mathbf{v}_{\alpha}}\overset{(3.56)}{=}\left(\delta_{\beta\alpha}-\bar{\xi}_{\alpha}\right)\mathbf{I}\ , \tag{6.22}$$

where $\delta_{\beta\alpha}$ is the KRONECKER delta with respect to the constituents, we obtain

$$\begin{aligned}\sum_{\alpha=1}^{n}\boldsymbol{\mathcal{F}}_{\mathbf{v}_{\alpha}}&=-\sum_{\alpha=1}^{n}\left(\sum_{\beta=1}^{n}\lambda^{\varepsilon}\left(\delta_{\beta\alpha}-\bar{\xi}_{\alpha}\right)\bar{\mathbf{T}}_{\beta}\right)-\underbrace{\sum_{\alpha=1}^{n}\lambda^{\varepsilon}_{,\mathbf{v}_{\alpha}}}_{\mathbf{0}}\sum_{\beta=1}^{n}\mathbf{u}_{\beta}\bar{\mathbf{T}}_{\beta}\\&=-\sum_{\alpha=1}^{n}\lambda^{\varepsilon}\bar{\mathbf{T}}_{\alpha}+\lambda^{\varepsilon}\underbrace{\sum_{\alpha=1}^{n}\bar{\xi}_{\alpha}}_{1}\sum_{\beta=1}^{n}\bar{\mathbf{T}}_{\beta}=\mathbf{0}\ ,\end{aligned} \tag{6.23}$$

where $(3.54)_1$ and (4.50) with [**A11**] have been used. Consequently, we observe that if [**A11**] is required, $\sum\boldsymbol{\mathcal{F}}_{\mathbf{v}_{\alpha}}$ vanishes automatically.

In a second step, we extend LIU's method published 1996 in [80] and valid for vector-valued functions to those including tensor-valued functions. Unfortunately, we cannot prevent additional assumptions, first on the skew-symmetric part of $\boldsymbol{\mathcal{F}}_{\mathbf{v}_{\alpha}}$ and, second, on the quantities $[\mathbf{B}_{\alpha},\ \mathcal{P}_{\mathbf{B}_{\alpha}}]$ and $[\bar{\mathbf{Z}}_{\alpha},\ \boldsymbol{\lambda}_{\alpha}^{Z}]$ arising in the term $(*)$ of (6.12) (see Theorem 1 below). First, we recall LIU's Lemma 1 ([80, Section 2]):

Lemma 1 (Liu's Lemma 1) *Let $\boldsymbol{F}(\mathbf{Q})$ be a (scalar-, vector- or tensor-valued) function defined on the space of second order tensors $\mathcal{L}(\mathcal{W})$ and suppose that $\boldsymbol{F}(\mathbf{Q})=\mathbf{0}$ for any $\mathbf{Q}\in\mathcal{Q}(\mathcal{W})$, where $\mathcal{Q}(\mathcal{W})$ is the group of orthogonal tensors on $\mathcal{W}$.*
Then the gradient of $\boldsymbol{F}$ at the identity tensor is symmetric, i. e., for any skew-symmetric $\mathbf{W}\in\mathcal{L}(\mathcal{W})$,[3]

$$\partial_{\mathbf{Q}}\boldsymbol{F}(\mathbf{I})\{\mathbf{W}\}=\mathbf{0}. \tag{6.25}$$

---

[3] The braces, $\{\cdot\}$, express linearity of the operator, i. e.,

$$\mathbf{A}\{\mathbf{V}+\lambda\mathbf{W}\}=\mathbf{A}\{\mathbf{V}\}+\lambda\mathbf{A}\{\mathbf{W}\} \tag{6.24}$$

□

We cite LIU's proof from [80]:

Proof. For the skew symmetric tensor $\mathbf{W} = -\mathbf{W}^{\mathrm{T}}$ and $0 < \varepsilon \ll 1$, the equality

$$(\mathbf{I} + \varepsilon\mathbf{W})(\mathbf{I} + \varepsilon\mathbf{W})^{\mathrm{T}} = \mathbf{I} + \varepsilon^2\mathbf{W}\mathbf{W}^{\mathrm{T}} \tag{6.26}$$

holds. Therefore, the tensor $(\mathbf{I} + \varepsilon\mathbf{W})$ is orthogonal to within second order terms $O(\varepsilon^2)$ in ε. This observation can be written as

$$\mathbf{I} + \varepsilon\mathbf{W} = \mathbf{Q_W} + O(\varepsilon^2) \ , \tag{6.27}$$

where $\mathbf{Q_W} \in \mathcal{Q}(\mathcal{W})$. $\boldsymbol{F}(\mathbf{Q_W}) = \mathbf{0}$ and $\boldsymbol{F}(\mathbf{I}) = \mathbf{0}$ are valid by assumption. If we now define the gradient $\partial_{\mathbf{Q}}\boldsymbol{F}$ as a linear transformation on $\mathcal{L}(\mathcal{W})$ we can use the above-mentioned results to conclude the proof with

$$\begin{aligned}
\boldsymbol{F}(\mathbf{I} + \varepsilon\mathbf{W}) - \boldsymbol{F}(\mathbf{I}) &=: \partial_{\mathbf{Q}}\boldsymbol{F}(\mathbf{I})\{\varepsilon\mathbf{W}\} + O(\varepsilon^2) \\
&= \boldsymbol{F}(\mathbf{Q_W} + O(\varepsilon^2)) - \boldsymbol{F}(\mathbf{Q_W}) \\
&= \partial_{\mathbf{Q}}\boldsymbol{F}(\mathbf{Q_W})\left\{O(\varepsilon^2)\right\} + O(\varepsilon^2) \\
&= O(\varepsilon^2) \ ,
\end{aligned}$$

which implies that $\partial_{\mathbf{Q}}\boldsymbol{F}(\mathbf{I})\{\mathbf{W}\}$ must vanish. □

LIU's second Lemma in [80] reads

Lemma 2 *Let $\mathbf{h}(\mathbf{A}, \mathbf{v})$ be an isotropic vector-valued function of a second rank tensor $\mathbf{A}$ and a vector $\mathbf{v}$. Then the function $\mathbf{h}$ satisfies the following relation*

$$\begin{aligned}
(\delta_{ij}h_k - \delta_{ik}h_j) &= \left(\frac{\partial h_i}{\partial v_j}v_k - \frac{\partial h_i}{\partial v_k}v_j\right) \\
&+ \left(\frac{\partial h_i}{\partial A_{jl}}A_{kl} + \frac{\partial h_i}{\partial A_{lj}}A_{lk} - \frac{\partial h_i}{\partial A_{kl}}A_{jl} - \frac{\partial h_i}{\partial A_{lk}}A_{lj}\right) \ .
\end{aligned} \tag{6.28}$$

□

We remark that a function $\mathbf{h}$ which depends on more than one second rank tensor or more than one vector, $\mathbf{h}(\vec{\mathbf{A}}, \ \vec{\mathbf{v}}) = \mathbf{h}(\mathbf{A}_1, \cdots, \mathbf{A}_R, \ \mathbf{v}_1, \cdots, \mathbf{v}_P)$, satisfies the identity

$$\begin{aligned}
(\delta_{ij}h_k - \delta_{ik}h_j) &= \sum_{\mathbf{v}}\left(\frac{\partial h_i}{\partial v_j}v_k - \frac{\partial h_i}{\partial v_k}v_j\right) \\
&+ \sum_{\mathbf{A}}\left(\frac{\partial h_i}{\partial A_{jl}}A_{kl} + \frac{\partial h_i}{\partial A_{lj}}A_{lk} - \frac{\partial h_i}{\partial A_{kl}}A_{jl} - \frac{\partial h_i}{\partial A_{lk}}A_{lj}\right) ,
\end{aligned} \tag{6.29}$$

where the symbols $\sum_{\mathbf{A}}$ and $\sum_{\mathbf{v}}$ stand for the summations over all tensors, $\mathbf{A}_I$, and all vectors, $\mathbf{v}_J$, upon which the function $\mathbf{h}$ depends. The proof of Lemma 2 and the extension (6.29) were also performed by LIU [80]. For our purpose, we expand Lemma 2 and relation (6.29) to tensor-valued functions.

Lemma 3 (Extension of Lemma 2)
Let $\mathbf{T}(\vec{\mathbf{A}},\ \vec{\mathbf{v}}) = \mathbf{T}(\mathbf{A}_1,\cdots,\mathbf{A}_R,\ \mathbf{v}_1,\cdots,\mathbf{v}_P)$ *be an isotropic tensor-valued function of* R *tensors,* $\mathbf{A}_I$*, and* P *vectors,* $\mathbf{v}_J$ *. Then, the function* $\mathbf{T}$ *satisfies the following relation*

$$
\begin{aligned}
(\delta_{ji}T_{kp} + \delta_{pj}T_{ik}) - (\delta_{ki}T_{jp} + \delta_{pk}T_{ij}) &= \sum_{J=1}^{P}\left(\frac{\partial T_{ip}}{\partial v_j^J}v_k^J - \frac{\partial T_{ip}}{\partial v_k^J}v_j^J\right) \\
&+ \sum_{I=1}^{R}\left(\frac{\partial T_{ip}}{\partial A_{jl}^I}A_{kl}^I + \frac{\partial T_{ip}}{\partial A_{lj}^I}A_{lk}^I - \frac{\partial T_{ip}}{\partial A_{kl}^I}A_{jl}^I - \frac{\partial T_{ip}}{\partial A_{lk}^I}A_{lj}^I\right). \qquad (6.30)
\end{aligned}
$$

□

The proof of this third Lemma parallels LIU's proof of Lemma 2 in [80].

Proof. Because $\mathbf{T}$ is isotropic, it must satisfy the following identity:

$$
\boldsymbol{F}(\mathbf{Q}) := \mathbf{T}(\mathbf{Q}\vec{\mathbf{A}}\mathbf{Q}^{\mathrm{T}},\ \mathbf{Q}\vec{\mathbf{v}}) - \mathbf{Q}\mathbf{T}(\vec{\mathbf{A}},\ \vec{\mathbf{v}})\mathbf{Q}^{\mathrm{T}} \equiv \mathbf{0}\ . \qquad (6.31)
$$

In the spirit of the proof of Lemma 1, we deduce from equation (6.31)

$$
\begin{aligned}
\varepsilon\partial_{\mathbf{Q}}\boldsymbol{F}(\mathbf{I})\{\mathbf{W}\} + O(2) &= \boldsymbol{F}(\mathbf{I}+\varepsilon\mathbf{W}) - \boldsymbol{F}(\mathbf{I}) \\
&= \Big(\mathbf{T}\left(\{\mathbf{I}+\varepsilon\mathbf{W}\}\vec{\mathbf{A}}\{\mathbf{I}+\varepsilon\mathbf{W}\}^{\mathrm{T}},\ \{\mathbf{I}+\varepsilon\mathbf{W}\}\vec{\mathbf{v}}\right) \\
&\qquad - \{\mathbf{I}+\varepsilon\mathbf{W}\}\mathbf{T}(\vec{\mathbf{A}},\ \vec{\mathbf{v}})\{\mathbf{I}+\varepsilon\mathbf{W}\}^{\mathrm{T}}\Big) \\
&\qquad - \Big(\mathbf{T}(\vec{\mathbf{A}},\ \vec{\mathbf{v}}) - \mathbf{T}(\vec{\mathbf{A}},\ \vec{\mathbf{v}})\Big) \\
&= \mathbf{T}\Big(\vec{\mathbf{A}} + \varepsilon^2\mathbf{W}\vec{\mathbf{A}}\mathbf{W}^{\mathrm{T}} + \varepsilon\Big\{\mathbf{W}\vec{\mathbf{A}} - \vec{\mathbf{A}}\mathbf{W}\Big\},\ \vec{\mathbf{v}} + \varepsilon\mathbf{W}\vec{\mathbf{v}}\Big) \\
&\qquad - \Big(\mathbf{T}(\vec{\mathbf{A}},\vec{\mathbf{v}}) + \varepsilon\mathbf{W}\mathbf{T}(\vec{\mathbf{A}},\ \vec{\mathbf{v}}) + \varepsilon\mathbf{T}(\vec{\mathbf{A}},\ \vec{\mathbf{v}})\mathbf{W}^{\mathrm{T}} \\
&\qquad + \varepsilon^2\mathbf{W}\mathbf{T}(\vec{\mathbf{A}},\ \vec{\mathbf{v}})\mathbf{W}^{\mathrm{T}}\Big) \\
&= \mathbf{T}(\vec{\mathbf{A}},\ \vec{\mathbf{v}}) + \varepsilon\sum_{I=1}^{R}\partial_{\mathbf{A}_I}\mathbf{T}\{\mathbf{W}\mathbf{A}_I - \mathbf{A}_I\mathbf{W}\} + \varepsilon\sum_{J=1}^{P}\partial_{\mathbf{v}_J}\mathbf{T}\{\mathbf{W}\mathbf{v}_J\} \\
&\qquad - \mathbf{T}(\vec{\mathbf{A}},\ \vec{\mathbf{v}}) - \varepsilon\mathbf{W}\mathbf{T}(\vec{\mathbf{A}},\ \vec{\mathbf{v}}) - \varepsilon\mathbf{T}(\vec{\mathbf{A}},\ \vec{\mathbf{v}})\mathbf{W}^{\mathrm{T}} + O(\varepsilon^2)\ , \qquad (6.32)
\end{aligned}
$$

where ε is a small number. Dividing by ε and subsequently letting ε tend to zero, then in view of Lemma 1 this yields,

$$\partial_{\mathbf{Q}} \boldsymbol{F}(\mathbf{I})\{\mathbf{W}\} = \sum_{I=1}^{R} \partial_{\mathbf{A}_I} \mathbf{T}\{\mathbf{W}\mathbf{A}_I - \mathbf{A}_I \mathbf{W}\} + \sum_{J=1}^{P} \partial_{\mathbf{v}_J} \mathbf{T}\{\mathbf{W}\mathbf{v}_J\} - \left\{\mathbf{W}\mathbf{T}(\vec{\mathbf{A}},\ \vec{\mathbf{v}}) + \mathbf{T}(\vec{\mathbf{A}},\ \vec{\mathbf{v}})\mathbf{W}^{\mathrm{T}}\right\} = \mathbf{0}\,. \tag{6.33}$$

If we now write the term in braces in the second row as

$$\mathbf{W}\mathbf{T}(\vec{\mathbf{A}},\ \vec{\mathbf{v}}) + \mathbf{T}(\vec{\mathbf{A}},\ \vec{\mathbf{v}})\mathbf{W}^{\mathrm{T}} = (\delta_{ji} T_{kp} + \delta_{pj} T_{ik})\, W_{jk}\ \mathbf{e}_i \otimes \mathbf{e}_p\ , \tag{6.34}$$

we obtain

$$\left\{\sum_{I=1}^{R}\left(\frac{\partial T_{ip}}{\partial A^I_{jl}} A^I_{kl} + \frac{\partial T_{ip}}{\partial A^I_{lj}} A^I_{lk}\right) + \sum_{J=1}^{P} \frac{\partial T_{ip}}{\partial v^J_j} v^J_k - (\delta_{ji} T_{kp} + \delta_{pj} T_{ik})\right\} W_{jk}\ \mathbf{e}_i \otimes \mathbf{e}_p = \mathbf{0}\ , \tag{6.35}$$

which must be satisfied for every skew-symmetric tensor $\mathbf{W}$. Therefore, the expression in braces in (6.35) must be symmetric in jk which can be written as

$$\left\{ (\delta_{ji} T_{kp} + \delta_{pj} T_{ik}) - (\delta_{ki} T_{jp} + \delta_{pk} T_{ij}) \right\} \mathbf{e}_i \otimes \mathbf{e}_p$$
$$= \left\{ \sum_{J=1}^{P} \left(\frac{\partial T_{ip}}{\partial v^J_j} v^J_k - \frac{\partial T_{ip}}{\partial v^J_k} v^J_j \right) + \sum_{I=1}^{R} \left(\frac{\partial T_{ip}}{\partial A^I_{jl}} A^I_{kl} + \frac{\partial T_{ip}}{\partial A^I_{lj}} A^I_{lk} - \frac{\partial T_{ip}}{\partial A^I_{kl}} A^I_{jl} - \frac{\partial T_{ip}}{\partial A^I_{lk}} A^I_{lj} \right) \right\} \mathbf{e}_i \otimes \mathbf{e}_p\ . \tag{6.36}$$

The proof of Lemma 2 is analogous. □

The auxiliary results spelled out in Lemmae 2 and 3 allow the formulation of the following theorem.

Theorem 1 (analogous to Theorem 1 of Liu [80])
Assume that the following properties hold [compare (6.14) and (6.18)]

(i) *For the M vector variables* $\mathbf{x}^J \in \left\{\nabla\theta,\ \vec{\nabla}\nu,\ \vec{\nabla}\rho\right\}$,[4]

[4] See Liu identities $(6.18)_{1,3,5}$.

$$
\begin{aligned}
(\boldsymbol{\mathcal{F}}_{\mathbf{x}^J})_{ij} + (\boldsymbol{\mathcal{F}}_{\mathbf{x}^J})_{ji} = \left(\frac{\partial \phi_i^{\rho\eta}}{\partial x_j^J} + \frac{\partial \phi_j^{\rho\eta}}{\partial x_i^J}\right) + \lambda^\varepsilon \left(\frac{\partial q_i}{\partial x_j^J} + \frac{\partial q_j}{\partial x_i^J}\right) \\
- \sum_{\beta=1}^{n} (\boldsymbol{\lambda}_\beta^v)_p \left(\frac{\partial \bar{T}_{ip}^\beta}{\partial x_j^J} + \frac{\partial \bar{T}_{jp}^\beta}{\partial x_i^J}\right) = 0.
\end{aligned} \tag{6.37}
$$

(ii) *For the vector variables* $\mathbf{v}_\alpha$, $(\alpha = 1, \ldots, n)$,

$$
\begin{aligned}
\textbf{[A12]} \quad (\boldsymbol{\mathcal{F}}_{\mathbf{v}_\alpha})_{ij} - (\boldsymbol{\mathcal{F}}_{\mathbf{v}_\alpha})_{ji} = \left(\frac{\partial \phi_i^{\rho\eta}}{\partial v_j^\alpha} - \frac{\partial \phi_j^{\rho\eta}}{\partial v_i^\alpha}\right) + \lambda^\varepsilon \left(\frac{\partial q_i}{\partial v_j^\alpha} - \frac{\partial q_j}{\partial v_i^\alpha}\right) \\
- \sum_{\beta=1}^{n} (\boldsymbol{\lambda}_\beta^v)_p \left(\frac{\partial \bar{T}_{ip}^\beta}{\partial v_j^\alpha} - \frac{\partial \bar{T}_{jp}^\beta}{\partial v_i^\alpha}\right) = 0.
\end{aligned}
$$

(iii) *For every tensor variable* $\breve{\mathbf{A}} \in \left\{\vec{\mathbf{D}},\ \vec{\mathbf{W}}\right\}^5$,

$$
(\boldsymbol{\mathcal{F}}_{\breve{\mathbf{A}}})_{ijk} = \frac{\partial \phi_i^{\rho\eta}}{\partial \breve{A}_{jk}} + \lambda^\varepsilon \frac{\partial q_i}{\partial \breve{A}_{jk}} - \sum_{\beta=1}^{n} \left(\boldsymbol{\lambda}_\beta^v\right)_p \left(\frac{\partial \bar{T}_{ip}^\beta}{\partial \breve{A}_{jk}}\right) = 0\ . \tag{6.38}
$$

(iv) *For every tensor variable* $\tilde{\mathbf{A}} \in \left\{\vec{\bar{\mathbf{Z}}},\ \vec{\mathbf{B}}\right\}$ *and the corresponding* $\mathcal{P}_{\tilde{\mathbf{A}}} \in \left\{\mathcal{P}_{\bar{\mathbf{Z}}_1}, \cdots, \mathcal{P}_{\bar{\mathbf{Z}}_n},\ \mathcal{P}_{\mathbf{B}_1}, \cdots, \mathcal{P}_{\mathbf{B}_n}\right\}^6$,

$$
(\boldsymbol{\mathcal{F}}_{\tilde{\mathbf{A}}})_{ijk} = \frac{\partial \phi_i^{\rho\eta}}{\partial \tilde{A}_{jk}} + \lambda^\varepsilon \frac{\partial q_i}{\partial \tilde{A}_{jk}} - \sum_{\beta=1}^{n} \left(\boldsymbol{\lambda}_\beta^v\right)_p \left(\frac{\partial \bar{T}_{ip}^\beta}{\partial \tilde{A}_{jk}}\right) = -u_i^\alpha \left(\mathcal{P}_{\tilde{\mathbf{A}}}\right)_{jk}\ , \tag{6.39}
$$

where the constituent-index α *of* u_i^α *has to match that of* $\tilde{\mathbf{A}}$.

(v) *Moreover, if* $\boldsymbol{\Omega}_\alpha = \mathbf{W}_\alpha$, *assume that* [compare (*) in (6.12)]

$$
\begin{aligned}
\textbf{[A13a]} \qquad & [\mathbf{B}_\alpha, \mathcal{P}_{\mathbf{B}_\alpha}] + \left[\bar{\mathbf{Z}}_\alpha,\ \boldsymbol{\lambda}_\alpha^Z\right] \\
& = -\bar{\rho}_\alpha \operatorname{skw}(\boldsymbol{\lambda}_\alpha^v \otimes \mathbf{u}_\alpha) \\
& \quad - \operatorname{skw}\left((\boldsymbol{\mathcal{F}}_{\mathbf{v}_\alpha})^T\right) + \rho^{-1} \bar{\rho}_\alpha \operatorname{skw}\left(\mathcal{P}_{\nabla\theta} \otimes \nabla\theta\right)\ ,
\end{aligned}
$$

[5] See Liu identities $(6.18)_{2,4}$.

[6] See Liu identities $(6.18)_{6,8}$.

but when $\boldsymbol{\Omega}_\alpha$ *is independent of* $\mathbf{W}_\alpha$*, that*

[A13b]
$$[\mathbf{B}_\alpha, \mathcal{P}_{\mathbf{B}_\alpha}] = -\bar{\rho}_\alpha \operatorname{skw}(\boldsymbol{\lambda}_\alpha^v \otimes \mathbf{u}_\alpha) - \operatorname{skw}\left((\boldsymbol{\mathcal{F}}_{\mathbf{v}_\alpha})^T\right) + \rho^{-1}\bar{\rho}_\alpha \operatorname{skw}\left(\mathcal{P}_{\nabla\theta} \otimes \nabla\theta\right),$$
$$\left[\bar{\mathbf{Z}}_\alpha, \boldsymbol{\lambda}_\alpha^Z\right] = \mathbf{0},$$

for all $\alpha = 1, \ldots, n$.

(vi) *Finally,*

[A14] *suppose that all* $\boldsymbol{\mathcal{F}}_{\mathbf{x}^J}$, $\mathbf{x}^J \in \left\{\nabla\theta,\ \nabla\vec{\rho},\ \nabla\vec{\nu}\right\}$ *are independent of* $\nabla\rho_1, \ldots, \nabla\rho_n$, $\nabla\nu_1, \ldots, \nabla\nu_n$ *and* $\nabla\theta$.

Then

$$\boldsymbol{\mathcal{F}}_{\nabla\theta} = \mathbf{0}, \quad \boldsymbol{\mathcal{F}}_{\nabla\rho_\alpha} = \mathbf{0}, \quad \boldsymbol{\mathcal{F}}_{\nabla\nu_\beta} = \mathbf{0}, \qquad \alpha = 1, \ldots, m\,, \quad \beta = 1, \ldots, n-1\,. \tag{6.40}$$

□

We remark that owing to the isotropy of $\boldsymbol{\mathcal{F}}$ the assumption of symmetry of all $\boldsymbol{\mathcal{F}}_{\mathbf{v}_\alpha}$ is justified for thermodynamic equilibrium (cf. SVENDSEN & HUTTER [115, page 2046]).

The assumptions in items (v) and (vi) do not follow from physical reasoning, but are necessary for the derivation of (6.40). Due to item (v), the terms identified by the asterisk in the residual inequality (6.12), together vanish identically.

Proof. Let us take the result of Lemma 3, apply it to the constituent CAUCHY stress tensor, $\bar{\mathbf{T}}_\alpha$, and multiply the resulting equation by

$$\boldsymbol{\lambda}_\alpha^v = (\boldsymbol{\lambda}_\alpha^v)_l\ \mathbf{e}_l\ . \tag{6.41}$$

If, for convenience, we omit the basis vector $\mathbf{e}_i$ and the bar in $\bar{\mathbf{T}}_\alpha$ and use the symmetry of $\mathbf{T}_\alpha$, we obtain via (6.36)

$$
\begin{aligned}
&\left\{\delta_{ij}(\boldsymbol{\lambda}^v_\alpha)_p T^\alpha_{pk} - \delta_{ik}(\boldsymbol{\lambda}^v_\alpha)_p T^\alpha_{pj}\right\} - \left\{(\boldsymbol{\lambda}^v_\alpha)_k T^\alpha_{ji} - (\boldsymbol{\lambda}^v_\alpha)_j T^\alpha_{ki}\right\} \\
&\quad = \sum_{J=1}^{N} (\boldsymbol{\lambda}^v_\alpha)_p \left(\frac{\partial T^\alpha_{pi}}{\partial x^J_j} x^J_k - \frac{\partial T^\alpha_{pi}}{\partial x^J_k} x^J_j \right) \\
&\qquad + \sum_{\breve{\mathbf{A}},\tilde{\mathbf{A}}} (\boldsymbol{\lambda}^v_\alpha)_p \left(\frac{\partial T^\alpha_{pi}}{\partial A_{jl}} A_{kl} + \frac{\partial T^\alpha_{pi}}{\partial A_{lj}} A_{lk} - \frac{\partial T^\alpha_{pi}}{\partial A_{kl}} A_{jl} - \frac{\partial T^\alpha_{pi}}{\partial A_{lk}} A_{lj} \right) ,
\end{aligned}
\tag{6.42}
$$

where N $(= M + n)$ is the number of vectors upon which the constituent CAUCHY stress tensors depend, i. e. $\{\nabla\theta,\ \vec{\nabla}\nu,\ \vec{\nabla}\rho,\ \vec{\mathbf{v}}\}$. Moreover, Lemma 2, applied to $\boldsymbol{\phi}^{\rho\eta}$ and $\mathbf{q}$, yields

$$
\begin{aligned}
&\left(\delta_{ij}\phi^{\rho\eta}_k - \delta_{ik}\phi^{\rho\eta}_j\right) \\
&\quad = \sum_{J=1}^{N} \left(\frac{\partial \phi^{\rho\eta}_i}{\partial x^J_j} x^J_k - \frac{\partial \phi^{\rho\eta}_i}{\partial x^J_k} x^J_j \right) \\
&\qquad + \sum_{\breve{\mathbf{A}}\tilde{\mathbf{A}}} \left(\frac{\partial \phi^{\rho\eta}_i}{\partial A_{jl}} A_{kl} + \frac{\partial \phi^{\rho\eta}_i}{\partial A_{lj}} A_{lk} - \frac{\partial \phi^{\rho\eta}_i}{\partial A_{kl}} A_{jl} - \frac{\partial \phi^{\rho\eta}_i}{\partial A_{lk}} A_{lj} \right)
\end{aligned}
\tag{6.43}
$$

and

$$
\begin{aligned}
&\lambda^\varepsilon \left(\delta_{ij} q_k - \delta_{ik} q_j\right) \\
&\quad = \sum_{J=1}^{N} \lambda^\varepsilon \left(\frac{\partial q_i}{\partial x^J_j} x^J_k - \frac{\partial q_i}{\partial x^J_k} x^J_j \right) \\
&\qquad + \sum_{\breve{\mathbf{A}}\tilde{\mathbf{A}}} \lambda^\varepsilon \left(\frac{\partial q_i}{\partial A_{jl}} A_{kl} + \frac{\partial q_i}{\partial A_{lj}} A_{lk} - \frac{\partial q_i}{\partial A_{kl}} A_{jl} - \frac{\partial q_i}{\partial A_{lk}} A_{lj} \right) .
\end{aligned}
\tag{6.44}
$$

If we now recall the definitions of $\mathbf{k}$ and $\boldsymbol{\mathcal{F}}$ (see (6.14), (6.16)), sum equation (6.42) over all constituents and subtract the result from corresponding sums of (6.43) and (6.44), we obtain

$$
\begin{aligned}
(\delta_{ij} k_k - \delta_{ik} k_j) &- \sum_{\alpha=1}^{n} \left((\boldsymbol{\lambda}^v_\alpha)_j (\bar{\mathbf{T}}_\alpha)_{ki} - (\boldsymbol{\lambda}^v_\alpha)_k (\bar{\mathbf{T}}_\alpha)_{ji} \right) \\
&= \sum_{J=1}^{N} \left(\mathcal{F}^{\mathbf{x}^J}_{ij} x^J_k - \mathcal{F}^{\mathbf{x}^J}_{ik} x^J_j \right) \\
&+ \sum_{\breve{\mathbf{A}}} \left(\mathcal{F}^{\breve{A}}_{ijl} \breve{A}_{kl} - \mathcal{F}^{\breve{A}}_{ikl} \breve{A}_{jl} + \mathcal{F}^{\breve{A}}_{ilj} \breve{A}_{lk} - \mathcal{F}^{\breve{A}}_{ilk} \breve{A}_{lj} \right) \\
&+ \sum_{\tilde{\mathbf{A}}} \left(\mathcal{F}^{\tilde{A}}_{ijl} \tilde{A}_{kl} - \mathcal{F}^{\tilde{A}}_{ikl} \tilde{A}_{jl} + \mathcal{F}^{\tilde{A}}_{ilj} \tilde{A}_{lk} - \mathcal{F}^{\tilde{A}}_{ilk} \tilde{A}_{lj} \right) ,
\end{aligned} \tag{6.45}
$$

where we employed the abbreviations

$$
\begin{aligned}
\mathcal{F}^{\mathbf{x}^J}_{ik} &:= (\boldsymbol{\mathcal{F}}_{\mathbf{x}^J})_{ik} , & \mathbf{x}^J &\in \{\nabla\theta,\ \vec{\mathbf{v}},\ \vec{\nabla}\nu,\ \vec{\nabla}\rho\}, \\
\mathcal{F}^{\breve{A}}_{ijk} &:= (\boldsymbol{\mathcal{F}}_{\breve{\mathbf{A}}})_{ijk} , & \breve{\mathbf{A}} &\in \left\{\vec{\mathbf{D}},\ \vec{\mathbf{W}}\right\} , \\
\mathcal{F}^{\tilde{A}}_{ijk} &:= (\boldsymbol{\mathcal{F}}_{\tilde{\mathbf{A}}})_{ijk} , & \tilde{\mathbf{A}} &\in \left\{\vec{\vec{\mathbf{Z}}},\ \vec{\mathbf{B}}\right\} .
\end{aligned} \tag{6.46}
$$

By using the Liu identities (6.4) (item (iii) of Theorem 1), we see that each member in the sum over $\breve{\mathbf{A}} \in \{\vec{\mathbf{D}},\ \vec{\mathbf{W}}\}$ vanishes by itself. Therefore, (6.45) reads

$$
\begin{aligned}
(\delta_{ij} k_k - \delta_{ik} k_j) &- \sum_{\alpha=1}^{n} \left((\boldsymbol{\lambda}^v_\alpha)_j (\bar{\mathbf{T}}_\alpha)_{ki} - (\boldsymbol{\lambda}^v_\alpha)_k (\bar{\mathbf{T}}_\alpha)_{ji} \right) \\
&= \sum_{J=1}^{N} \left(\mathcal{F}^{\mathbf{x}^J}_{ij} x^J_k - \mathcal{F}^{\mathbf{x}^J}_{ik} x^J_j \right) \\
&+ \sum_{\tilde{\mathbf{A}}} \left(\mathcal{F}^{\tilde{A}}_{ijl} \tilde{A}_{kl} - \mathcal{F}^{\tilde{A}}_{ikl} \tilde{A}_{jl} + \mathcal{F}^{\tilde{A}}_{ilj} \tilde{A}_{lk} - \mathcal{F}^{\tilde{A}}_{ilk} \tilde{A}_{lj} \right) .
\end{aligned} \tag{6.47}
$$

The first part of the second sum on the right-hand side of (6.47) can be expressed as

$$\sum_{\tilde{\mathbf{A}}} \left(\mathcal{F}^{\tilde{A}}_{ijl} \tilde{A}_{kl} - \mathcal{F}^{\tilde{A}}_{ikl} \tilde{A}_{jl} \right) = - \sum_{\beta=1}^{n} \left(u_i^\beta (\boldsymbol{\lambda}^Z_\beta)_{jl} \bar{Z}^\beta_{kl} - u_i^\beta (\boldsymbol{\lambda}^Z_\beta)_{kl} \bar{Z}^\beta_{jl} \right)$$

$$- \sum_{\beta=1}^{n} \left(u_i^\beta (\mathcal{P}_{\mathbf{B}_\beta})_{jl} B^\beta_{kl} - u_i^\beta (\mathcal{P}_{\mathbf{B}_\beta})_{kl} B^\beta_{jl} \right)$$

$$= - \sum_{\beta=1}^{n} u_i^\beta \left(\left[\boldsymbol{\lambda}^Z_\beta , \bar{\mathbf{Z}}_\beta \right]_{jk} + \left[\mathcal{P}_{\mathbf{B}_\beta}, \mathbf{B}_\beta \right]_{jk} \right) \tag{6.48}$$

via the symmetry of $\bar{\mathbf{Z}}_\beta$, $\mathbf{B}_\beta$, $\boldsymbol{\lambda}^Z_\beta$ and $(6.1)_8$ & (6.7) (item (iv) of Theorem 1). Due to the symmetry of $\bar{\mathbf{Z}}_\beta$, $\boldsymbol{\lambda}^Z_\beta$, $\mathbf{B}_\beta$ and $\mathcal{P}_{\mathbf{B}_\beta}$ the same result follows also for the second part of the second sum; we thus obtain

$$\sum_{\tilde{\mathbf{A}}} \left(\mathcal{F}^{\tilde{A}}_{ijl} \tilde{A}_{kl} - \mathcal{F}^{\tilde{A}}_{ikl} \tilde{A}_{jl} + \mathcal{F}^{\tilde{A}}_{ilj} \tilde{A}_{lk} - \mathcal{F}^{\tilde{A}}_{ilk} \tilde{A}_{lj} \right)$$

$$= -2 \sum_{\beta=1}^{n} u_i^\beta \left(\left[\boldsymbol{\lambda}^Z_\beta , \bar{\mathbf{Z}}_\beta \right]_{jk} + \left[\mathcal{P}_{\mathbf{B}_\beta}, \mathbf{B}_\beta \right]_{jk} \right) . \tag{6.49}$$

In item (v) of Theorem 1 both assumptions, [**A13a**] and [**A13b**], allow us to write (6.49) in the form

$$\sum_{\tilde{\mathbf{A}}} \left(\mathcal{F}^{\tilde{A}}_{ijl} \tilde{A}_{kl} - \mathcal{F}^{\tilde{A}}_{ikl} \tilde{A}_{jl} + \mathcal{F}^{\tilde{A}}_{ilj} \tilde{A}_{lk} - \mathcal{F}^{\tilde{A}}_{ilk} \tilde{A}_{lj} \right)$$

$$= -2 \sum_{\alpha=1}^{n} u_i^\alpha \Big\{ \left(\operatorname{skw}(\boldsymbol{\mathcal{F}}_{\mathbf{v}_\alpha})^{\mathrm{T}} \right)_{jk} + \bar{\rho}_\alpha \left(\operatorname{skw}(\boldsymbol{\lambda}^v_\alpha \otimes \mathbf{u}_\alpha) \right)_{jk}$$

$$- \rho^{-1} \bar{\rho}_\alpha \left(\operatorname{skw} \left(\mathcal{P}_{\nabla\theta} \otimes \nabla\theta \right) \right)_{jk} \Big\}$$

$$\overset{(3.57)}{=} -2 \sum_{\alpha=1}^{n} u_i^\alpha \left\{ \left(\operatorname{skw}(\boldsymbol{\mathcal{F}}_{\mathbf{v}_\alpha})^{\mathrm{T}} \right)_{jk} + \bar{\rho}_\alpha \left(\operatorname{skw}(\boldsymbol{\lambda}^v_\alpha \otimes \mathbf{u}_\alpha) \right)_{jk} \right\}$$

$$\overset{(\mathrm{ii})}{=} -2 \sum_{\alpha=1}^{n} u_i^\alpha \left\{ \bar{\rho}_\alpha \left(\operatorname{skw}(\boldsymbol{\lambda}^v_\alpha \otimes \mathbf{u}_\alpha) \right)_{jk} \right\}$$

$$\overset{[\mathbf{A11}]}{=} 0 . \tag{6.50}$$

With these manipulations (6.47) reduces to

$$(\delta_{ij}k_k - \delta_{ik}k_j) - \underbrace{\sum_{\alpha=1}^{n}\Big((\boldsymbol{\lambda}_\alpha^v)_j(\bar{\mathbf{T}}_\alpha)_{ki} - (\boldsymbol{\lambda}_\alpha^v)_k(\bar{\mathbf{T}}_\alpha)_{ji}\Big)}_{(*)}$$

$$= \sum_{J=1}^{N}\Big(\mathcal{F}_{ij}^{\mathbf{x}^J} x_k^J - \mathcal{F}_{ik}^{\mathbf{x}^J} x_j^J\Big) \ . \qquad (6.51)$$

This equation is one major result of the proof. It will be of use later on. Liu [80] derived a similar equation [80, eqn.(11)], but compared to (6.51) the additional term (∗) does not occur. Liu [80] is only dealing with single-material bodies, so this term is exclusively due to the mixture theory approach pursued here.

We further exploit (6.51) by choosing special values for i, j and k. We obtain

- $(i,\ j,\ k) = (1,\ 2,\ 3)$:

$$\begin{aligned}&-\sum_{\alpha=1}^{n}\Big((\boldsymbol{\lambda}_\alpha^v)_2(\bar{\mathbf{T}}_\alpha)_{31} - (\boldsymbol{\lambda}_\alpha^v)_3(\bar{\mathbf{T}}_\alpha)_{21}\Big)\\ &\quad = \sum_{J=1}^{M}\Big(\mathcal{F}_{12}^{\mathbf{x}^J} x_3^J - \mathcal{F}_{13}^{\mathbf{x}^J} x_2^J\Big) + \sum_{\alpha=1}^{n}\big(\mathcal{F}_{12}^{\mathbf{v}_\alpha} v_3^\alpha - \mathcal{F}_{13}^{\mathbf{v}_\alpha} v_2^\alpha\big) \ ,\end{aligned} \qquad (6.52a)$$

- $(i,\ j,\ k) = (2,\ 3,\ 1)$:

$$\begin{aligned}&-\sum_{\alpha=1}^{n}\Big((\boldsymbol{\lambda}_\alpha^v)_3(\bar{\mathbf{T}}_\alpha)_{12} - (\boldsymbol{\lambda}_\alpha^v)_1(\bar{\mathbf{T}}_\alpha)_{32}\Big)\\ &\quad = \sum_{J=1}^{M}\Big(\mathcal{F}_{23}^{\mathbf{x}^J} x_1^J - \mathcal{F}_{21}^{\mathbf{x}^J} x_3^J\Big) + \sum_{\alpha=1}^{n}\big(\mathcal{F}_{23}^{\mathbf{v}_\alpha} v_1^\alpha - \mathcal{F}_{21}^{\mathbf{v}_\alpha} v_3^\alpha\big) ,\end{aligned} \qquad (6.52b)$$

- $(i,\ j,\ k) = (3,\ 1,\ 2)$:

$$\begin{aligned}&-\sum_{\alpha=1}^{n}\Big((\boldsymbol{\lambda}_\alpha^v)_1(\bar{\mathbf{T}}_\alpha)_{23} - (\boldsymbol{\lambda}_\alpha^v)_2(\bar{\mathbf{T}}_\alpha)_{13}\Big)\\ &\quad = \sum_{J=1}^{M}\Big(\mathcal{F}_{31}^{\mathbf{x}^J} x_2^J - \mathcal{F}_{32}^{\mathbf{x}^J} x_1^J\Big) + \sum_{\alpha=1}^{n}\big(\mathcal{F}_{31}^{\mathbf{v}_\alpha} v_2^\alpha - \mathcal{F}_{32}^{\mathbf{v}_\alpha} v_1^\alpha\big) .\end{aligned} \qquad (6.52c)$$

If we now sum all three equations of (6.52), use the symmetry of all $\bar{\mathbf{T}}_\alpha$ and all $\boldsymbol{\mathcal{F}}_{\mathbf{v}_\alpha}$ (item (ii) in Theorem 1 or [**A12**]) and the skew-symmetry of $\boldsymbol{\mathcal{F}}_{\mathbf{x}^J}$ with $\mathbf{x}^J \in \{\nabla\theta,\ \vec{\nabla}\rho,\ \vec{\nabla}\nu\}$, see (6.3), it is observed that

$$2\sum_{J=1}^{M}\left(\mathcal{F}_{13}^{\mathbf{x}^J}\mathbf{x}_2^J+\mathcal{F}_{21}^{\mathbf{x}^J}\mathbf{x}_3^J+\mathcal{F}_{32}^{\mathbf{x}^J}\mathbf{x}_1^J\right)=2\sum_{J=1}^{M}\begin{pmatrix}\mathcal{F}_{32}^{\mathbf{x}^J}\\ \mathcal{F}_{13}^{\mathbf{x}^J}\\ \mathcal{F}_{21}^{\mathbf{x}^J}\end{pmatrix}\cdot\underbrace{\begin{pmatrix}x_1^J\\ x_2^J\\ x_3^J\end{pmatrix}}_{\mathbf{x}^J}=\mathbf{0}. \tag{6.53}$$

In item (vi) of Theorem 1 we assumed that $\boldsymbol{\mathcal{F}}_{\mathbf{x}^J}$, $\mathbf{x}^J\in\{\nabla\theta,\ \vec{\nabla\rho},\ \vec{\nabla\nu}\}$, are independent of $\nabla\rho_1,\ldots,\nabla\rho_n,\ \nabla\nu_1,\ \ldots,\nabla\nu_n$ and $\nabla\theta$. Therefore, since $\mathbf{x}^J$ in (6.53) are arbitrary, we necessarily have

$$\begin{pmatrix}\mathcal{F}_{32}^{\mathbf{x}^J}\\ \mathcal{F}_{13}^{\mathbf{x}^J}\\ \mathcal{F}_{21}^{\mathbf{x}^J}\end{pmatrix}=\mathbf{0}\ , \tag{6.54}$$

and with the skew-symmetry of $\boldsymbol{\mathcal{F}}_{\mathbf{x}^J}$ we obtain

$$\boldsymbol{\mathcal{F}}_{\mathbf{x}^J}=\mathbf{0},\qquad \mathbf{x}^J\in\left\{\nabla\theta,\ \vec{\nabla\rho},\ \vec{\nabla\nu}\right\}\ . \tag{6.55}$$

This completes the proof. □

The restriction (6.55) is not the only consequence of equation (6.51). When choosing yet another set of indices a representation for $\mathbf{k}$ can be deduced. To this end, we first incorporate (6.55) into (6.51) and obtain

$$\begin{aligned}(\delta_{ij}k_k-\delta_{ik}k_j)&-\sum_{\alpha=1}^{n}\left((\boldsymbol{\lambda}_\alpha^v)_j(\bar{\mathbf{T}}_\alpha)_{ki}-(\boldsymbol{\lambda}_\alpha^v)_k(\bar{\mathbf{T}}_\alpha)_{ji}\right)\\ &=\sum_{\alpha=1}^{n}\left(\mathcal{F}_{ij}^{\mathbf{v}_\alpha}v_k^\alpha-\mathcal{F}_{ik}^{\mathbf{v}_\alpha}v_j^\alpha\right)\ .\end{aligned} \tag{6.56}$$

If we choose the indices

$$(i,\ j,\ k)=\left\{\begin{matrix}1,\ 1,\ 2\\ 1,\ 1,\ 3\\ 2,\ 2,\ 1\\ 2,\ 2,\ 3\\ 3,\ 3,\ 1\\ 3,\ 3,\ 2\end{matrix}\right\}\ , \tag{6.57}$$

take into account assumptions [**A11**], [**A12**], use the fact that λ^ε depends only on constitutive variables and employ the result (4.50) of the principle of objectivity for $\sum\boldsymbol{\mathcal{F}}_{\mathbf{v}_\alpha}$ we obtain, after a lengthy calculation (see Appendix (B.4)),

$$\mathbf{k}=-\tfrac{1}{2}\lambda^\varepsilon\sum_\alpha\left\{\left\{\bar{\mathbf{T}}_\alpha-\mathrm{tr}\left(\bar{\mathbf{T}}_\alpha\right)\mathbf{I}\right\}\mathbf{u}_\alpha+(\lambda^\varepsilon)^{-1}\left\{\boldsymbol{\mathcal{F}}_{\mathbf{v}_\alpha}-\mathrm{tr}\left(\boldsymbol{\mathcal{F}}_{\mathbf{v}_\alpha}\right)\mathbf{I}\right\}\mathbf{u}_\alpha\right\}, \tag{6.58}$$

or

$$\mathbf{k} = -\tfrac{1}{2}\lambda^{\varepsilon} \sum_{\alpha} \mathbf{K}_{\alpha}(\mathbb{S})\ \mathbf{u}_{\alpha}\ , \tag{6.59}$$

with

$$\mathbf{K}_{\alpha} = \{\bar{\mathbf{T}}_{\alpha} - \operatorname{tr}(\bar{\mathbf{T}}_{\alpha})\,\mathbf{I}\} + (\lambda^{\varepsilon})^{-1}\{\boldsymbol{\mathcal{F}}_{\mathbf{v}_{\alpha}} - \operatorname{tr}(\boldsymbol{\mathcal{F}}_{\mathbf{v}_{\alpha}})\,\mathbf{I}\}\ . \tag{6.60}$$

So, $\mathbf{k}$ is given by a combination of the diffusion velocities with tensorial coefficients which are constitutive quantities.

From the definition (6.16) of $\mathbf{k}$ and [**A11**], i. e.

$$\mathbf{k} = \boldsymbol{\phi}^{\rho\eta} + \lambda^{\varepsilon}\left[\mathbf{q} + \sum \bar{\mathbf{T}}_{\alpha}\mathbf{u}_{\alpha}\right]\ , \tag{6.61}$$

and the result (6.58), (6.59) we can now deduce a representation for $\boldsymbol{\phi}^{\rho\eta}$. If we write the first sum on the right-hand side of (6.58) in the form

$$\begin{aligned} &-\lambda^{\varepsilon}\sum \tfrac{1}{2}\Big\{\bar{\mathbf{T}}_{\alpha} - \operatorname{tr}(\bar{\mathbf{T}}_{\alpha})\,\mathbf{I}\Big\}\mathbf{u}_{\alpha} \\ ,\qquad &= -\lambda^{\varepsilon}\sum\Big\{\tfrac{1}{2}(\bar{\mathbf{T}}_{\alpha} - \tfrac{1}{3}\operatorname{tr}(\bar{\mathbf{T}}_{\alpha})\mathbf{I}) - \tfrac{1}{3}\operatorname{tr}(\bar{\mathbf{T}}_{\alpha})\mathbf{I}\Big\}\mathbf{u}_{\alpha} \end{aligned} \tag{6.62}$$

we obtain via the definition of $\mathbf{k}$, (6.61),

$$\boldsymbol{\phi}^{\rho\eta} = \begin{cases} -\lambda^{\varepsilon}\Big\{\mathbf{q} + \sum\{\,(\bar{\mathbf{T}}_{\alpha} - \tfrac{1}{3}\operatorname{tr}(\bar{\mathbf{T}}_{\alpha})\mathbf{I}) + \tfrac{1}{3}\operatorname{tr}(\bar{\mathbf{T}}_{\alpha})\mathbf{I}\}\mathbf{u}_{\alpha}\Big\}, & \text{if } \mathbf{k} = \mathbf{0}, \\ -\lambda^{\varepsilon}\mathbf{q} - \lambda^{\varepsilon}\sum\Big\{\,\tfrac{3}{2}\{\bar{\mathbf{T}}_{\alpha} - \tfrac{1}{3}\operatorname{tr}(\bar{\mathbf{T}}_{\alpha})\mathbf{I}\}\mathbf{u}_{\alpha} & \\ \qquad + \tfrac{1}{2}(\lambda^{\varepsilon})^{-1}\{\boldsymbol{\mathcal{F}}_{\mathbf{v}_{\alpha}} - \operatorname{tr}(\boldsymbol{\mathcal{F}}_{\mathbf{v}_{\alpha}})\mathbf{I}\}\mathbf{u}_{\alpha}\Big\}, & \text{if } \mathbf{k} \neq \mathbf{0}. \end{cases} \tag{6.63}$$

For simple, single-material bodies the extra entropy flux is collinear with the heat flux vector (cf. Svendsen et al. [116]). In (6.63) we observe that this is no longer the case, if we have multiple constituents with differing velocities. Thus, the present mixture model is, as expected, not in agreement with the postulate of Clausius & Duhem. The additional contributions to the entropy flux vector are, first, due to the deviatoric Cauchy stress tensors $\left(\bar{\mathbf{T}}_{\alpha} - \tfrac{1}{3}\operatorname{tr}(\bar{\mathbf{T}}_{\alpha})\,\mathbf{I}\right)$ and, second, due to the vector-valued one-form $\boldsymbol{\mathcal{F}}_{\mathbf{v}_{\alpha}}$ ($\alpha = 1, \ldots, n$). Thus, the flux of entropy is not only due to heat-flux but results also from shear-stresses (more precisely: deviatoric stresses) within the constituents and, provided $\mathbf{k} \neq \mathbf{0}$, a contribution which is related to the constituent velocity differences. In the context of the kinetic theory these additional terms can be interpreted as 'higher moments' (cf. Truesdell and Muncaster [120]).

Note also that $\mathbf{k}$ and $(\boldsymbol{\phi}^{\rho\eta} + \lambda^{\varepsilon}\mathbf{q})$ are both vectors that can be written as vectorial combinations of $\mathbf{u}_{\alpha}$. This implies that both vanish identically provided $\mathbf{u}_{\alpha} = \mathbf{0}$ for all components K_{α},

$$\mathbf{k}\,\Big|_{\forall\,\mathbf{u}_\alpha=\mathbf{0}} = \mathbf{0}\;, \qquad (\boldsymbol{\phi}^{\rho\eta} + \lambda^{\varepsilon}\mathbf{q})\,\Big|_{\forall\,\mathbf{u}_\alpha=\mathbf{0}} = \mathbf{0}\;. \tag{6.64}$$

This is obvious from (6.58)-(6.59).

With the help of the result (6.58) we also intend to reduce the number of dependencies of the LAGRANGE multiplier λ^{ε}. If we account for relation (6.17), apply the exterior derivative to (6.58) and use the product rule of differentiation, we obtain

$$\begin{aligned}
\mathrm{d}\mathbf{k} &\overset{(6.17)}{=} \boldsymbol{\mathcal{F}} + \mathrm{d}\lambda^{\varepsilon}\Big\{\mathbf{q} + \sum \bar{\mathbf{T}}_\alpha \mathbf{u}_\alpha\Big\} + \lambda^{\varepsilon} \sum \bar{\mathbf{T}}_\alpha (\mathrm{d}\mathbf{u}_\alpha) \\
&= -\mathrm{d}\lambda^{\varepsilon}\Big\{\sum \tfrac{1}{2}\{\bar{\mathbf{T}}_\alpha - \mathrm{tr}(\bar{\mathbf{T}}_\alpha)\mathbf{I}\}\mathbf{u}_\alpha\Big\} \\
&\quad -\tfrac{1}{2}\lambda^{\varepsilon}\sum\Big\{\{\bar{\mathbf{T}}_\alpha - \mathrm{tr}(\bar{\mathbf{T}}_\alpha)\mathbf{I}\}(\mathrm{d}\mathbf{u}_\alpha) + \mathrm{d}\big(\bar{\mathbf{T}}_\alpha - \mathrm{tr}(\bar{\mathbf{T}}_\alpha)\mathbf{I}\big)\mathbf{u}_\alpha\Big\} \\
&\quad -\tfrac{1}{2}\sum\Big\{\{\boldsymbol{\mathcal{F}}_{\mathbf{v}_\alpha} - \mathrm{tr}(\boldsymbol{\mathcal{F}}_{\mathbf{v}_\alpha})\mathbf{I}\}(\mathrm{d}\mathbf{u}_\alpha) + \mathrm{d}\big(\boldsymbol{\mathcal{F}}_{\mathbf{v}_\alpha} - \mathrm{tr}(\boldsymbol{\mathcal{F}}_{\mathbf{v}_\alpha})\mathbf{I}\big)\mathbf{u}_\alpha\Big\}\,.
\end{aligned} \tag{6.65}$$

Unfortunately, we cannot reduce, on the basis of (6.65), the dependencies of the function λ^{ε} because no constitutive law has yet been specified for the constituent CAUCHY stress tensors and the one-form $\boldsymbol{\mathcal{F}}_{\mathbf{v}_\alpha}$.

However, (6.65) can be used to find a representation for $\boldsymbol{\mathcal{F}}_{\dot\theta}$ or $\mathcal{P}_{\nabla\theta}$, respectively (see $(6.18)_7$); indeed if we specialize the exterior derivatives in (6.65) to the derivative with respect to $\dot\theta$, we obtain

$$\begin{aligned}
\mathbf{k}_{,\dot\theta} &= \boldsymbol{\mathcal{F}}_{\dot\theta} + \lambda^{\varepsilon}_{,\dot\theta}\Big\{\mathbf{q} + \sum \bar{\mathbf{T}}_\alpha \mathbf{u}_\alpha\Big\} \\
&= -\lambda^{\varepsilon}_{,\dot\theta}\sum\Big\{\tfrac{1}{2}\big(\bar{\mathbf{T}}_\alpha - \mathrm{tr}(\bar{\mathbf{T}}_\alpha)\mathbf{I}\big)\mathbf{u}_\alpha\Big\} \\
&\quad -\lambda^{\varepsilon}\sum\Big\{\tfrac{1}{2}\big(\bar{\mathbf{T}}_\alpha - \mathrm{tr}\,(\bar{\mathbf{T}}_\alpha)\,\mathbf{I}\big)_{,\dot\theta}\mathbf{u}_\alpha\Big\} \\
&\quad -\sum \tfrac{1}{2}\big(\boldsymbol{\mathcal{F}}_{\mathbf{v}_\alpha} - \mathrm{tr}\,(\boldsymbol{\mathcal{F}}_{\mathbf{v}_\alpha})\,\mathbf{I}\big)_{,\dot\theta}\mathbf{u}_\alpha\;,
\end{aligned} \tag{6.66}$$

which, via $(6.18)_7$, yields the expression

$$\begin{aligned}
\mathcal{P}_{\nabla\theta} &= -\lambda^{\varepsilon}_{,\dot\theta}\Big\{\mathbf{q} + \sum \bar{\mathbf{T}}_\alpha \mathbf{u}_\alpha\Big\} \\
&\quad -\lambda^{\varepsilon}_{,\dot\theta}\sum\Big\{\tfrac{1}{2}\big(\bar{\mathbf{T}}_\alpha - \mathrm{tr}(\bar{\mathbf{T}}_\alpha)\mathbf{I}\big)\mathbf{u}_\alpha\Big\} \\
&\quad -\lambda^{\varepsilon}\sum\Big\{\tfrac{1}{2}\big(\bar{\mathbf{T}}_\alpha - \mathrm{tr}\,(\bar{\mathbf{T}}_\alpha)\,\mathbf{I}\big)_{,\dot\theta}\mathbf{u}_\alpha\Big\} \\
&\quad -\sum \tfrac{1}{2}\big(\boldsymbol{\mathcal{F}}_{\mathbf{v}_\alpha} - \mathrm{tr}\,(\boldsymbol{\mathcal{F}}_{\mathbf{v}_\alpha})\,\mathbf{I}\big)_{,\dot\theta}\mathbf{u}_\alpha\,.
\end{aligned} \tag{6.67}$$

If we again use relation (6.62) and write

$$\sum \bar{\mathbf{T}}_\alpha \mathbf{u}_\alpha = \sum \left\{ \left(\bar{\mathbf{T}}_\alpha - \tfrac{1}{3}\,\mathrm{tr}(\bar{\mathbf{T}}_\alpha)\mathbf{I} \right) + \tfrac{1}{3}\,\mathrm{tr}(\bar{\mathbf{T}}_\alpha) \right\} \mathbf{u}_\alpha \,, \tag{6.68}$$

we obtain

$$\begin{aligned} \mathcal{P}_{\nabla\theta} = &-\lambda^{\varepsilon}_{,\dot\theta}\, \mathbf{q} - \tfrac{3}{2}\lambda^{\varepsilon}_{,\dot\theta} \sum \left(\bar{\mathbf{T}}_\alpha - \tfrac{1}{3}\,\mathrm{tr}(\bar{\mathbf{T}}_\alpha)\mathbf{I} \right) \mathbf{u}_\alpha \\ &-\tfrac{1}{2}\lambda^{\varepsilon} \sum \left(\bar{\mathbf{T}}_\alpha - \mathrm{tr}(\bar{\mathbf{T}}_\alpha)\mathbf{I} \right)_{,\dot\theta} \mathbf{u}_\alpha \\ &-\tfrac{1}{2} \sum \left(\boldsymbol{\mathcal{F}}_{\mathbf{v}_\alpha} - \mathrm{tr}\left(\boldsymbol{\mathcal{F}}_{\mathbf{v}_\alpha} \right) \mathbf{I} \right)_{,\dot\theta} \mathbf{u}_\alpha \,. \end{aligned} \tag{6.69}$$

In their frictional, single-material theory, SVENDSEN et al. [116] obtained the result $\mathcal{P}_{\nabla\theta} = -\lambda^{\varepsilon}_{,\dot\theta}\mathbf{q}$ which can be recovered from (6.69) if, for instance, the number of constituents is set to one or if there is no constituent velocity difference.

6.3 Integrability Conditions

In this section we aim to translate the LIU identities for the quantities (see (6.1))

$$\mathcal{P}_{x_I}, \qquad x_I \in \mathbb{S}\backslash\{\theta,\ \nabla\theta,\ \rho,\ \nu\} \tag{6.70}$$

to conditions on constitutive quantities. For this purpose, we shall explore the integrability conditions for the above terms; this is necessary if we wish $\mathcal{P} = \sum_{I=1}^{K} \mathcal{P}_{x_I} \mathrm{d}x_I$ to be in the form of a total differential and thus form the *general* GIBBS *relation.* In doing so, we are following the procedure of SVENDSEN & HUTTER [115]. Accordingly, the constitutive quantities, and in particular $\mathrm{d}(\rho\eta)$ and $\mathrm{d}(\rho\varepsilon)$ which arise in the definition of $\mathcal{P}$, see (5.13), are exact one-forms; i. e. they satisfy the relations

$$\mathrm{d}^2(\rho\eta) = 0 \quad \text{and} \quad \mathrm{d}^2(\rho\varepsilon) = 0 \,. \tag{6.71}$$

These restrictions on the derivatives of $\rho\eta$ and $\rho\varepsilon$ correspond to their own integrability conditions. The requirement $\mathrm{d}^2(\rho\eta) = 0$ is equivalent to

$$(\rho\eta)_{,x_I x_J} - (\rho\eta)_{,x_J x_I} = 0, \qquad \forall x_I, x_J \in \mathbb{S} \,. \tag{6.72}$$

An analogous result also holds for $\rho\varepsilon$. Application of the exterior derivative $\mathrm{d}(\cdot)$ to

$$\mathcal{P} := \mathrm{d}(\rho\eta) - \lambda^{\varepsilon}\mathrm{d}(\rho\varepsilon) \,, \tag{6.73}$$

and use of (6.71), results in

$$0 = \mathrm{d}^2(\rho\eta) = \mathrm{d}\lambda^\varepsilon \wedge \mathrm{d}(\rho\varepsilon) + \mathrm{d}\mathcal{P}\ . \tag{6.74}$$

To prove this, we apply the exterior derivative to the definition of $\mathcal{P}$ and then obtain

$$0 = \mathrm{d}^2(\rho\eta) = \mathrm{d}\big(\lambda^\varepsilon \mathrm{d}(\rho\varepsilon)\big) + \mathrm{d}\mathcal{P}\ . \tag{6.75}$$

The first term on the right-hand side of (6.75) can be written as in (6.72). This yields

$$\begin{aligned}
&\big(\lambda^\varepsilon(\rho\varepsilon)_{,x_I}\big)_{,x_J} - \big(\lambda^\varepsilon(\rho\varepsilon)_{,x_J}\big)_{,x_I} \\
&= \lambda^\varepsilon{}_{,x_J}(\rho\varepsilon)_{,x_I} + \lambda^\varepsilon(\rho\varepsilon)_{,x_I x_J} - \lambda^\varepsilon{}_{,x_I}(\rho\varepsilon)_{,x_J} - \lambda^\varepsilon(\rho\varepsilon)_{,x_J x_I} \\
&= \lambda^\varepsilon \underbrace{\Big\{(\rho\varepsilon)_{,x_I x_J} - (\rho\varepsilon)_{,x_J x_I}\Big\}}_{\mathbf{0}} + \lambda^\varepsilon{}_{,x_J}(\rho\varepsilon)_{,x_I} - \lambda^\varepsilon{}_{,x_I}(\rho\varepsilon)_{,x_J}\ ,
\end{aligned} \tag{6.76}$$

for all $x_I, x_J \in \mathbb{S}$. With the definition of the wedge product (see (2.18)) we observe that

$$\mathrm{d}\lambda^\varepsilon \wedge \mathrm{d}(\rho\varepsilon), \tag{6.77}$$

is equivalent to

$$\lambda^\varepsilon{}_{,x_J}(\rho\varepsilon)_{,x_I} - \lambda^\varepsilon{}_{,x_I}(\rho\varepsilon)_{,x_J} \qquad \forall x_I, x_J \in \mathbb{S}\ , \tag{6.78}$$

and thus the desired relation (6.74) emerges. It can also be written as

$$\begin{gathered}
\mathcal{P}_{x_I,x_J} - \mathcal{P}_{x_J,x_I} = \lambda^\varepsilon{}_{,x_I}(\rho\varepsilon)_{,x_J} - \lambda^\varepsilon{}_{,x_J}(\rho\varepsilon)_{,x_I}\ , \\
x_I, x_J \in \mathbb{S} \quad (I < J)\ .
\end{gathered} \tag{6.79}$$

This can be interpreted as integrability conditions for the one-form $\mathcal{P}$.

From (6.1) we know that $\mathcal{P}_{x_I} = 0$ for $x_I \in \big\{\dot{\theta},\ \nabla\vec{\rho},\ \nabla\vec{\nu},\ \vec{\mathbf{D}},\ \vec{\mathbf{W}}\big\}$ and thus (6.79) splits into the following statements

$$\text{(i)} \qquad \mathcal{P}_{x_I,x_J} - \mathcal{P}_{x_J,x_I} = \lambda^\varepsilon{}_{,x_I}(\rho\varepsilon)_{,x_J} - \lambda^\varepsilon{}_{,x_J}(\rho\varepsilon)_{,x_I}\ , \tag{6.80}$$

$$x_I, x_J \in \big\{\theta,\ \nabla\theta,\ \vec{\rho},\ \vec{\nu},\ \vec{\mathbf{v}},\ \vec{\mathbf{B}},\ \vec{\vec{\mathbf{Z}}}\big\} \quad (I < J)\ ,$$

$$\text{(ii)} \qquad \mathcal{P}_{x_I,x_J} = \lambda^\varepsilon{}_{,x_I}(\rho\varepsilon)_{,x_J} - \lambda^\varepsilon{}_{,x_J}(\rho\varepsilon)_{,x_I}\ , \tag{6.81}$$

$$x_I \in \big\{\theta,\ \nabla\theta,\ \vec{\rho},\ \vec{\nu},\ \vec{\mathbf{v}},\ \vec{\mathbf{B}},\ \vec{\vec{\mathbf{Z}}}\big\}\ ,$$

$$x_J \in \big\{\dot{\theta},\ \nabla\vec{\rho},\ \nabla\vec{\nu},\ \vec{\mathbf{D}},\ \vec{\mathbf{W}}\big\}\ ,$$

$$\text{(iii)} \qquad 0 = \lambda^\varepsilon{}_{,x_I}(\rho\varepsilon)_{,x_J} - \lambda^\varepsilon{}_{,x_J}(\rho\varepsilon)_{,x_I}\ , \tag{6.82}$$

$$x_I, x_J \in \big\{\dot{\theta},\ \nabla\vec{\rho},\ \nabla\vec{\nu},\ \vec{\mathbf{D}},\ \vec{\mathbf{W}}\big\} \quad (I < J)\ .$$

We are not able to derive any further restrictions for constitutive quantities from equations (6.80) to (6.82), because within the present developments λ^ε still depends on the entire set of constitutive variabes, $\mathbb{S}$. To make progress in the thermodynamic analysis, we now drastically reduce the variable dependence of λ^ε by imposing the assumption

[**A15**] $$\lambda^\varepsilon = \hat{\lambda}^\varepsilon(\theta,\ \dot{\theta})\ .$$

It is motivated by many earlier results, e. g. by Müller [97] and many others for materials for which [**A15**] has been proved. We have stated before (see Section 4.2) that the dependence of λ^ε on $\dot{\theta}$ is necessary for the linearised heat equation to be hyperbolic.

[**A15**] neglects contributions of all constituent quantities except θ and $\dot{\theta}$ which are common to all constituents. This is fairly well justified, because λ^ε is a Lagrange multiplier for a mixture balance equation which describes the mixture as a whole. Consequently, contributions of single constituent variables should not arise, rather only mixture quantities, e. g. apart from θ and $\dot{\theta}$, the mixture mass density, ρ, the mixture stretching tensor, $\mathbf{D}$, etc. If we, in addition, think of λ^ε as a quantity that describes only those effects that are related to the mixture behaving as a single body, we can require λ^ε to have the same properties as its analogue in thermodynamic theories of single material bodies (cf. Müller [97], Hutter [58]). There, it is shown that λ^ε is a universal function of $\dot{\theta}$ and θ, the terminology 'universal' meaning 'independent of the material behaviour'. If λ^ε is proved to be an universal function of θ and $\dot{\theta}$ then the coldness $\hat{\lambda}^\varepsilon(\theta,\ \dot{\theta})$ define a derived concept, not given a priori. [**A15**] destroys this property and assigns to $\hat{\lambda}^\varepsilon(\theta,\ \dot{\theta})$ a similar role as is assigned to the absolute temperature in the Clausius-Duhem inequality. So, from a conceptual point of view, one might equally have assumed [**A15**] ab initio. This way, many computations would have turned out to be simpler. This was not done because it was our intention to possibly prove [**A15**].

With this assumption at hand, we can further reduce (6.80)-(6.82) to the forms

$$
\begin{aligned}
&\text{(i)} \qquad \mathcal{P}_{x_I,x_J} = \mathcal{P}_{x_J,x_I}, \\
&\qquad\qquad x_I, x_J \in \left\{\nabla\theta,\ \vec{\rho},\ \vec{\nu},\ \vec{\mathbf{v}},\ \vec{\mathbf{B}},\ \vec{\bar{\mathbf{Z}}}\right\} \quad (I<J)\ , \\
&\text{(ii)} \qquad (\rho\eta)_{,x_I\dot{\theta}} = \lambda^{\varepsilon}(\rho\varepsilon)_{,x_I\dot{\theta}}\ , \\
&\qquad\quad (\rho\eta)_{,x_I\theta} - \mathcal{P}_{\theta,x_I} = \lambda^{\varepsilon}(\rho\varepsilon)_{,x_I\theta}\ , \\
&\qquad\qquad x_I \in \left\{\nabla\theta, \vec{\rho},\ \vec{\nu},\ \vec{\mathbf{v}},\ \vec{\mathbf{B}},\ \vec{\bar{\mathbf{Z}}}\right\}\ , \\
&\text{(iii)} \qquad \varepsilon_{,x_I} = 0, \\
&\qquad\qquad x_I \in \left\{\vec{\nabla}\rho,\ \vec{\nabla}\nu,\ \vec{\mathbf{D}},\ \vec{\mathbf{W}}\right\}\ ,
\end{aligned}
\tag{6.83}
$$

where for (ii) the definition of $\mathcal{P}$, (6.73), has been used. From the last of these relations we immediately obtain

$$
\varepsilon = \hat{\varepsilon}(\mathbb{S}_R)\ , \tag{6.84}
$$

$$
\mathbb{S}_R := \left\{\theta,\ \dot{\theta},\ \nabla\theta,\ \vec{\rho},\ \vec{\nu},\ \vec{\mathbf{v}},\ \vec{\mathbf{B}},\ \vec{\bar{\mathbf{Z}}}\right\} = \mathbb{S}\setminus\{\vec{\nabla}\rho, \vec{\nabla}\nu, \vec{\mathbf{D}}, \vec{\mathbf{W}}\}\ , \tag{6.85}
$$

which, with the help of

$$
\mathcal{P}_{x_I} = (\rho\eta)_{,x_I} - \lambda^{\varepsilon}(\rho\varepsilon)_{,x_I} \tag{6.86}
$$

and the fact that $\mathcal{P}_{x_I}$ has the same dependence as ε, see $(6.1)_{3,5,7,9}$, leads to the same dependence for the entropy, viz.,

$$
\eta = \hat{\eta}\left(\theta,\ \dot{\theta},\ \nabla\theta,\ \vec{\rho},\ \vec{\nu},\ \vec{\mathbf{v}},\ \vec{\mathbf{B}},\ \vec{\bar{\mathbf{Z}}}\right)\ . \tag{6.87}
$$

If we consider the definition of the one-form $\mathcal{P}$, along with the Liu identities (6.1) for $\mathcal{P}$, we obtain

$$
\begin{aligned}
\mathcal{P} := \mathrm{d}(\rho\eta) - \lambda^{\varepsilon}\mathrm{d}(\rho\varepsilon) &= \sum_{I=1}^{K} \mathcal{P}_{x_I}\mathrm{d}x_I \\
&= \mathcal{P}_{\theta}(\mathrm{d}\theta) + \mathcal{P}_{\nabla\theta}(\mathrm{d}\nabla\theta) + \sum_{\alpha=1}^{m} \bar{l}^{\rho}_{\alpha}(\mathrm{d}\rho_{\alpha}) + \sum_{\alpha=1}^{n-1}(l^{\nu}_{\alpha}+s)(\mathrm{d}\nu_{\alpha}) \\
&\quad + \sum \bar{\rho}_{\alpha}\boldsymbol{\lambda}^{v}_{\alpha}\cdot(\mathrm{d}\mathbf{v}_{\alpha}) + \sum \mathcal{P}_{\mathbf{B}_{\alpha}}\cdot(\mathrm{d}\mathbf{B}_{\alpha}) + \sum \boldsymbol{\lambda}^{Z}_{\alpha}\cdot(\mathrm{d}\bar{\mathbf{Z}}_{\alpha})\ ,
\end{aligned}
\tag{6.88}
$$

with $\bar{l}^{\rho}_{\alpha}$, l^{ν}_{α} and s defined in (5.19), (5.20) and (5.22). Relation (6.88) is regarded as the *general* Gibbs *relation*. Using definitions $(3.64)_{2,3}$, we write

$$
\begin{aligned}
\mathrm{d}(\rho\varepsilon) &= \mathrm{d}(\rho\varepsilon_I) + \mathrm{d}(\rho\varepsilon_D) \\
&\overset{(3.57)}{=} \mathrm{d}(\rho\varepsilon_I) + \mathrm{d}\left(\sum_{\alpha=1}^{n} \tfrac{1}{2}\bar{\rho}_\alpha \mathbf{u}_\alpha \cdot \mathbf{u}_\alpha\right) \\
&= \mathrm{d}(\rho\varepsilon_I) + \sum_{\alpha=1}^{n} \left\{\bar{\rho}_\alpha \mathbf{u}_\alpha \cdot \mathrm{d}(\mathbf{v}_\alpha) + \tfrac{1}{2}(\mathbf{u}_\alpha \cdot \mathbf{u}_\alpha)\mathrm{d}(\bar{\rho}_\alpha)\right\} ,
\end{aligned}
\tag{6.89}
$$

which, when substituted into

$$
\mathrm{d}(\rho\eta) - \lambda^\varepsilon \mathrm{d}(\rho\varepsilon) = \sum_{I=1}^{K} \mathcal{P}_{x_I} \mathrm{d}x_I \ , \tag{6.90}
$$

yields

$$
\begin{aligned}
&\mathrm{d}(\rho\eta) - \lambda^\varepsilon \mathrm{d}(\rho\varepsilon_I) \\
&\quad = \sum_{I=1}^{K} \mathcal{P}_{x_I} \mathrm{d}x_I + \lambda^\varepsilon \mathrm{d}(\rho\varepsilon_D) \\
&\quad = \sum_{I=1}^{K} \mathcal{P}_{x_I} \mathrm{d}x_I + \lambda^\varepsilon \sum_{\alpha=1}^{n} \left\{\bar{\rho}_\alpha \mathbf{u}_\alpha \cdot \mathrm{d}(\mathbf{v}_\alpha) + \tfrac{1}{2}(\mathbf{u}_\alpha \cdot \mathbf{u}_\alpha)\mathrm{d}(\bar{\rho}_\alpha)\right\} .
\end{aligned}
\tag{6.91}
$$

Therefore, using (6.88), this expression takes the form

$$
\begin{aligned}
\mathrm{d}(\rho\eta) - \lambda^\varepsilon \mathrm{d}(\rho\varepsilon_I) &= \mathcal{P}_\theta(\mathrm{d}\theta) + \mathcal{P}_{\nabla\theta}(\mathrm{d}\nabla\theta) + \sum_{\alpha=1}^{m} \underset{\alpha}{\bar{l}}{}^{\rho}_{\mathrm{I}}(\mathrm{d}\rho_\alpha) \\
&\quad + \sum_{\alpha=1}^{n-1} \left(\underset{\alpha}{l}{}^{\nu}_{\mathrm{I}} + s\right)(\mathrm{d}\nu_\alpha) + \sum \bar{\rho}_\alpha \left(\boldsymbol{\lambda}^v_\alpha + \lambda^\varepsilon \mathbf{u}_\alpha\right) \cdot (\mathrm{d}\mathbf{v}_\alpha) \\
&\quad + \sum \mathcal{P}_{\mathbf{B}_\alpha} \cdot (\mathrm{d}\mathbf{B}_\alpha) + \sum \boldsymbol{\lambda}^Z_\alpha \cdot (\mathrm{d}\bar{\mathbf{Z}}_\alpha) \ ,
\end{aligned}
\tag{6.92}
$$

in which we have used the definitions (5.19), (5.20), (5.22) with the aid of which the quantities

$$
\begin{aligned}
&\underset{\alpha}{l}{}^{\rho}_{\mathrm{I}} := l^\rho_\alpha + \tfrac{1}{2}\lambda^\varepsilon(\mathbf{u}_\alpha \cdot \mathbf{u}_\alpha), \\
&\underset{\alpha}{l}{}^{\nu}_{\mathrm{I}} := l^\nu_\alpha + \tfrac{1}{2}\lambda^\varepsilon \rho_\alpha(\mathbf{u}_\alpha \cdot \mathbf{u}_\alpha) - \mathsf{k}, \\
&\mathsf{k} := \tfrac{1}{2}\lambda^\varepsilon \rho_n(\mathbf{u}_n \cdot \mathbf{u}_n)
\end{aligned}
\tag{6.93}
$$

have been defined. Let us now employ the identity

$$\mathrm{d}(\rho\eta) - \lambda^\varepsilon \mathrm{d}(\rho\varepsilon_I) = -\lambda^\varepsilon \mathrm{d}\big(\rho\varepsilon_I - (\lambda^\varepsilon)^{-1}\rho\eta\big) + (\lambda^\varepsilon)^{-1}\,(\rho\eta)\,\mathrm{d}\lambda^\varepsilon \tag{6.94}$$

to write (6.92) as

$$\begin{aligned} -\mathrm{d}\big(\rho\varepsilon_I - \frac{1}{\lambda^\varepsilon}\rho\eta\big) = +\frac{1}{\lambda^\varepsilon}\Big\{ & \Big(\mathcal{P}_\theta - (\lambda^\varepsilon)^{-1}\rho\eta\,(\lambda^\varepsilon)_{,\theta}\Big)(\mathrm{d}\theta) \\ & - (\lambda^\varepsilon)^{-1}\,\rho\eta\,(\lambda^\varepsilon)_{,\dot\theta}\,(\mathrm{d}\dot\theta) + \mathcal{P}_{\nabla\theta}(\mathrm{d}\nabla\theta) \\ & + \sum_{\alpha=1}^{m} \underset{\alpha}{\bar{l}^\rho_{\mathrm{I}}}(\mathrm{d}\rho_\alpha) + \sum_{\alpha=1}^{n-1}\Big(\underset{\alpha}{l^\nu_{\mathrm{I}}} + s\Big)(\mathrm{d}\nu_\alpha) \\ & + \sum \bar\rho_\alpha\,(\boldsymbol{\lambda}^v_\alpha + \lambda^\varepsilon \mathbf{u}_\alpha)\cdot(\mathrm{d}\mathbf{v}_\alpha) \\ & + \sum \mathcal{P}_{\mathbf{B}_\alpha}\cdot(\mathrm{d}\mathbf{B}_\alpha) + \sum \lambda^Z_\alpha\cdot(\mathrm{d}\bar{\mathbf{Z}}_\alpha)\Big\}\,. \end{aligned} \tag{6.95}$$

If we specialize $\mathrm{d}(\cdot)$ in (6.95) to derivatives with respect to the constitutive variables, the following set of equations is obtained

$$\begin{aligned} -\big(\rho\varepsilon_I - \frac{1}{\lambda^\varepsilon}\rho\eta\big)_{,\theta} &= \frac{1}{\lambda^\varepsilon}\mathcal{P}_\theta - \frac{1}{(\lambda^\varepsilon)^2}\,\rho\eta\,\lambda^\varepsilon{}_{,\theta}\,, \\ -\big(\rho\varepsilon_I - \frac{1}{\lambda^\varepsilon}\rho\eta\big)_{,\dot\theta} &= -\frac{1}{(\lambda^\varepsilon)^2}\,\rho\eta\,\lambda^\varepsilon{}_{,\dot\theta}\,, \\ -\big(\rho\varepsilon_I - \frac{1}{\lambda^\varepsilon}\rho\eta\big)_{,\nabla\theta} &= \frac{1}{\lambda^\varepsilon}\mathcal{P}_{\nabla\theta}\,, \\ -\big(\rho\varepsilon_I - \frac{1}{\lambda^\varepsilon}\rho\eta\big)_{,\rho_\alpha} &= \frac{1}{\lambda^\varepsilon}\,\underset{\alpha}{\bar{l}^\rho_{\mathrm{I}}}\,, && \alpha = 1,\dots,m\,, \\ -\big(\rho\varepsilon_I - \frac{1}{\lambda^\varepsilon}\rho\eta\big)_{,\nu_\alpha} &= \frac{1}{\lambda^\varepsilon}\Big(\underset{\alpha}{l^\nu_{\mathrm{I}}} + s\Big)\,, && \alpha = 1,\dots,n-1\,, \\ -\big(\rho\varepsilon_I - \frac{1}{\lambda^\varepsilon}\rho\eta\big)_{,\mathbf{v}_\alpha} &= \frac{1}{\lambda^\varepsilon}\bar\rho_\alpha(\boldsymbol{\lambda}^v_\alpha + \lambda^\varepsilon\mathbf{u}_\alpha)\,, && \alpha = 1,\dots,n\,, \\ -\big(\rho\varepsilon_I - \frac{1}{\lambda^\varepsilon}\rho\eta\big)_{,\mathbf{B}_\alpha} &= \frac{1}{\lambda^\varepsilon}\mathcal{P}_{\mathbf{B}_\alpha}\,, && \alpha = 1,\dots,n\,, \\ -\big(\rho\varepsilon_I - \frac{1}{\lambda^\varepsilon}\rho\eta\big)_{,\bar{\mathbf{Z}}_\alpha} &= \frac{1}{\lambda^\varepsilon}\boldsymbol{\lambda}^Z_\alpha\,, && \alpha = 1,\dots,n\,. \end{aligned} \tag{6.96}$$

From these relations we observe that the LAGRANGE multipliers for the momenta $\boldsymbol{\lambda}^v_\alpha$ ($\alpha = 1,\dots,n$), those for the masses λ^ρ_α ($\alpha = 1,\dots,m$) and volume fractions λ^ν_α ($\alpha = 1,\dots,n-1$), hidden in the definitions of $\underset{\alpha}{\bar{l}^\rho_{\mathrm{I}}}$ and $\underset{\alpha}{l^\nu_{\mathrm{I}}}$, see (5.19), (5.20), depend on the constitutive variables, because they are determined by the 'inner' part of a HELMHOLTZ *free energy*-like quantity, defined as

$$
\begin{aligned}
\Psi^G &:= \varepsilon - (\lambda^\varepsilon)^{-1}\eta = \Psi_I^G + \Psi_D^G \,, \\
\Psi_I^G &:= \varepsilon_I - (\lambda^\varepsilon)^{-1}\eta \,, \qquad \Psi_D^G := \varepsilon_D \,.
\end{aligned}
\tag{6.97}
$$

Moreover, in view of $(6.96)_5$, (6.93) and (5.20), λ_α^ν ($\alpha = 1, \ldots, n-1$) also depend on the constraint variable, s, due to saturation. It is also not correct to call Ψ^G the HELMHOLTZ free energy because we have not made the assumption $\lambda^\varepsilon = (\theta)^{-1}$ see [**A15**]. This is the reason for us to use the superscript G as the identifier of this distinction. On the other hand, we can regard Ψ_I^G as a *potential* because in (6.96), λ^ε is at our disposal to ascertain that $\mathrm{d}\left(\rho\varepsilon_I - \frac{1}{\lambda^\varepsilon}\rho\eta\right)$ is the *total derivative* of a well defined function. In this way, we can regard λ^ε as an *integrating denominator*.

In applying assumption [**A11**] to $(6.96)_6$ we observe that its right-hand side vanishes; so Ψ_I^G becomes independent of all $\mathbf{v}_\alpha$. Thus, [**A11**] effectively says that Ψ_I^G is unaffected by any $\mathbf{v}_\alpha$, i. e.[7]

$$
\Psi_I^G = \hat{\Psi}_I^G(\theta,\ \dot{\theta},\ \nabla\theta,\ \vec{\rho},\ \vec{\nu},\ \vec{\mathbf{B}},\ \vec{\vec{\mathbf{Z}}}) \,. \tag{6.98}
$$

For single material bodies this assumption is a consequence of the principle of objectivity, however for mixtures, [**A11**] goes beyond the principle of objectivity, because not only

$$
\sum (\Psi_I^G)_{,\mathbf{v}_\alpha} = 0 \,, \tag{6.99}
$$

for the sum over the constituents must hold, but a fortiori

$$
(\Psi_I^G)_{,\mathbf{v}_\alpha} = 0 \,, \qquad \alpha = 1, \ldots, n \,, \tag{6.100}
$$

for each constituent individually.

We now introduce the following definitions:

1. The *(true) partial thermodynamic pressures*

$$
\bar{p}_\alpha^G := \rho\rho_\alpha(\Psi_I^G)_{,\rho_\alpha} \,, \qquad \alpha = 1, \ldots, m \,, \tag{6.101}
$$

2. the *configuration pressures*

$$
\beta_\alpha^G := \rho(\Psi_I^G)_{,\nu_\alpha} \,, \qquad \alpha = 1, \ldots, n-1 \,, \tag{6.102}
$$

3. the *saturation pressure*

$$
\varsigma := \frac{s}{\lambda^\varepsilon} \overset{(5.22)}{=} -\frac{l_n^\nu}{\lambda^\varepsilon} \underset{(5.20)}{\overset{(5.19)}{=}} \rho_n \mathbf{u}_n \cdot \mathbf{v}_n - \frac{1}{\lambda^\varepsilon}(\rho_n\lambda_n^\rho + \lambda_n^\nu) \,. \tag{6.103}
$$

[7] In fact, a relation like $(6.96)_6$ arises in many mixture theories. It is the simplifiction of this relation, which is the primary motivation for assumption [**A11**].

The first two are functionally known, once the HELMHOLTZ-like free energy is determined as a function of ρ_α and ν_α. If such dependencies are not present, these pressures vanish and thus do not enter the field equations. For the thermodynamic pressures this is e. g. the case for each volume preserving constituent. The configuration pressure should be accounted for more often than one might think; for instance, in soil mechanics any dependence on void ratio corresponds here to a dependence of Ψ_I^G on the volume fractions. Finally, the saturation pressure is a field related to the nth LAGRANGE multiplier λ_n^ρ, λ_n^ν and λ^ε as indicated in (6.103).

The above definitions allow us to derive from $(6.96)_{4,5}$ that

$$-(\lambda^\varepsilon)^{-1} \underset{\alpha}{l}{}^{\rho}_{\mathrm{I}} = \Psi_I^G + \rho_\alpha^{-1}\, p_\alpha^G, \qquad \alpha = 1, \dots, m\ , \tag{6.104}$$

and

$$\begin{aligned} -(\lambda^\varepsilon)^{-1} \underset{\alpha}{l}{}^{\nu}_{\mathrm{I}} &= \frac{s}{\lambda^\varepsilon} + \rho_{,\nu_\alpha} \Psi_I^G + \rho(\Psi_I^G)_{,\nu_\alpha} \\ &= \varsigma + (\rho_\alpha - \rho_n)\Psi_I^G + \beta_\alpha^G, \qquad \alpha = 1, \dots, n-1\ , \end{aligned} \tag{6.105}$$

in which we also used the auxiliary result

$$\rho_{,\nu_\alpha} = \rho_\alpha - \rho_n\ , \qquad \alpha = 1, \dots, n-1\ . \tag{6.106}$$

The set of $(m+n-1)$ relations in $(6.96)_{4,5}$ which, besides s, includes the $(m+n-1)$ LAGRANGE multipliers for the constituent masses, λ_α^ρ, and volume fractions, λ_α^ν, confirms that the field s (or ς) must be regarded as independent of the constitutive variables, and thus is considered as an independent constraint field.

Following SVENDSEN & HUTTER [115], we also introduce the *'inner' parts of the* GIBBS *free energy-like quantities*, $\underset{\alpha}{\mu}{}^{G}_{\mathrm{I}}$ $(\alpha = 1, \dots, n)$, corresponding to Ψ_I^G. As pointed out before, only when $\lambda^\varepsilon = \theta^{-1}$ we are allowed to call Ψ_I^G the 'inner' part of the HELMHOLTZ free energy, and $\underset{\alpha}{\mu}{}^{G}_{\mathrm{I}}$ the *'inner' parts of the constituent* GIBBS *free energies* (chemical potentials, free enthalpies); nevertheless, all satisfy the requirements of potentials. We define $\underset{\alpha}{\mu}{}^{G}_{\mathrm{I}}$ as

$$\underset{\alpha}{\mu}{}^{G}_{\mathrm{I}} := -(\lambda^\varepsilon)^{-1} \underset{\alpha}{l}{}^{\rho}_{\mathrm{I}} \overset{(6.104)}{=} \Psi_I^G + (\rho_\alpha)^{-1} p_\alpha^G\ , \tag{6.107}$$

where the right-hand side of (6.107) is formally the same as the usual definition of the GIBBS free energy of a viscous fluid (cf. HUTTER [60] or any other book on thermodynamics treating single fluids). If we consider relations (6.101) to (6.105) and the identity

$$\underset{\alpha}{l}{}^{\nu}_{\mathrm{I}} = \rho_\alpha \underset{\alpha}{l}{}^{\rho}_{\mathrm{I}} + \lambda_\alpha^\nu - \mathsf{k}\ , \tag{6.108}$$

which is deducible from (6.93) and (5.20), we obtain for the above variant of the GIBBS free energies

$$\bar{\rho}_\alpha \mu^G_{\stackrel{\alpha}{\mathrm{I}}} = \begin{cases} \rho_\alpha (\rho \Psi^G_I)_{,\rho_\alpha} \stackrel{(6.108)}{=} \nu_\alpha \Big\{ (\rho \Psi^G_I)_{,\nu_\alpha} + \iota_\alpha + \varsigma_I \Big\} , & \alpha = 1, \dots, m \, , \\ \nu_\alpha \Big\{ (\rho \Psi^G_I)_{,\nu_\alpha} + \iota_\alpha + \varsigma_I \Big\} , & \alpha = m+1, \dots, n-1 \, , \\ \nu_\alpha (\iota_\alpha + \varsigma_I) \, , & \alpha = n \, , \end{cases} \tag{6.109}$$

where the abbreviations

$$\iota_\alpha := (\lambda^\varepsilon)^{-1} \lambda^\nu_\alpha \tag{6.110}$$

and

$$\varsigma_I := \varsigma - \tfrac{1}{2} \rho_n (\mathbf{u}_n \cdot \mathbf{u}_n) \tag{6.111}$$

have been used. Whilst ι_α depends on the LAGRANGE multipliers λ^ε and λ^ν_α, ς_I also has a dependence on the diffusive kinetic energy of constituent K_n.

We continue with the integrability conditions on $\boldsymbol{\phi}^{\rho\eta}$, $\mathbf{q}$ and $\bar{\mathbf{T}}_\alpha$. In the language of differentiable forms, we require $\mathrm{d}\boldsymbol{\phi}^{\rho\eta}$, $\mathrm{d}\mathbf{q}$ and $\mathrm{d}\bar{\mathbf{T}}_\alpha$ to be exact one-forms, i. e.

$$\mathrm{d}^2 \left(\boldsymbol{\phi}^{\rho\eta} \right) = \mathbf{0}, \qquad \mathrm{d}^2 \left(\mathbf{q} \right) = \mathbf{0}, \qquad \mathrm{d}^2 \left(\bar{\mathbf{T}}_\alpha \right) = \mathbf{0} \, . \tag{6.112}$$

These relations must be satisfied to make $\boldsymbol{\mathcal{F}}$, defined in (5.14), an exact differential. With the help of (5.14), we can transform $(6.112)_1$ into[8]

$$\mathbf{0} = \mathrm{d}^2 \boldsymbol{\phi}^{\rho\eta} = \mathrm{d}\boldsymbol{\mathcal{F}} - (\mathrm{d}\lambda^\varepsilon) \wedge (\mathrm{d}\mathbf{q}) + \sum (\mathrm{d}\boldsymbol{\lambda}^v_\alpha) \wedge (\mathrm{d}\bar{\mathbf{T}}_\alpha) \, , \tag{6.113}$$

which can also be written as

$$\begin{aligned} \boldsymbol{\mathcal{F}}_{\mathbf{x}_I,\mathbf{x}_J} - \boldsymbol{\mathcal{F}}_{\mathbf{x}_J,\mathbf{x}_I} = & \Big(\mathbf{q}_{,\mathbf{x}_I} (\lambda^\varepsilon)_{,\mathbf{x}_J} - \mathbf{q}_{,\mathbf{x}_J} (\lambda^\varepsilon)_{,\mathbf{x}_I} \Big) \\ & - \sum \Big\{ (\bar{\mathbf{T}}_\alpha)_{,\mathbf{x}_I} (\boldsymbol{\lambda}^v_\alpha)_{,\mathbf{x}_J} - (\bar{\mathbf{T}}_\alpha)_{,\mathbf{x}_J} (\boldsymbol{\lambda}^v_\alpha)_{,\mathbf{x}_I} \Big\}, \end{aligned} \tag{6.114}$$

for $\mathbf{x}_I, \mathbf{x}_J \in \mathbb{S}$ with $(I < J)$. Taking into account [**A11**], [**A15**], $\mathbf{u}_\alpha = \hat{\mathbf{u}}_\alpha(\vec{\rho}, \vec{\nu}, \vec{\mathbf{v}})$ and (see $(6.2)_{2,4}$ and (6.4)),

$$\boldsymbol{\mathcal{F}}_{\mathbf{x}_I} = 0, \qquad \mathbf{x}_I \in \left\{ \nabla\theta, \, \vec{\nabla\rho}, \, \vec{\nabla\nu}, \, \vec{\mathbf{D}}, \, \vec{\mathbf{W}} \right\} \tag{6.115}$$

we obtain

[8] The derivation of (6.113) is similar to that of (6.74).

$$
\begin{aligned}
&\boldsymbol{\mathcal{F}}_{\mathbf{x}_I,\mathbf{x}_J} = \boldsymbol{\mathcal{F}}_{\mathbf{x}_J,\mathbf{x}_I}\,, \\
&\qquad\qquad \mathbf{x}_I,\ \mathbf{x}_J\ \in \mathbb{S}\backslash\{\theta,\ \dot{\theta},\ \vec{\rho},\ \vec{\nu},\ \vec{\mathbf{v}}\}\,, \\
&\boldsymbol{\mathcal{F}}_{\mathbf{x}_I,\mathbf{x}_J} - \boldsymbol{\mathcal{F}}_{\mathbf{x}_J,\mathbf{x}_I} = \lambda^\varepsilon \sum \left\{(\bar{\mathbf{T}}_\alpha)_{,\mathbf{x}_I} (\mathbf{u}_\alpha)_{,\mathbf{x}_J} - (\bar{\mathbf{T}}_\alpha)_{,\mathbf{x}_J} (\mathbf{u}_\alpha)_{,\mathbf{x}_I}\right\}, \\
&\qquad\qquad \mathbf{x}_I,\ \mathbf{x}_J\ \in \left\{\vec{\rho},\ \vec{\nu},\ \vec{\mathbf{v}}\right\}\,, \\
&\boldsymbol{\mathcal{F}}_{\mathbf{x}_I,\mathbf{x}_J} = (\lambda^\varepsilon)_{,\mathbf{x}_I} \left\{\mathbf{q}_{,\mathbf{x}_J} + \sum \left\{(\bar{\mathbf{T}}_\alpha)_{,\mathbf{x}_J} \mathbf{u}_\alpha\right\}\right\}, \\
&\qquad\qquad \mathbf{x}_J\ \in \left\{\nabla\theta,\ \vec{\nabla\rho},\ \vec{\nabla\nu},\ \vec{\mathbf{D}},\ \vec{\mathbf{W}}\right\}\,, \\
&\qquad\qquad \mathbf{x}_I\ \in \left\{\theta,\ \dot{\theta}\right\}\,, \\
&\boldsymbol{\mathcal{F}}_{\mathbf{x}_I,\mathbf{x}_J} = \lambda^\varepsilon \sum \left\{(\bar{\mathbf{T}}_\alpha)_{,\mathbf{x}_J} (\mathbf{u}_\alpha)_{,\mathbf{x}_I}\right\}, \\
&\qquad\qquad \mathbf{x}_J\ \in \left\{\nabla\theta,\ \vec{\nabla\rho},\ \vec{\nabla\nu},\ \vec{\mathbf{D}},\ \vec{\mathbf{W}}\right\}\,, \\
&\qquad\qquad \mathbf{x}_I\ \in \left\{\vec{\rho},\ \vec{\nu},\ \vec{\mathbf{v}}\right\}\,.
\end{aligned}
\tag{6.116}
$$

We listed these integrability conditions on $\boldsymbol{\mathcal{F}}$ for the sake of completeness. The above relations have no direct impact on the model, but of course, we are not allowed to violate them.

It is worthwhile to highlight the key results obtained via the exploitation of the integrability conditions which follow from the LIU identities.

(i) On the basis of [**A11**], i. e. $\boldsymbol{\lambda}^v_\alpha = -\lambda^\varepsilon \mathbf{u}_\alpha$ and [**A15**], i. e. $\lambda^\varepsilon = \hat{\lambda}^\varepsilon(\theta,\ \dot{\theta})$, it was shown that with the aid of (6.83) and earlier relations the internal energy ε, entropy η and HELMHOLTZ-like free energy Ψ^G_I cannot depend on $\vec{\nabla\rho}$, $\vec{\nabla\nu}$, $\vec{\mathbf{D}}$ and $\vec{\mathbf{W}}$ and are thus only functions of $\mathbb{S}_R = \left\{\theta,\ \dot{\theta},\ \nabla\theta,\ \vec{\rho},\ \vec{\nu},\ \vec{\mathbf{v}},\ \vec{\mathbf{B}},\ \vec{\vec{\mathbf{Z}}}\right\}$. Of these thermodynamic potentials, Ψ^G, see (6.97) and (6.98), is particularly significant since [**A11**], in conjunction with $(6.96)_6$, also rules out a dependence of Ψ^G_I on any $\mathbf{v}_\alpha$: $\Psi^G_I = \Psi^G_I(\mathbb{S}_R \setminus \{\vec{\mathbf{v}}\})$.

(ii) The constituent thermodynamic pressures, $\bar{p}^G_\alpha$, constituent configuration pressures, β^G_α and the inner parts of the constituent GIBBS free energies, $\mu^G_{\alpha\mathrm{I}}$ are all density weighted derivatives of the inner part of the HELMHOLTZ-like free energy Ψ^G_I, see (6.101), (6.102), (6.107). They are, hence, equally functions of $\mathbb{S}_R \setminus \{\vec{\mathbf{v}}\}$.

(iii) All LAGRANGE parameters except λ^ν_n can be expressed in terms of thermodynamic quantities. To see this, let us combine the definitions (5.19) and $(6.93)_1$ with (6.104) to obtain

$$l^{\rho}_{\alpha} := \lambda^{\rho}_{\alpha} + \boldsymbol{\lambda}^{v}_{\alpha} \cdot \mathbf{v}_{\alpha} = -\lambda^{\varepsilon}\left(\Psi^{G}_{I} + \rho^{-1}_{\alpha} p^{G}_{\alpha} - \tfrac{1}{2}\mathbf{u}_{\alpha}\cdot\mathbf{u}_{\alpha}\right) , \qquad (\alpha = 1,\ldots,n) , \tag{6.117}$$

or with [**A11**],

$$\lambda^{\rho}_{\alpha} = \lambda^{\varepsilon}\left(\mathbf{u}_{\alpha}\cdot\mathbf{v}_{\alpha} - \Psi^{G}_{I} - \rho^{-1}_{\alpha} p^{G}_{\alpha} + \tfrac{1}{2}\mathbf{u}_{\alpha}\cdot\mathbf{u}_{\alpha}\right) , \qquad (\alpha = 1,\ldots,m) . \tag{6.118}$$

Similarly, from a combination of (5.20) and $(6.93)_{2,3}$ with (6.104) we deduce

$$\begin{aligned}\lambda^{\nu}_{\alpha} &= (-\rho_{\alpha} l^{\rho}_{\alpha})\,\mathrm{H}(m-\alpha) - \lambda^{\varepsilon}\,\big(\,\varsigma + (\rho_{\alpha}-\rho_{n})\Psi^{G}_{I} \\ &\quad + \tfrac{1}{2}(\rho_{\alpha}\mathbf{u}_{\alpha}\cdot\mathbf{u}_{\alpha} - \rho_{n}\mathbf{u}_{n}\cdot\mathbf{u}_{n})\,\big) \\ &= \lambda^{\varepsilon}\boldsymbol{\iota}_{\alpha} , \qquad (\alpha = 1,\ldots,n-1) ,\end{aligned} \tag{6.119}$$

where H is the HEAVISIDE step function. It is seen that with [**A11**] and the above formulae (6.118), (6.119), all LAGRANGE multipliers, except λ^{ν}_{n}, are determined as functions of the universal coldness function $\hat{\lambda}^{\varepsilon}(\theta,\ \dot{\theta})$, the HELMHOLTZ-like free energy, the densities ρ_{α}, and diffusion velocities $\mathbf{u}_{\alpha}$.

(iv) The only undetermined field is a quantity proportional to $-\lambda^{\nu}_{n}$, defined in (5.22) as $s = -l^{\nu}_{n}$, or in (6.103) as $\varsigma = s/\lambda^{\varepsilon}$ and referred to as saturation pressure, see e. g. (6.119). This scalar variable replaces as a new free field the n-th volume fraction, that is lost as an independent field through the saturation condition.

(v) Inner parts of the constituent GIBBS-like free energies, defined in $(6.107)_1$, have their analogue, $\boldsymbol{\iota}_{\alpha} := \lambda^{\nu}_{\alpha}/\lambda^{\varepsilon}$ $(\alpha \le n-1)$, and later will be shown to occur together only with the specific mass and volume fraction production rate densities, respectively. Thus, they play only a role when chemical reactions or fragmentations or internal phase changes occur and non-trivial volume fraction production rates arise. Therefore, in processes where these are absent, they are 'dormant' variables.

(vi) It is straightforward to deduce from $(6.96)_1$ and definition (6.97) that

$$\eta = -\frac{1}{(1/\lambda^{\varepsilon})_{,\theta}}\left\{\Psi^{G}_{I,\theta} + \frac{\mathcal{P}_{\theta}}{\rho\lambda^{\varepsilon}}\right\} , \tag{6.120}$$

which reduces to the classical relation when $\dot{\theta}$ is not among the independent constitutive variables,

$$\eta = -\Psi^G_{I,\theta} \qquad \text{if} \quad \dot{\theta} \notin \mathbb{S} \ , \tag{6.121}$$

since in this case $\mathcal{P}_\theta = 0$, see e. g. (5.30). Else $\mathcal{P}_\theta$ must be determined, (see (vi),(7.36) and (7.86) for its equilibrium value).

(vii) It is equally straightforward to derive from $(6.96)_{2,3,7,8}$, definition (6.97) and (6.120) the formulae

$$\eta = -\frac{1}{(1/\lambda^\varepsilon)_{,\dot{\theta}}} \Psi^G_{I,\dot{\theta}} \ , \tag{6.122}$$

$$\mathcal{P}_\theta = -\rho\lambda^\varepsilon \left\{ (1/\lambda^\varepsilon)_{,\theta}\eta + \Psi^G_{I,\theta} \right\} \ , \tag{6.123}$$

$$\mathcal{P}_{\nabla\theta} = -\rho\lambda^\varepsilon \Psi^G_{I,\nabla\theta} \ , \tag{6.124}$$

$$\mathcal{P}_{\mathbf{B}_\alpha} = \begin{cases} -\rho\lambda^\varepsilon \Psi^G_{I,\mathbf{B}_\alpha} - \lambda^\varepsilon \Psi^G_I \nu_\alpha \frac{\partial \rho_\alpha}{\partial \mathbf{B}_\alpha} \ , & \text{if } \alpha = 1,\dots,m \ , \\ -\rho\lambda^\varepsilon \Psi^G_{I,\mathbf{B}_\alpha} \ , & \text{if } \alpha = m+1,\dots,n \ , \end{cases} \tag{6.125}$$

$$\mathcal{P}_{\bar{\mathbf{Z}}_\alpha} = \boldsymbol{\lambda}^Z_\alpha = -\rho\lambda^\varepsilon \Psi^G_{I,\bar{\mathbf{Z}}_\alpha} \ , \qquad \alpha = 1,\dots,n \ , \tag{6.126}$$

provided $\dot{\theta}$ is an independent constitutive variable. These formulae complete the evaluation of the integrability conditions (6.96). They are interesting because: *first*, they show that non-equilibrium entropy and $\mathcal{P}_\theta$, $\mathcal{P}_{\nabla\theta}$, $\mathcal{P}_{\mathbf{B}_\alpha}$ and $\boldsymbol{\lambda}^Z_\alpha$ $(\alpha = 1,\dots,n)$ are all derivable from the HELMHOLTZ-like free energy, Ψ^G_I. *Second*, combining (6.120) with (6.122) yields an expression for $\mathcal{P}_\theta$, namely

$$\mathcal{P}_\theta = \rho\lambda^\varepsilon (1/\lambda^\varepsilon)_{,\theta} \left\{ \frac{\Psi^G_{I,\dot{\theta}}}{(1/\lambda^\varepsilon)_{,\dot{\theta}}} - \frac{\Psi^G_{I,\theta}}{(1/\lambda^\varepsilon)_{,\theta}} \right\} \ . \tag{6.127}$$

Here, the term in braces is the dynamic minus the static entropy. *Third*, if Ψ^G_I does not depend on $\dot{\theta}$, then $\mathcal{P}_\theta = 0$, and the entropy follows from (6.120). Moreover, in this case $\boldsymbol{\mathcal{F}}_{\dot{\theta}} = \mathcal{P}_{\nabla\theta} = \mathbf{0}$ (see (6.5)), and the HELMHOLTZ-like free energy is independent of $\nabla\theta$.

All these results show a structure which a posteriori support expectations that one might have wished to guess but which are nevertheless surprising. Totally new and surprising is perhaps only (6.127).

Chapter 7
Thermodynamic Analysis II Residual Inequality, Thermodynamic Equilibrium, Isotropic Expansion

Abstract After the full exploitation of the LIU identities in the preceding chapter, we draw in this chapter some (but not all) inferences which follow from the condition that the entropy production density assumes its minimum value in thermodynamic equilibrium. This requirement implies that $\partial\pi^{\rho\eta}/\partial \mathsf{n}_I \,|_{\rm E} = 0$, where $\pi^{\rho\eta}$ is the entropy production density and n_I are those independent variables which vanish in equilibrium. The evaluation of this condition first requires $\pi^{\rho\eta}$ to be expressed in an appropriate form. Choosing for n_I, in turn, the variables $\mathbf{v}_\alpha$, $\dot{\theta}$, $\nabla\theta$ and $\mathbf{D}_\alpha$, which are the constituent velocities, the time rates of change of the temperature, the temperature gradient and constituent stretchings, allows evaluation of the equilibrium representations of the constituent interaction forces, entropy, heat flux vector and constituent stress tensors, which exhibit a clear structure of their dependences on (i) a thermodynamic potential (HELMHOLTZ-like free energy) and thermodynamic, configuration and saturation pressures, (ii) extra entropy flux, (iii) frictional effects via their production terms and (iv) interaction rate densities of constituent mass and volume fractions. It becomes very clear how the various equilibrium terms are affected if simplifying assumptions are made about the functional dependencies of the above mentioned production terms.
The remainder of the chapter deals with quasi-linear expansions of the constitutive relations for the interaction rate densities of mass, volume fraction and for the extra entropy flux vector about a state in which these quantities are linear in the thermodynamic non-equilibrium variables and in the constituent mass and volume fraction gradients. The formulae for the constituent interaction forces, entropy, equilibrium heat flux and constituent stresses are slightly simplified thereby.

7.1 Residual Entropy Inequality in Final Form

In Section 6.1 we have already found one form of the residual entropy inequality, namely (6.12). Now, we wish to incorporate into the entropy inequality the results from the last two sections, i. e. the extra entropy flux $\mathbf{k}$, (6.16),

L. Schneider, K. Hutter, *Solid-Fluid Mixtures of Frictional Materials in Geophysical and Geotechnical Context*, Advances in Geophysical and Environmental Mechanics and Mathematics, DOI 10.1007/978-3-642-02968-4_7,

the pressures p^G_α and β^G_β ($\alpha = 1,\dots,m,\ \beta = 1,\dots,n-1$), (6.101) and (6.102), and the potentials, Ψ^G_I,(6.97) and $\underset{\alpha}{\mu}{}^G_{\mathrm{I}}$, (6.107). Recalling [**A11**] and the definition (6.16) of $\mathbf{k}$ yields

$$\mathbf{k} = \phi^{\rho\eta} + \lambda^\varepsilon \left\{\mathbf{q} + \sum \bar{\mathbf{T}}_\alpha \mathbf{u}_\alpha\right\} , \tag{7.1}$$

and its connection to $\boldsymbol{\mathcal{F}}$ (see, (6.17)) implies the following relations:

$$\begin{aligned}
&\mathbf{k}_{,x_I} = 0, \qquad x_I \in \left\{\nabla\theta,\ \vec{\nabla\rho},\ \vec{\nabla\nu},\ \vec{\mathbf{D}},\ \vec{\mathbf{W}}\right\} ,\\
&\mathbf{k}_{,\theta} = \boldsymbol{\mathcal{F}}_\theta + {\lambda^\varepsilon}_{,\theta} \left\{\mathbf{q} + \sum \bar{\mathbf{T}}_\gamma \mathbf{u}_\gamma\right\} ,\\
&\mathbf{k}_{,\dot\theta} = \boldsymbol{\mathcal{F}}_{\dot\theta} + {\lambda^\varepsilon}_{,\dot\theta} \left\{\mathbf{q} + \sum \bar{\mathbf{T}}_\gamma \mathbf{u}_\gamma\right\} ,\\
&\mathbf{k}_{,\rho_\alpha} = \boldsymbol{\mathcal{F}}_{\rho_\alpha} + \lambda^\varepsilon \sum_{\gamma=1}^n \bar{\mathbf{T}}_\gamma (\mathbf{u}_\gamma)_{,\rho_\alpha} \overset{(3.68)_1}{=} \boldsymbol{\mathcal{F}}_{\rho_\alpha} - \nu_\alpha \lambda^\varepsilon \rho^{-1} \mathbf{T}_I \mathbf{u}_\alpha ,\\
&\mathbf{k}_{,\nu_\beta} = \boldsymbol{\mathcal{F}}_{\nu_\beta} + \lambda^\varepsilon \sum_{\gamma=1}^n \bar{\mathbf{T}}_\gamma (\mathbf{u}_\gamma)_{,\nu_\beta} = \boldsymbol{\mathcal{F}}_{\nu_\beta} - \lambda^\varepsilon \mathbf{T}_I \big(\xi_\beta \mathbf{u}_\beta - \xi_n \mathbf{u}_n\big) , \qquad (7.2)\\
&\mathbf{k}_{,\mathbf{v}_\gamma} = \boldsymbol{\mathcal{F}}_{\mathbf{v}_\gamma} + \lambda^\varepsilon \sum_{\alpha=1}^n \bar{\mathbf{T}}_\alpha (\mathbf{u}_\alpha)_{,\mathbf{v}_\gamma} \overset{(4.48)_1}{=} \boldsymbol{\mathcal{F}}_{\mathbf{v}_\gamma} + \lambda^\varepsilon \big\{\bar{\mathbf{T}}_\gamma - \bar{\xi}_\gamma \mathbf{T}_I\big\} ,\\
&\mathbf{k}_{,\mathbf{B}_\gamma} = \boldsymbol{\mathcal{F}}_{\mathbf{B}_\gamma} \overset{(6.18)_8}{=} -(\mathbf{u}_\gamma \otimes \mathcal{P}_{\mathbf{B}_\gamma}) \overset{(6.96)_7}{=} \lambda^\varepsilon \rho (\mathbf{u}_\gamma \otimes (\Psi^G_I)_{,\mathbf{B}_\gamma}) ,\\
&\mathbf{k}_{,\bar{\mathbf{Z}}_\gamma} = \boldsymbol{\mathcal{F}}_{\bar{\mathbf{Z}}_\gamma} \overset{(6.18)_6}{=} -(\mathbf{u}_\gamma \otimes \boldsymbol{\lambda}^Z_\gamma) \overset{(6.96)_8}{=} \lambda^\varepsilon \rho (\mathbf{u}_\gamma \otimes (\Psi^G_I)_{,\bar{\mathbf{Z}}_\gamma}) ,\\
&\qquad \text{for} \quad \alpha = 1,\dots,m, \quad \beta = 1,\dots,n-1 \quad \text{and} \quad \gamma = 1,\dots,n .
\end{aligned}$$

The first of equations (7.2) follows from (6.17), $(6.18)_{2,4}$ and (6.40). Equations $(7.2)_{4,5,6}$ are derived from (6.17) and one of the identities

$$\begin{aligned}
(\mathbf{u}_\gamma)_{,\rho_\alpha} &= -\,(\rho)^{-1} \nu_\alpha \mathbf{u}_\alpha ,\\
(\mathbf{u}_\gamma)_{,\nu_\beta} &= -\,\big(\xi_\beta \mathbf{u}_\beta - \xi_n \mathbf{u}_n\big) ,\\
(\mathbf{u}_\alpha)_{,\mathbf{v}_\gamma} &= (\delta_{\alpha\gamma} - \bar{\xi}_\gamma)\mathbf{I} ,
\end{aligned} \tag{7.3}$$

where relations $(7.3)_{1,2}$ are derived in Appendix B.5 whilst the derivation of $(7.3)_3$ is straightforward.

With relations (7.2) we are in the position to replace $\boldsymbol{\mathcal{F}}_{\mathbf{x}_I}$, ($\mathbf{x}_I \in \{\theta,\ \vec{\rho},\ \vec{\nu},\ \vec{\mathbf{v}}\}$) in (6.12) by derivatives of $\mathbf{k}$. If we also apply [**A11**] and use either [**A13a**] or [**A13b**] we obtain the following alternative form of the residual entropy inequality:

$$
\begin{aligned}
\pi^{\rho\eta} = \quad & \mathcal{P}_\theta\,(\dot{\theta}) - \Big\{\mathbf{k}_{,\theta} - (\lambda^\varepsilon)_{,\theta}\left\{\mathbf{q} + \sum \bar{\mathbf{T}}_\alpha \mathbf{u}_\alpha\right\}\Big\} \cdot (\nabla\theta) \\
& + \sum_{\alpha=1}^{m} \Big\{\nu_\alpha \big(\boldsymbol{\Gamma}^* - l_\alpha^\rho \mathbf{I}\big)\mathbf{u}_\alpha - \mathbf{k}_{,\rho_\alpha} - \nu_\alpha \lambda^\varepsilon \rho^{-1} \mathbf{T}_I\ \mathbf{u}_\alpha \\
& \qquad\qquad - \rho^{-1}\nu_\alpha\, \mathrm{skw}\,(\mathcal{P}_{\nabla\theta} \otimes \nabla\theta)\, \mathbf{u}_\alpha\Big\} \cdot (\nabla\rho_\alpha) \\
& + \sum_{\alpha=1}^{n-1} \Big\{\big(\rho_\alpha \boldsymbol{\Gamma}^* - l_\alpha^\nu \mathbf{I}\big)\mathbf{u}_\alpha - \mathbf{k}_{,\nu_\alpha} - \lambda^\varepsilon \mathbf{T}_I \big(\xi_\alpha \mathbf{u}_\alpha - \xi_n \mathbf{u}_n\big) \\
& \qquad\qquad + s\mathbf{v} - \mathbf{s}^* - \mathrm{skw}\,(\mathcal{P}_{\nabla\theta} \otimes \nabla\theta)\,\big(\xi_\alpha \mathbf{u}_\alpha - \xi_n \mathbf{u}_n\big)\Big\} \cdot (\nabla\nu_\alpha) \\
& + \sum_{\alpha=1}^{n} \Big\{\nu_\alpha \big(\rho_\alpha \boldsymbol{\Gamma}^* - l_\alpha^\nu \mathbf{I}\big) + \bar{\rho}_\alpha \lambda^\varepsilon (\mathbf{u}_\alpha \otimes \mathbf{u}_\alpha) + \langle \mathbf{B}_\alpha\ ,\ \mathcal{P}_{\mathbf{B}_\alpha}\rangle \\
& \qquad\qquad - \mathbf{k}_{,\mathbf{v}_\alpha} + \lambda^\varepsilon \big(\bar{\mathbf{T}}_\alpha - \bar{\xi}_\alpha \mathbf{T}_I\ \big)\Big\} \cdot (\mathbf{D}_\alpha) \\
& + \sum_{\alpha=1}^{n} \boldsymbol{\lambda}_\alpha^Z \cdot \bar{\boldsymbol{\Phi}}_\alpha \\
& + \sum_{\alpha=1}^{n} \left\{\boldsymbol{\lambda}_\alpha^v \cdot \bar{\mathbf{m}}_\alpha^i + l_\alpha^\rho \bar{\rho}_\alpha c_\alpha + \lambda_\alpha^\nu \bar{n}_\alpha\right\} \\
\geqslant 0. &
\end{aligned}
\tag{7.4}
$$

In comparison to (6.12) the contribution indicated in (6.12) by the asterisk is now gone and $\mathcal{F}_{x_I}$ are replaced by the derivatives of $\mathbf{k}$ with respect to $\{\theta,\ \rho_\alpha,\ \nu_\alpha,\ \mathbf{v}_\alpha\}$. By incorporating $(\Psi_I^G)_{,\mathbf{B}_\alpha}$, $(\Psi_I^G)_{,\bar{\mathbf{Z}}_\alpha}$, the free enthalpies $\mu_{\alpha\mathrm{I}}^G$, the true thermodynamic and the configuration pressures, p_α^G and β_α^G, respectively into (7.4), we obtain, after cumbersome calculations, the following inequality (see Appendix B.6)

$$
\begin{aligned}
\pi^{\rho\eta} = \quad & \mathcal{P}_\theta\ (\dot{\theta}) - \Big\{ \mathbf{k},_\theta - (\lambda^\varepsilon),_\theta \{\mathbf{q} + \sum \bar{\mathbf{T}}_\alpha \mathbf{u}_\alpha\} \Big\} \cdot (\nabla\theta) \\
& + \lambda^\varepsilon \sum_{\alpha=1}^{m} \Big\{ \nu_\alpha \big(\boldsymbol{\Delta}_D^{*\alpha} + \rho_\alpha^{-1} p_\alpha^G \mathbf{I}\big) \mathbf{u}_\alpha - (\lambda^\varepsilon)^{-1} \mathbf{k},_{\rho_\alpha} \\
& \qquad - \nu_\alpha\ (\rho\lambda^\varepsilon)^{-1}\, \mathrm{skw}\,(\mathcal{P}_{\nabla\theta} \otimes \nabla\theta)\, \mathbf{u}_\alpha \Big\} \cdot (\nabla\rho_\alpha) \\
& + \lambda^\varepsilon \sum_{\alpha=1}^{n-1} \Big\{ \big\{\rho_\alpha \boldsymbol{\Delta}_D^{*\alpha} + \zeta_\alpha \mathbf{I} - \tfrac{1}{2}\rho_n (\mathbf{u}_n \cdot \mathbf{u}_n) \mathbf{I}\big\} \mathbf{u}_\alpha - \mathbf{c} - (\lambda^\varepsilon)^{-1} \mathbf{k},_{\nu_\alpha} \\
& \qquad - (\lambda^\varepsilon)^{-1}\, \mathrm{skw}\,(\mathcal{P}_{\nabla\theta} \otimes \nabla\theta)\, \big(\xi_\alpha \mathbf{u}_\alpha - \xi_n \mathbf{u}_n\big) \Big\} \cdot (\nabla\nu_\alpha) \\
& + \lambda^\varepsilon \sum_{\alpha=1}^{n} \Big\{ \bar{\rho}_\alpha \big(\boldsymbol{\Delta}_D^{*\alpha} + \mathbf{u}_\alpha \otimes \mathbf{u}_\alpha\big) + \big(\bar{\zeta}_\alpha - \tfrac{1}{2}\nu_\alpha \rho_n (\mathbf{u}_n \cdot \mathbf{u}_n)\big) \mathbf{I} \\
& \qquad - 2\rho\ \mathrm{sym}\,\big((\Psi_I^G),_{\mathbf{B}_\alpha}\big) \mathbf{B}_\alpha - (\lambda^\varepsilon)^{-1} \mathbf{k},_{\mathbf{v}_\alpha} + \bar{\mathbf{T}}_\alpha \Big\} \cdot (\mathbf{D}_\alpha) \\
& - \lambda^\varepsilon \sum_{\alpha=1}^{n} \rho (\Psi_I^G),_{\bar{\mathbf{Z}}_\alpha} \cdot \bar{\boldsymbol{\Phi}}_\alpha \\
& - \lambda^\varepsilon \sum_{\alpha=1}^{n} \Big\{ \mathbf{u}_\alpha \cdot \bar{\mathbf{m}}_\alpha^i + \bar{\rho}_\alpha \big(\mu_{\alpha \mathrm{I}}^G + \tfrac{1}{2}\mathbf{u}_\alpha \cdot \mathbf{u}_\alpha\big) c_\alpha - \boldsymbol{\iota}_\alpha \bar{n}_\alpha \Big\} \\
\geqslant 0 \quad & ,
\end{aligned}
\tag{7.5}
$$

in which the following definitions have been used:

$$
\boldsymbol{\Delta}_D^{*\alpha} := \rho^{-1} \big\{ \mathbf{T}_D - (\lambda^\varepsilon)^{-1}\, \mathrm{sym}\,(\mathcal{P}_{\nabla\theta} \otimes \nabla\theta) \big\} - \Psi_D^G \mathbf{I} + \tfrac{1}{2}(\mathbf{u}_\alpha \cdot \mathbf{u}_\alpha)\mathbf{I}\ , \tag{7.6}
$$

$$
\zeta_\alpha := \begin{cases} \beta_\alpha^G - \rho_n \Psi_I^G + \varsigma, & \alpha = 1, \dots, n-1, \\ -\rho_n \Psi_I^G + \varsigma, & \alpha = n, \end{cases} \tag{7.7}
$$

$$
\mathbf{c} := \big(\rho_n \boldsymbol{\Delta}_D^{*n} + \zeta_n \mathbf{I}\big) \mathbf{u}_n\ , \tag{7.8}
$$

with $\mathbf{T}_D$ given in (3.68). For convenience, we also recall definitions (6.97), (6.101), (6.102) and (6.103), i. e.

$$
\begin{aligned}
& \Psi_I^G := \varepsilon_I - (\lambda^\varepsilon)^{-1}\eta\ , && \bar{p}_\alpha^G := \rho\rho_\alpha (\Psi_I^G),_{\rho_\alpha}\ , \\
& \beta_\alpha^G := \rho (\Psi_I^G),_{\nu_\alpha}\ , && \varsigma := \tfrac{s}{\lambda^\varepsilon}\ , \\
& \boldsymbol{\iota}_\alpha := \tfrac{\lambda_\alpha^\nu}{\lambda^\varepsilon}\ ,
\end{aligned}
\tag{7.9}
$$

and also recall (6.120), (6.122)-(6.126). Needless to say, that all quantities in (7.6)-(7.9) and (6.120), (6.122)-(6.126) are known once Ψ_I^G, ς and $\mathbf{u}_\alpha$ are prescribed, where $\boldsymbol{\iota}_\alpha$ follow, from (6.110) and (6.119). Moreover, (7.5) has the appropriate form from which thermodynamic equilibrium properties can be deduced.

7.2 Mixture Thermodynamic Equilibrium

We say that a process describes a *thermodynamic equilibrium* if no entropy is produced in the course of such a process. In the present situation the mixture entropy production rate density, $\pi^{\rho\eta}$, vanishes if the following conditions hold:

- the *non-equilibrium variables*, n, vanish, i. e.

$$\mathsf{n} := \left\{\dot{\theta},\ \nabla\theta,\ \vec{\mathbf{v}},\ \vec{\mathbf{D}},\ \vec{\mathbf{W}}\right\} = \mathbf{0}, \tag{7.10}$$

- the constitutive quantities of the frictional production rate densities, $\bar{\boldsymbol{\Phi}}_\alpha$ $(\alpha = 1, \ldots, n)$, are zero in thermodynamic equilibrium,

$$\lim_{\mathsf{n}\to 0} \bar{\boldsymbol{\Phi}}_\alpha \quad =: \quad \bar{\boldsymbol{\Phi}}_\alpha\big|_{\mathrm{E}} \quad = \quad \mathbf{0}, \qquad \alpha = 1, \ldots, n\ , \tag{7.11}$$

- the interaction rate densities for mass, c_α, and volume fraction, n_α $(\alpha = 1, \ldots, n)$, also vanish in thermodynamic equilibrium,

$$\lim_{\mathsf{n}\to 0} c_\alpha \ =: \ c_\alpha\big|_{\mathrm{E}} = 0, \quad \lim_{\mathsf{n}\to 0} n_\alpha \ =: \ n_\alpha\big|_{\mathrm{E}} = 0, \quad \alpha = 1, \ldots, n\,. \tag{7.12}$$

The complement to the set of non-equilibrium variables in $\mathbb{S}$, n, is that of the *equilibrium variables*

$$\mathsf{e} := \left\{\theta,\ \vec{\rho},\ \vec{\nu},\ \vec{\nabla\rho},\ \vec{\nabla\nu},\ \vec{\mathbf{B}},\ \vec{\bar{\mathbf{Z}}}\right\}. \tag{7.13}$$

Without loss of generality, it is reasonable to decompose all constitutive quantities, denoted by $\hat{\mathsf{C}}(\mathsf{e},\mathsf{n})$, into an equilibrium part, $\hat{\mathsf{C}}\big|_{\mathrm{E}}$, and a non-equilibrium part, $\hat{\mathsf{C}}\big|_{\mathrm{N}}$, i. e.

$$\begin{aligned}
&\hat{\mathsf{C}}\ (\mathsf{e},\ \mathsf{n}) = \hat{\mathsf{C}}\ \big|_{\mathrm{E}}(\mathsf{e}) + \hat{\mathsf{C}}\ \big|_{\mathrm{N}}(\mathsf{e},\ \mathsf{n})\ ,\\
&\hat{\mathsf{C}}\ \big|_{\mathrm{E}}(\mathsf{e}) := \lim_{\mathsf{n}\to 0} \hat{\mathsf{C}}\ (\mathsf{e},\ \mathsf{n})\ ,\\
&\hat{\mathsf{C}}\ \big|_{\mathrm{N}}(\mathsf{e},\ \mathsf{n}) := \hat{\mathsf{C}}(\mathsf{e},\ \mathsf{n}) - \hat{\mathsf{C}}\ \big|_{\mathrm{E}}(\mathsf{e})\ .
\end{aligned} \tag{7.14}$$

In the present section our aim is to find representations for the equilibrium parts of the constitutive quantities. To this end, we exploit the fact that under conditions (i) to (iii) above, $\pi^{\rho\eta}$ assumes its minimum value. This property of $\pi^{\rho\eta}$ is obvious, because, first, $\pi^{\rho\eta}$ satisfies the original entropy inequality

$$\pi^{\rho\eta} \geqslant 0 \ , \tag{7.15}$$

and, second, the right-hand side of (7.5) vanishes if conditions (i) to (iii) are applied to it, i. e.

$$\pi^{\rho\eta}\big|_{\mathrm{E}} = 0 \ . \tag{7.16}$$

For the proof of the latter statement, use has to be made of the facts

$$\mathbf{k}\big|_{\mathrm{E}} = \mathbf{0} \ , \qquad \mathbf{k}_{,\rho_\alpha}\big|_{\mathrm{E}} = \mathbf{0} \ , \qquad \mathbf{k}_{,\nu_\alpha}\big|_{\mathrm{E}} = \mathbf{0} \ , \tag{7.17}$$

results which directly follow from (6.58) or (6.59) and (B.43), (B.44). Thus, *necessary conditions* for $\pi^{\rho\eta}$ having a minimum at $\mathsf{n} = \mathbf{0}$, are

$$\pi^{\rho\eta}{}_{,\mathsf{n}_I}\big|_{\mathrm{E}} = 0 \qquad \forall\, \mathsf{n}_I \in \mathsf{n} \ , \tag{7.18}$$

as well as

$$\left\{\pi^{\rho\eta}{}_{,\mathsf{n}_I\mathsf{n}_J}\big|_{\mathrm{E}}\right\} \quad \text{is non-negative definite} \quad \mathsf{n}_I, \mathsf{n}_J \in \mathsf{n} \ . \tag{7.19}$$

In preparation for the exploitation of relation (7.18), we collect some auxiliary results:

- We have already mentioned that the vector-valued extra entropy flux $\mathbf{k}$ vanishes in thermodynamic equilibrium (see (7.17)). In addition, from $(7.2)_1$ we observe that $\mathbf{v}_\alpha$ $(\alpha = 1, \ldots, n)$ are the only vector-valued constitutive variables which $\mathbf{k}$ depends upon, i. e.

$$\mathbf{k} = \hat{\mathbf{k}}\big(\theta,\ \dot{\theta},\ \vec{\rho},\ \vec{\nu},\ \vec{\mathbf{v}},\ \vec{\mathbf{B}},\ \vec{\vec{\mathbf{Z}}}\big) = \hat{\mathbf{k}}\,(\mathbb{S}_R) \ . \tag{7.20}$$

  Thus, in every term of the isotropic representation of $\mathbf{k}$, $\vec{\mathbf{v}}$ must be present at least once[1] and consequently,

$$\mathbf{k}_{,\mathbf{y}_I}\big|_{\mathrm{E}} = \mathbf{0} \ , \quad \mathbf{k}_{,\mathbf{y}_I\mathbf{y}_J}\big|_{\mathrm{E}} = \mathbf{0} \ , \quad \mathbf{y}_I,\ \mathbf{y}_J \in \mathbb{S}_R \backslash \{\vec{\mathbf{v}}\} \ , \tag{7.21}$$

  but in general

$$\mathbf{k}_{,\mathbf{v}_\alpha}\big|_{\mathrm{E}} \neq \mathbf{0} \ , \quad \mathbf{k}_{,\mathbf{v}_\alpha\mathbf{y}_I}\big|_{\mathrm{E}} \neq \mathbf{0} \ , \quad \mathbf{y}_I \in \mathbb{S}_R \backslash \{\vec{\mathbf{v}}\} \ , \ \alpha = 1, \ldots, n. \tag{7.22}$$

- The interaction supply rate densities for momentum, $\mathbf{m}_\alpha$ $(\alpha = 1, \ldots, n)$ are allowed to be non-zero in thermodynamic equilibrium, but it follows from (4.6) and $(4.7)_2$ that

$$\sum \bar{\mathbf{m}}^i_\alpha\big|_{\mathrm{E}} = \sum \bar{\mathbf{m}}_\alpha\big|_{\mathrm{E}} = \mathbf{0} \ . \tag{7.23}$$

[1] This is confirmed by equation (6.58).

- For convenience, we recall relations (3.66), (3.64) and (6.98):

$$\mathbf{T}_D = -\sum \bar{\rho}_\alpha \mathbf{u}_\alpha \otimes \mathbf{u}_\alpha = \hat{\mathbf{T}}_D(\vec{\rho},\ \vec{\nu},\ \vec{\mathbf{v}}) \ , \tag{7.24}$$

$$\Psi_D^G = \varepsilon_D = \tfrac{1}{2}\sum \bar{\xi}_\alpha \mathbf{u}_\alpha \cdot \mathbf{u}_\alpha = \hat{\Psi}_D(\vec{\rho},\ \vec{\nu},\ \vec{\mathbf{v}}) \ , \tag{7.25}$$

$$\Psi_I^G = \hat{\Psi}_I^G(\theta,\ \dot{\theta},\ \nabla\theta,\ \vec{\rho},\ \vec{\nu},\ \vec{\mathbf{B}},\ \vec{\bar{\mathbf{Z}}}) \ . \tag{7.26}$$

- The true thermodynamic pressures, p_α^G ($\alpha = 1,\dots,m$), and the configuration pressures β_α^G ($\alpha = 1,\dots,n-1$), possess the same dependencies as Ψ_I^G. If we consider relations (7.24), (7.25) and the definitions for $\boldsymbol{\Delta}_D^{*\alpha}$ and c (see (7.6) and (7.8)), it is observed that

$$\boldsymbol{\Delta}_D^{*\alpha}\big|_{\mathrm{E}} = \mathbf{0} \qquad \text{and} \qquad \mathsf{c}\big|_{\mathrm{E}} = \mathbf{0}. \tag{7.27}$$

We start the exploitation of condition (7.18) for $\pi^{\rho\eta}$ having a minimum at $\mathsf{n} = 0$ with the evaluation of

$$\pi^{\rho\eta}{}_{,\mathbf{v}_\beta}\big|_{\mathrm{E}} \stackrel{!}{=} 0 \ . \tag{7.28}$$

With the help of $(7.3)_3$, we obtain from (7.5)

$$\begin{aligned}
\pi^{\rho\eta}{}_{,\mathbf{v}_\beta}\big|_{\mathrm{E}} &= \lambda^\varepsilon\big|_{\mathrm{E}} \sum_{\alpha=1}^{m} \Big\{ (\delta_{\alpha\beta} - \bar{\xi}_\beta)\, \nu_\alpha\, (\rho_\alpha)^{-1} p_\alpha^G\big|_{\mathrm{E}} \mathbf{I} - (\lambda^\varepsilon)^{-1}\big|_{\mathrm{E}} \mathbf{k}_{,\rho_\alpha \mathbf{v}_\beta}\big|_{\mathrm{E}} \Big\} \nabla\rho_\alpha \\
&+ \lambda^\varepsilon\big|_{\mathrm{E}} \sum_{\alpha=1}^{n-1} \Big\{ (\delta_{\alpha\beta} - \bar{\xi}_\beta)\zeta_\alpha\big|_{\mathrm{E}} \mathbf{I} - \mathsf{c}_{,\mathbf{v}_\beta}\big|_{\mathrm{E}} - (\lambda^\varepsilon)^{-1}\big|_{\mathrm{E}} \mathbf{k}_{,\nu_\alpha \mathbf{v}_\beta}\big|_{\mathrm{E}} \Big\} \nabla\nu_\alpha \\
&- \lambda^\varepsilon\big|_{\mathrm{E}} \sum_{\alpha=1}^{n} \rho\, (\Psi_I^G)_{,\bar{\mathbf{Z}}_\alpha}\big|_{\mathrm{E}} (\bar{\boldsymbol{\Phi}}_\alpha)_{,\mathbf{v}_\beta}\big|_{\mathrm{E}} \\
&- \lambda^\varepsilon\big|_{\mathrm{E}} \sum_{\alpha=1}^{n} (\delta_{\alpha\beta} - \bar{\xi}_\beta)\bar{\mathbf{m}}_\alpha^i\big|_{\mathrm{E}} \\
&- \lambda^\varepsilon\big|_{\mathrm{E}} \sum_{\alpha=1}^{n} \Big\{ \bar{\rho}_\alpha\, \mu_{\mathrm{I}\alpha}^G\big|_{\mathrm{E}}\, (c_\alpha)_{,\mathbf{v}_\beta}\big|_{\mathrm{E}} - \boldsymbol{\iota}_\alpha\big|_{\mathrm{E}} (\bar{n}_\alpha)_{,\mathbf{v}_\beta}\big|_{\mathrm{E}} \Big\} \\
&\stackrel{!}{=} 0 \ .
\end{aligned} \tag{7.29}$$

This equation can be used to evaluate $\bar{\mathbf{m}}_\alpha^i\big|_{\mathrm{E}}$. Taking into account that

$$-\lambda^{\varepsilon}\big|_{\mathrm{E}}\sum_{\alpha=1}^{n}(\delta_{\alpha\beta}-\bar{\xi}_{\beta})\bar{\mathbf{m}}_{\alpha}^{i}\big|_{\mathrm{E}}\overset{(7.23)}{=}-\lambda^{\varepsilon}\big|_{\mathrm{E}}\,\bar{\mathbf{m}}_{\beta}^{i}\big|_{\mathrm{E}}+\lambda^{\varepsilon}\big|_{\mathrm{E}}\bar{\xi}_{\beta}\underbrace{\sum_{\alpha=1}^{n}\bar{\mathbf{m}}_{\alpha}^{i}\big|_{\mathrm{E}}}_{0}\,, \tag{7.30}$$

we obtain

$$\begin{aligned}
\bar{\mathbf{m}}_{\beta}^{i}\big|_{\mathrm{E}} &= \bar{\mathbf{m}}_{\beta}\big|_{\mathrm{E}}\\
&= \sum_{\alpha=1}^{m}\Big\{\big(\delta_{\alpha\beta}-\bar{\xi}_{\beta}\big)\;\nu_{\alpha}\;(\rho_{\alpha})^{-1}p_{\alpha}^{G}\big|_{\mathrm{E}}\;\mathbf{I}-(\lambda^{\varepsilon})^{-1}\big|_{\mathrm{E}}\;\mathbf{k}_{,\rho_{\alpha}\mathbf{v}_{\beta}}\big|_{\mathrm{E}}\Big\}\nabla\rho_{\alpha}\\
&\quad+\sum_{\alpha=1}^{n-1}\Big\{\big(\delta_{\alpha\beta}-\bar{\xi}_{\beta}\big)\zeta_{\alpha}\big|_{\mathrm{E}}\mathbf{I}-\mathbf{c}_{,\mathbf{v}_{\beta}}\big|_{\mathrm{E}}-(\lambda^{\varepsilon})^{-1}\big|_{\mathrm{E}}\;\mathbf{k}_{,\nu_{\alpha}\mathbf{v}_{\beta}}\big|_{\mathrm{E}}\Big\}\nabla\nu_{\alpha}\\
&\quad-\sum_{\alpha=1}^{n}\rho\left(\Psi_{I}^{G}\right)_{,\bar{\mathbf{Z}}_{\alpha}}\big|_{\mathrm{E}}\left(\bar{\mathbf{\Phi}}_{\alpha}\right)_{,\mathbf{v}_{\beta}}\big|_{\mathrm{E}}\\
&\quad-\sum_{\alpha=1}^{n}\Big\{\bar{\rho}_{\alpha}\mu_{\alpha\,\mathrm{I}}^{G}\big|_{\mathrm{E}}\;(c_{\alpha})_{,\mathbf{v}_{\beta}}\big|_{\mathrm{E}}-\boldsymbol{\iota}_{\alpha}\big|_{\mathrm{E}}(\bar{n}_{\alpha})_{,\mathbf{v}_{\beta}}\big|_{\mathrm{E}}\Big\}\,.
\end{aligned} \tag{7.31}$$

This result is significant for several reasons.

First, we observe that the frictional constitutive quantities, $\bar{\mathbf{\Phi}}_{\alpha}$ ($\alpha = 1,\ldots,n$), influence the equilibrium quantities, $\bar{\mathbf{m}}_{\beta}^{i}\big|_{\mathrm{E}}$, directly, via the term involving $\left(\bar{\mathbf{\Phi}}_{\alpha}\right)_{,\mathbf{v}_{\beta}}\big|_{\mathrm{E}}$, but this term is not likely to be significant, so that frictional effects will eventually influence the equilibrium interaction forces via the constitutive variables $\bar{\mathbf{Z}}_{\alpha}$ ($\alpha = 1,\ldots,n$). The argument why we will eventually assume $\left(\bar{\mathbf{\Phi}}_{\alpha}\right)_{,\mathbf{v}_{\beta}}\big|_{\mathrm{E}} = \mathbf{0}$ can be given as follows: In an expansion of the isotropic tensor function $\bar{\mathbf{\Phi}}_{\alpha}(\mathbb{S})$ non-vanishing contributions of $\left(\bar{\mathbf{\Phi}}_{\alpha}\right)_{,\mathbf{v}_{\beta}}\big|_{\mathrm{E}}$ can only come from symmetrized dyads $\mathbf{a}\otimes\mathbf{v}_{\beta}$ and scalar invariants of $\mathbb{S}$ which are linear in $\mathbf{v}_{\beta}$, where $\mathbf{a}$ is any vector formed with the set $\mathbb{S}$, see [**A8**]. Such candidates are only $\{\mathbf{a}\} = \{\vec{\nabla\rho},\ \vec{\nabla\nu},\ \vec{\mathbf{B}}\vec{\nabla\rho},\ \vec{\mathbf{B}}\vec{\nabla\nu},\ \vec{\vec{\mathbf{Z}}}\vec{\nabla\rho},\ \vec{\vec{\mathbf{Z}}}\vec{\nabla\nu}\}$ and higher order vector valued products of $\vec{\mathbf{B}}$, $\vec{\vec{\mathbf{Z}}}$ with $\vec{\nabla\rho}$ and $\vec{\nabla\nu}$ and the first invariants $\mathrm{I}_{\mathrm{sym}(\mathbf{a}\otimes\mathbf{v}_{\alpha})}$. In the most simple cases where $(\vec{\nabla\rho},\ \vec{\nabla\nu}) \notin \mathbb{S}$, we have $\{\mathbf{a}\} = \{\mathbf{0}\}$ and consequently $\left(\bar{\mathbf{\Phi}}_{\alpha}\right)_{,\mathbf{v}_{\beta}}\big|_{\mathrm{E}} = \mathbf{0}$. This suggests that $\left(\bar{\mathbf{\Phi}}_{\alpha}\right)_{,\mathbf{v}_{\beta}}\big|_{\mathrm{E}}$ is weak when $(\vec{\nabla\rho},\ \vec{\nabla\nu}) \in \mathbb{S}$. We may then write (7.31) without the frictional term

$$\bar{\mathbf{m}}^i_\beta\big|_{\mathrm{E}} = \bar{\mathbf{m}}_\beta\big|_{\mathrm{E}} = \sum_{\alpha=1}^{m}\Big\{\left(\delta_{\alpha\beta} - \bar{\xi}_\beta\right)\nu_\alpha\left(\rho_\alpha\right)^{-1} p^G_\alpha\big|_{\mathrm{E}}\mathbf{I} - (\lambda^\varepsilon)^{-1}\big|_{\mathrm{E}}\mathbf{k},_{\rho_\alpha \mathbf{v}_\beta}\big|_{\mathrm{E}}\Big\}\nabla\rho_\alpha$$

$$+\sum_{\alpha=1}^{n-1}\Big\{\left(\delta_{\alpha\beta} - \bar{\xi}_\beta\right)\zeta_\alpha\big|_{\mathrm{E}}\mathbf{I} - \mathbf{c},_{\mathbf{v}_\beta}\big|_{\mathrm{E}} - (\lambda^\varepsilon)^{-1}\big|_{\mathrm{E}}\,\mathbf{k},_{\nu_\alpha \mathbf{v}_\beta}\big|_{\mathrm{E}}\Big\}\nabla\nu_\alpha$$

$$-\sum_{\alpha=1}^{n}\Big\{\bar{\rho}_\alpha\mu^G_{\alpha\mathrm{I}}\big|_{\mathrm{E}}\,(c_\alpha),_{\mathbf{v}_\beta}\big|_{\mathrm{E}} - \iota_\alpha\big|_{\mathrm{E}}(\bar{n}_\alpha),_{\mathbf{v}_\beta}\big|_{\mathrm{E}}\Big\}\,. \tag{7.32}$$

Second, for the special case that all constituents are density preserving and no mass- and volume fraction interaction rate densities are present (first and fourth line on the right-hand side of (7.31) are absent), consideration of $\nabla\nu_\alpha$ $(\alpha = 1,\ldots,n)$ is essential for the description of $\bar{\mathbf{m}}^i_\beta\big|_{\mathrm{E}}$ $(\beta = 1,\ldots,n)$. However, if we would disregard all $\nabla\nu_\alpha$ in the constitutive law [**A8**], $\bar{\mathbf{m}}^i_\beta\big|_{\mathrm{E}}$ would not be zero in that case because the pressure like variable ζ_α contains the saturation pressure, see (7.7).

Third, and conceptually significant, the equilibrium interaction supply rate densities of momenta are given as a result of thermodynamic equilibrium requirements even though they have purely mechanical significance.[2] There is no freedom for their choice.

If $\mathsf{n}_I = \dot{\theta}$, we derive from (7.18) and (7.5)

$$\pi^{\rho\eta},_{\dot{\theta}}\big|_{\mathrm{E}} = \mathcal{P}_\theta\big|_{\mathrm{E}} - \lambda^\varepsilon\big|_{\mathrm{E}}\sum_{\alpha=1}^{n}\rho\left(\Psi^G_I\right),_{\bar{\mathbf{Z}}_\alpha}\big|_{\mathrm{E}}\cdot(\bar{\mathbf{\Phi}}_\alpha),_{\dot{\theta}}\big|_{\mathrm{E}}$$

$$-\lambda^\varepsilon\big|_{\mathrm{E}}\sum_{\alpha=1}^{n}\Big\{\bar{\rho}_\alpha\mu^G_{\alpha\mathrm{I}}\big|_{\mathrm{E}}\,(c_\alpha),_{\dot{\theta}}\big|_{\mathrm{E}} - \iota_\alpha\big|_{\mathrm{E}}(\bar{n}_\alpha),_{\dot{\theta}}\big|_{\mathrm{E}}\Big\}$$

$$\stackrel{!}{=} 0\,. \tag{7.33}$$

Here, we used the relations

$$\mathbf{k},_{\rho_\alpha}\big|_{\mathrm{E}} = \mathbf{0},\qquad \mathbf{k},_{\nu_\alpha}\big|_{\mathrm{E}} = \mathbf{0},\qquad \mathbf{k},_{\rho_\alpha\dot{\theta}}\big|_{\mathrm{E}} = \mathbf{0},\qquad \mathbf{k},_{\nu_\alpha\dot{\theta}}\big|_{\mathrm{E}} = \mathbf{0} \tag{7.34}$$

and

$$\mathbf{c}\big|_{\mathrm{E}} = \mathbf{0},\qquad \mathbf{c},_{\dot{\theta}}\big|_{\mathrm{E}} = \mathbf{0}\,. \tag{7.35}$$

---

[2] We emphasise that this is a consequence of the Second Law of Thermodynamics. Purely mechanical reasoning, that is sometimes used to 'derive' expressions for the interaction force, must be regarded as a priori estimates. They must be complemented by an additional term whose structure follows from the Second Law of Thermodynamics.

The identities (7.34) follow from (7.21), and (7.35) is obtained from the definition of $\mathbf{c}$ (see (7.8)). Relation (7.33) allows determination of $\mathcal{P}_\theta\big|_{\mathrm{E}}$ and yields

$$\begin{aligned}\mathcal{P}_\theta\big|_{\mathrm{E}} &= \lambda^\varepsilon\big|_{\mathrm{E}} \sum_{\alpha=1}^{n} \rho \left(\Psi_I^G\right),_{\bar{\mathbf{z}}_\alpha}\big|_{\mathrm{E}} \cdot (\bar{\mathbf{\Phi}}_\alpha),_{\dot{\theta}}\big|_{\mathrm{E}} \\ &\quad + \lambda^\varepsilon\big|_{\mathrm{E}} \sum_{\alpha=1}^{n} \left\{ \bar{\rho}_\alpha \mu^G_{\underset{\alpha}{\mathrm{I}}}\big|_{\mathrm{E}} \, (c_\alpha),_{\dot{\theta}}\big|_{\mathrm{E}} - \boldsymbol{\iota}_\alpha\big|_{\mathrm{E}} (\bar{n}_\alpha),_{\dot{\theta}}\big|_{\mathrm{E}} \right\} ,\end{aligned} \tag{7.36}$$

which is a revealing by-product of our thermodynamic analysis. Indeed, with (6.120), or

$$\eta = -\frac{1}{(1/\lambda^\varepsilon)_{,\theta}} \left\{ \Psi^G_{I,\theta} + \frac{\mathcal{P}_\theta}{\rho\lambda^\varepsilon} \right\} , \tag{7.37}$$

the entropy in thermostatic equilibrium takes the form

$$\begin{aligned}\eta\big|_{\mathrm{E}} &= \frac{-1}{(1/\lambda^\varepsilon)_{,\theta}\big|_{\mathrm{E}}} \left\{ (\Psi^G_{I,\theta})\big|_{\mathrm{E}} + \frac{\mathcal{P}_\theta\big|_{\mathrm{E}}}{\rho\lambda^\varepsilon\big|_{\mathrm{E}}} \right\} \\ &= \frac{-1}{(1/\lambda^\varepsilon)_{,\theta}\big|_{\mathrm{E}}} \Bigg\{ (\Psi^G_{I,\theta})\big|_{\mathrm{E}} + \sum_{\alpha=1}^{n} \left(\Psi_I^G\right),_{\bar{\mathbf{z}}_\alpha}\big|_{\mathrm{E}} \cdot (\bar{\mathbf{\Phi}}_\alpha),_{\dot{\theta}}\big|_{\mathrm{E}} \\ &\qquad + \sum_{\alpha=1}^{n} \frac{1}{\rho} \left\{ \bar{\rho}_\alpha \mu^G_{\underset{\alpha}{\mathrm{I}}}\big|_{\mathrm{E}} \, (c_\alpha),_{\dot{\theta}}\big|_{\mathrm{E}} - \boldsymbol{\iota}_\alpha\big|_{\mathrm{E}} (\bar{n}_\alpha),_{\dot{\theta}}\big|_{\mathrm{E}} \right\} \Bigg\} .\end{aligned} \tag{7.38}$$

Moreover, we remark that when $\dot{\theta} \notin \mathbb{S}$, one has $\mathcal{P}_\theta = 0$, see (6.12), $\lambda^\varepsilon = \lambda^\varepsilon(\theta) = 1/\theta$, so that (7.37) implies the classical relation $\eta = -\Psi^G_{I,\theta}$. Alternatively, for $\dot{\theta} \in \mathbb{S}$ the right-hand side of (7.38) is known when the thermodynamic quantities Ψ_I^G, $\bar{\mathbf{\Phi}}_\alpha$, $\mu^G_{\underset{\alpha}{\mathrm{I}}}$ and $\boldsymbol{\iota}_\alpha$ are prescribed.

If we specialize n_I to $\nabla\theta$ we obtain from (7.18) and (7.5)

$$\begin{aligned}\pi^{\rho\eta},_{\nabla\theta}\big|_{\mathrm{E}} &= -\big(\underbrace{\mathbf{k},_{\theta}\big|_{\mathrm{E}}}_{0} - (\lambda^\varepsilon),_{\theta}\big|_{\mathrm{E}} \mathbf{q}\big|_{\mathrm{E}}\big) - \lambda^\varepsilon\big|_{\mathrm{E}} \sum_{\alpha=1}^{n} \rho \left(\Psi_I^G\right),_{\mathbf{z}_\alpha}\big|_{\mathrm{E}} \cdot (\bar{\mathbf{\Phi}}_\alpha),_{\nabla\theta}\big|_{\mathrm{E}} \\ &\quad - \lambda^\varepsilon\big|_{\mathrm{E}} \sum_{\alpha=1}^{n} \left\{ \bar{\rho}_\alpha \, \mu^G_{\underset{\alpha}{\mathrm{I}}}\big|_{\mathrm{E}} \, (c_\alpha),_{\nabla\theta}\big|_{\mathrm{E}} - \boldsymbol{\iota}_\alpha\big|_{\mathrm{E}} (\bar{n}_\alpha),_{\nabla\theta}\big|_{\mathrm{E}} \right\} \\ &\overset{!}{=} 0 ,\end{aligned} \tag{7.39}$$

where (7.21) has been used to derive

$$\mathbf{k}_{,\theta}\big|_{\mathrm{E}} = \mathbf{0}, \qquad \mathbf{k}_{,\rho_\alpha \nabla\theta}\big|_{\mathrm{E}} = \mathbf{0}, \quad \text{and} \quad \mathbf{k}_{,\nu_\alpha \nabla\theta}\big|_{\mathrm{E}} = \mathbf{0}\,. \tag{7.40}$$

From equation (7.39) we obtain

$$(\lambda^\varepsilon)_{,\theta}\big|_{\mathrm{E}}\, \mathbf{q}\big|_{\mathrm{E}} = \lambda^\varepsilon\big|_{\mathrm{E}} \sum_{\alpha=1}^{n} \rho \left(\Psi_I^G\right)_{,\bar{\mathbf{Z}}_\alpha}\big|_{\mathrm{E}} \cdot (\bar{\boldsymbol{\Phi}}_\alpha)_{,\nabla\theta}\big|_{\mathrm{E}}$$

$$+ \lambda^\varepsilon\big|_{\mathrm{E}} \sum_{\alpha=1}^{n} \left\{ \bar{\rho}_\alpha \, \underset{\alpha}{\mu}{}^G_{\mathrm{I}}\big|_{\mathrm{E}}\, (c_\alpha)_{,\nabla\theta}\big|_{\mathrm{E}} - \boldsymbol{\iota}_\alpha\big|_{\mathrm{E}} (\bar{n}_\alpha)_{,\nabla\theta}\big|_{\mathrm{E}} \right\}. \tag{7.41}$$

We observe that the equilibrium mixture energy flux, $\mathbf{q}\big|_{\mathrm{E}}$, reduces to zero if either, first, the interaction rate densities for mass and volume fraction and the frictional production rate densities depend at least quadratically on the temperature gradient, second, they do not depend on it or, third, the interaction terms for mass, volume fraction and friction are not present at all. Without making these assumptions, the energy flux vector does not vanish a priori in mixture thermodynamic equilibrium.

At last, from

$$\pi^{\rho\eta}{}_{,\mathbf{D}_\alpha}\big|_{\mathrm{E}} = \mathbf{0}\,, \tag{7.42}$$

we deduce a constitutive relation for the constituent equilibrium CAUCHY stress tensors $\mathbf{T}_\beta$ ($\beta = 1, \ldots, n$). With the help of inequality (7.5) we obtain

$$\pi^{\rho\eta}{}_{,\mathbf{D}_\beta}\big|_{\mathrm{E}} = \lambda^\varepsilon\big|_{\mathrm{E}} \left\{ \bar{\zeta}_\beta \mathbf{I} - 2\rho \,\mathrm{sym}\left((\Psi_I^G)_{,\mathbf{B}_\beta} \right) \mathbf{B}_\beta - (\lambda^\varepsilon)^{-1} \mathbf{k}_{,\mathbf{v}_\beta} + \bar{\mathbf{T}}_\beta \right\}\big|_{\mathrm{E}}$$

$$- \lambda^\varepsilon\big|_{\mathrm{E}} \sum \rho \left(\Psi_I^G\right)_{,\bar{\mathbf{Z}}_\alpha}\big|_{\mathrm{E}} (\bar{\boldsymbol{\Phi}}_\alpha)_{,\mathbf{D}_\beta}\big|_{\mathrm{E}}$$

$$- \lambda^\varepsilon\big|_{\mathrm{E}} \sum \left\{ \bar{\rho}_\alpha \, \underset{\alpha}{\mu}{}^G_{\mathrm{I}}\big|_{\mathrm{E}}\, (c_\alpha)_{,\mathbf{D}_\beta}\big|_{\mathrm{E}} - \boldsymbol{\iota}_\alpha\big|_{\mathrm{E}} (\bar{n}_\alpha)_{,\mathbf{D}_\beta}\big|_{\mathrm{E}} \right\}$$

$$\overset{!}{=} \mathbf{0}\,, \tag{7.43}$$

where again use has been made of $(7.21)_2$ and (7.8) to rule out dependencies on $\mathbf{k}_{,\rho_\alpha \mathbf{D}_\alpha}\big|_{\mathrm{E}}$, $\mathbf{k}_{,\nu_\alpha \mathbf{D}_\alpha}\big|_{\mathrm{E}}$ and $\mathbf{c}_{,\mathbf{D}_\alpha}\big|_{\mathrm{E}}$, respectively. Equation (7.43) can be rewritten to yield the constitutive relations for the partial stresses

$$\bar{\mathbf{T}}_\beta\big|_{\mathrm{E}} = -\bar{\zeta}_\beta\big|_{\mathrm{E}} \mathbf{I} + 2\rho \;\mathrm{sym}\left((\Psi_I^G)_{,\mathbf{B}_\beta} \right)\big|_{\mathrm{E}} \mathbf{B}_\beta + (\lambda^\varepsilon)^{-1}\big|_{\mathrm{E}} \mathbf{k}_{,\mathbf{v}_\beta}\big|_{\mathrm{E}}$$

$$+ \sum \rho \left(\Psi_I^G\right)_{,\bar{\mathbf{Z}}_\alpha}\big|_{\mathrm{E}} (\bar{\boldsymbol{\Phi}}_\alpha)_{,\mathbf{D}_\beta}\big|_{\mathrm{E}}$$

$$+ \sum \left\{ \bar{\rho}_\alpha \, \underset{\alpha}{\mu}{}^G_{\mathrm{I}}\big|_{\mathrm{E}}\, (c_\alpha)_{,\mathbf{D}_\beta}\big|_{\mathrm{E}} - \boldsymbol{\iota}_\alpha\big|_{\mathrm{E}} (\bar{n}_\alpha)_{,\mathbf{D}_\beta}\big|_{\mathrm{E}} \right\} \tag{7.44}$$

which consist of five different terms. The first contains the configuration and saturation pressures, $\underset{\alpha}{\beta}{}^G$ and ς, through the combination $\bar{\zeta}_\beta\big|_{\mathrm{E}}$ (see (7.7)).

The true thermodynamic pressure, p_α^G, does not arise explicitly in (7.44) but with the help of (6.104), (6.105) and (6.108), i. e.

$$\begin{aligned} p_\alpha^G &= \beta_\alpha^G + \varsigma - \rho_n \, \Psi_I^G + \iota_\alpha - (\lambda^\varepsilon)^{-1}\mathsf{k} \\ &= \zeta_\alpha + \iota_\alpha - (\lambda^\varepsilon)^{-1}\mathsf{k}, \end{aligned} \qquad \alpha = 1, \ldots, m \,, \qquad (7.45)$$

we could replace ζ_α by p_α^G, ι_α and $(\lambda^\varepsilon)^{-1}\mathsf{k}$ at least for the first m constituents.

The second part of (7.44) which explicitly involves the constituent left CAUCHY-GREEN tensor, $\mathbf{B}_\beta$, describes the elastic contribution to the equilibrium constituent CAUCHY stress tensor, $\bar{\mathbf{T}}_\beta\big|_{\mathrm{E}}$.[3] For mature debris flows it is known that elastic effects are far from being dominant, but elastic contributions are nevertheless often used to describe shear stresses in thermodynamic equilibrium ('heap problem'). We are inclined to think, however, that the above shear stresses are less due to elastic and more to frictional effects which are thought to be describable e. g. by means of the hypo-plastic theory. The fourth term of constitutive relation (7.44) is exclusively due to frictional effects and has already been introduced by SVENDSEN et al. [116].

It is evident from (7.31), (7.36), (7.37), (7.41) and (7.44) that $\bar{\mathbf{m}}_\beta^i\big|_{\mathrm{E}}$, $\mathcal{P}_\theta\big|_{\mathrm{E}}$, $\eta\big|_{\mathrm{E}}$, $\mathbf{q}\big|_{\mathrm{E}}$ and $\bar{\mathbf{T}}_\beta\big|_{\mathrm{E}}$ contain in general terms which are due to the HELMHOLTZ-like free energy, the saturation pressure, elastic deformation, extra entropy flux, frictional effects and mass and volume fraction interaction rate densities. There is obviously some structure in these formulae, but this structure appears to be somewhat hidden in the overwhelming complexity of the notation. In the remainder of this section we shall attempt to shed light on this structure.

In relations (7.31), (7.36), (7.41) and (7.44) the quantity $\lambda^\varepsilon\big|_{\mathrm{E}}$ arises which, owing to

$$\lambda^\varepsilon = \lambda^\varepsilon\big|_{\mathrm{E}}(\theta) + \lambda^\varepsilon\big|_{\mathrm{N}}(\theta,\, \dot{\theta}), \qquad \lambda^\varepsilon\big|_{\mathrm{E}}(\theta) := \lim_{\mathrm{n}\to 0}(\lambda^\varepsilon) \,, \qquad (7.46)$$

only depends on the temperature. If we again stress the connection of λ^ε to its equivalent variable in single-material theories, it is reasonable to assume that

[**A16**] $$\lambda^\varepsilon\big|_{\mathrm{E}}(\theta) = \theta^{-1} \,,$$

[3] This dependence is not necessarily a collinearity of $\mathbf{B}_\beta$ and $\bar{\mathbf{T}}_\beta\big|_{\mathrm{E}}$, since in (7.44) $\mathbf{B}_\beta$ is premultiplied with a second rank tensor $2\rho \, \mathrm{sym}\,\big((\Psi_I^G)_{,\mathbf{B}_\beta}\big)\big|_{\mathrm{E}}$, which in most cases is not proportional to the unit tensor.

where θ is now identified with the KELVIN temperature. In this case $\Psi^G\big|_{\mathrm{E}}$ reduces to

$$\Psi^G\big|_{\mathrm{E}} = \varepsilon\big|_{\mathrm{E}} - \theta\eta\big|_{\mathrm{E}} =: \Psi \ , \tag{7.47}$$

in which the latter quantity represents the equilibrium part of the HELMHOLTZ-like free energy. It will be called HELMHOLTZ free energy. However we shall keep the notation $\Psi^G\big|_{\mathrm{E}}$ for it. Under assumption [**A16**] we recover from (7.31)

$$\begin{aligned}
\bar{\mathbf{m}}^i_\beta\big|_{\mathrm{E}} = &\sum_{\alpha=1}^{m} \left\{ \left(\delta_{\alpha\beta} - \bar{\xi}_\beta\right) \nu_\alpha \left(\rho_\alpha\right)^{-1} p^G_\alpha\big|_{\mathrm{E}}\, \mathbf{I} - \theta\, \mathbf{k}_{,\rho_\alpha \mathbf{v}_\beta}\big|_{\mathrm{E}} \right\} \nabla\rho_\alpha \\
&+ \sum_{\alpha=1}^{n-1} \left\{ \left(\delta_{\alpha\beta} - \bar{\xi}_\beta\right)\zeta_\alpha\big|_{\mathrm{E}} \mathbf{I} - \mathbf{c}_{,\mathbf{v}_\beta}\big|_{\mathrm{E}} - \theta\, \mathbf{k}_{,\nu_\alpha \mathbf{v}_\beta}\big|_{\mathrm{E}} \right\} \nabla\nu_\alpha \\
&+ \sum_{\alpha=1}^{n} \rho \left(\Psi^G_I\right)_{,\bar{\mathbf{Z}}_\alpha}\big|_{\mathrm{E}} \left(\bar{\boldsymbol{\Phi}}_\alpha\right)_{,\mathbf{v}_\beta}\big|_{\mathrm{E}} \\
&- \sum \left\{ \bar{\rho}_\alpha \mu^G_{\alpha\mathrm{I}}\big|_{\mathrm{E}} \left(c_\alpha\right)_{,\mathbf{v}_\beta}\big|_{\mathrm{E}} - \iota_\alpha\big|_{\mathrm{E}} (\bar{n}_\alpha)_{,\mathbf{v}_\beta}\big|_{\mathrm{E}} \right\} ,
\end{aligned} \tag{7.48}$$

which is observed to be at least structurally in agreement with the result of SVENDSEN & HUTTER [115, eqn.(8.9)]. The representations for $\mathbf{q}\big|_{\mathrm{E}}$ and $\bar{\mathbf{T}}_\alpha\big|_{\mathrm{E}}$, ((7.41) and (7.44)), do not considerably alter under assumption [**A16**].

It is at this point worthwhile to pause and to summarize what has been attained. We have expressed the equilibrium interaction forces (7.31), the equilibrium entropy (7.38), the equilibrium energy flux (7.41) and the equilibrium partial stress tensors (7.44) in terms of clearly identifiable contributions, all of which were obtained from the exploitation of the entropy principle, in particular the so-called LIU identities and the inferences deduced from them. A first set of these is generally of direct thermodynamic origin and involves a thermodynamic potential, derivatives of it with respect to the true constituent densities and the volume fraction densities, the saturation pressure as well as certain derivatives of the extra entropy flux vector. In the expression for the interaction forces (7.31) these terms also depend explicitly on $\nabla\rho_\alpha$ and $\nabla\nu_\alpha$ (first two lines in (7.31)). In the expression for the entropy (7.38) they only involve $\Psi^G_{I,\theta}\big|_{\mathrm{E}}$, in the heat flux (7.41) they are not present at all, and in the expression for the equilibrium constituent stress tensors (7.44) they comprise the pressure like quantity $\bar{\zeta}_\alpha$, the elastic contributions and the extra entropy flux, stated in the first line of (7.44). The remaining contributions to the above mentioned equilibrium quantities are equally of thermodynamic origin, but they only exist if the production rate densities of the frictional tensorial variable, $\bar{\boldsymbol{\Phi}}_\alpha$ and the constituent mass, c_α and volume fractions, n_α do not vanish. They all appear as sums of products of derivatives of $\bar{\boldsymbol{\Phi}}_\alpha$, c_α and n_α with prefactors which are the same in the expressions of $\bar{\mathbf{m}}^i_\beta\big|_{\mathrm{E}}$, $\eta\big|_{\mathrm{E}}$, $\mathbf{q}\big|_{\mathrm{E}}$ and $\bar{\mathbf{T}}_\alpha\big|_{\mathrm{E}}$. The prefactors are

$\rho\left(\Psi^G_I\right),_{\bar{\mathbf{Z}}_\alpha}\big|_{\mathrm{E}}$	for the friction term $\bar{\mathbf{\Phi}}_\alpha$	
$\bar{\rho}_\alpha \mu^G_{\alpha \mathrm{I}}\big|_{\mathrm{E}}$	for the mass production c_α	(7.49)
$\boldsymbol{\iota}_\alpha\big|_{\mathrm{E}}$	for the volume fraction production n_α .	

They are known once and for all when the inner HELMHOLTZ-like free energy is known as a function of its variables, and the constituent free energies, see (6.109), and the parameters $\boldsymbol{\iota}_\alpha$, see (6.119), are known. The above factors, in turn, are multiplied

for $\bar{\mathbf{m}}^i_\beta\big|_{\mathrm{E}}$	with $\left(\bar{\mathbf{\Phi}}_\alpha\right),_{\mathbf{v}_\beta}\big|_{\mathrm{E}}$,	$(c_\alpha),_{\mathbf{v}_\beta}\big|_{\mathrm{E}}$,	$(\bar{n}_\alpha),_{\mathbf{v}_\beta}\big|_{\mathrm{E}}$	
for $\eta\big|_{\mathrm{E}}$	with $\left(\bar{\mathbf{\Phi}}_\alpha\right),_{\dot{\theta}}\big|_{\mathrm{E}}$,	$(c_\alpha),_{\dot{\theta}}\big|_{\mathrm{E}}$,	$(\bar{n}_\alpha),_{\dot{\theta}}\big|_{\mathrm{E}}$	(7.50)
for $\mathbf{q}\big|_{\mathrm{E}}$	with $\left(\bar{\mathbf{\Phi}}_\alpha\right),_{\nabla\theta}\big|_{\mathrm{E}}$,	$(c_\alpha),_{\nabla\theta}\big|_{\mathrm{E}}$,	$(\bar{n}_\alpha),_{\nabla\theta}\big|_{\mathrm{E}}$	
for $\bar{\mathbf{T}}_\alpha\big|_{\mathrm{E}}$	with $\left(\bar{\mathbf{\Phi}}_\alpha\right),_{\mathbf{D}_\beta}\big|_{\mathrm{E}}$,	$(c_\alpha),_{\mathbf{D}_\beta}\big|_{\mathrm{E}}$,	$(\bar{n}_\alpha),_{\mathbf{D}_\beta}\big|_{\mathrm{E}}$	

and subsequently summed to reveal the corresponding representations arising in (7.31), (7.38), (7.41) and (7.44). The two lists (7.49) and (7.50) disclose the thermodynamic structure of the various contributions to the equilibrium quantities $\bar{\mathbf{m}}^i_\beta\big|_{\mathrm{E}}$, $\eta\big|_{\mathrm{E}}$, $\mathbf{q}\big|_{\mathrm{E}}$ and $\bar{\mathbf{T}}_\alpha\big|_{\mathrm{E}}$ particularly clearly: The constituent equilibrium interaction forces receive contributions exclusively via derivatives of $\bar{\mathbf{\Phi}}_\alpha$, c_α and n_α with respect to the constituent velocities. Analogously, the equilibrium entropy is directly affected only by corresponding derivatives with respect to $\dot{\theta}$, the equilibrium heat flux vector by corresponding derivatives with respect to $\nabla\theta$ and the equilibrium constituent stresses by those with respect to $\mathbf{D}_\beta$. This demonstrates, on the one hand, that contributions to $\bar{\mathbf{m}}^i_\beta\big|_{\mathrm{E}}$ and $\operatorname{div}\left(\bar{\mathbf{T}}_\alpha\big|_{\mathrm{E}}\right)$ are not likely interchangeable. On the other hand, the list (7.50) can serve as a help when explicitly parameterising constitutive relations for $\mathbf{\Phi}_\alpha$, c_α and n_α. For instance, we may have reason to assume that thermal effects are insignificant. Then, it may be justified to postulate that

$$\left(\bar{\mathbf{\Phi}}_\alpha,\ c_\alpha,\ n_\alpha\right),_{\dot{\theta}} = 0\ , \quad \left(\bar{\mathbf{\Phi}}_\alpha,\ c_\alpha,\ n_\alpha\right),_{\nabla\theta} = \mathbf{0}\ . \tag{7.51}$$

This would yield

$$\eta\big|_{\mathrm{E}} = -\Psi^G_{I,\theta} \qquad \text{and} \qquad \mathbf{q}\big|_{\mathrm{E}} = \mathbf{0}\ , \tag{7.52}$$

agreeing with the classical relations. Moreover, we have already provided reasons to assume $\bar{\mathbf{\Phi}}_{\alpha,\mathbf{v}_\beta} = \mathbf{0}$, see discussion following (7.31); in this case the contribution of the frictional variable to the interaction force would vanish, whilst that to the constituent stresses would still be present via $\left(\bar{\mathbf{\Phi}}_\alpha\right),_{\mathbf{D}_\beta}\big|_{\mathrm{E}}$.

We shall see that these terms, when properly parameterised, will give rise to the thermodynamic justification of hypo-plasticity. In any case, keeping the full generality of the constitutive dependences of $\bar{\boldsymbol{\Phi}}_\alpha$, c_α and n_α brought, via (7.49) and (7.50), light into a deeper understanding of the structure of the formulae of the constitutive quantities as imposed by the entropy principle. It is now much easier to understand the role played by ad-hoc simplifications of certain constitutive relations for $\bar{\boldsymbol{\Phi}}_\alpha$, c_α and n_α than it would have been if such simplifications had been introduced at the outset.

7.3 'Isotropic' Expansions for the Interaction Supply Rate Densities of Mass and Volume Fraction and the Extra Entropy Flux

To further inspect the constitutive relations of the last section we perform an '*isotropic*' *expansion* (cf. SVENDSEN & HUTTER [115]) of $\vec{c}$, $\vec{n}$ and $\mathbf{k}$ about the state $\boldsymbol{\mathcal{Y}} = \mathbf{0}$, where we define $\boldsymbol{\mathcal{Y}}$ as

$$\boldsymbol{\mathcal{Y}} := \left\{\dot{\theta},\ \nabla\theta,\ \vec{\nabla\rho},\ \vec{\nabla\nu},\ \vec{\mathbf{v}},\ \vec{\mathbf{D}},\ \vec{\mathbf{W}}\right\} . \tag{7.53}$$

It is observed that, *first*, the set of non-equilibrium variables n is a subset of $\boldsymbol{\mathcal{Y}}$. Thus the state $\boldsymbol{\mathcal{Y}} = \mathbf{0}$ is a subspace of the space describing mixture thermodynamic equilibrium because $\vec{\nabla\rho}$ and $\vec{\nabla\nu}$ are also set to zero. In other words, the set $\mathbb{S}_{\boldsymbol{\mathcal{Y}}}$ of elements $\boldsymbol{\mathcal{Y}}$ is the direct sum of $\mathbb{S}_{\mathsf{n}}$ of the non-equilibrium elements n plus $\mathbb{S}_{\vec{\nabla\rho}} \cup \mathbb{S}_{\vec{\nabla\nu}}$,

$$\mathbb{S}_{\boldsymbol{\mathcal{Y}}} = \mathbb{S}_{\mathsf{n}} \oplus \left(\mathbb{S}_{\vec{\nabla\rho}} \cup \mathbb{S}_{\vec{\nabla\nu}}\right) \tag{7.54}$$

Second, all vector-valued variables of the set of constitutive variables, $\mathbb{S}$, are included and thus all vector-valued, isotropic constitutive quantities and therefore also $\mathbf{k}$ must vanish at $\boldsymbol{\mathcal{Y}} = \mathbf{0}$.

In this expansion only the most simple dependencies of the above variables are mentioned explicitly, where the higher non-linear contributions are subsumed in residual quantities. By neglecting the higher non-linear contributions, we obtain an expansion about the state $\boldsymbol{\mathcal{Y}} = \mathbf{0}$ which, due to the additional requirements $\nabla\rho_\alpha = \mathbf{0}$ $(\alpha = 1, \ldots, m)$ and $\nabla\nu_\alpha = \mathbf{0}$ $(\alpha = 1, \ldots, n-1)$, is found to be more restrictive than an expansion about the mixture thermodynamic equilibrium.

Interaction rate densities c_α and n_α

A linear representation of the scalars c_α and n_α in terms of $\boldsymbol{\mathcal{Y}}$ would only involve $\dot\theta$ and $\mathrm{tr}(\mathbf{D}_\beta)$, which is too simple and likely unrealistic. So, at least a quadratic expansion is necessary. A minimal form of an 'isotropic' expansion of c_α thus reads as follows

$$
\begin{aligned}
c_\alpha = c_\alpha^{\dot\theta}(\dot\theta) &+ \Big\{\sum_{\beta=1}^{m} c_{\alpha\beta}^{\theta\rho}(\nabla\rho_\beta) + \sum_{\beta=1}^{n-1} c_{\alpha\beta}^{\theta\nu}(\nabla\nu_\beta)\Big\}\cdot(\nabla\theta) \\
&+ \sum_{\beta=1}^{n}\Big\{\sum_{\gamma=1}^{m} c_{\alpha\beta\gamma}^{\mathbf{v}\rho}(\nabla\rho_\gamma) + \sum_{\gamma=1}^{n-1} c_{\alpha\beta\gamma}^{\mathbf{v}\nu}(\nabla\nu_\gamma)\Big\}\cdot(\mathbf{v}_\beta) \\
&+ \sum_{\beta=1}^{n}\Big\{c_{\alpha\beta}^{\mathrm{D}}\mathbf{I} + \mathbf{C}_{\alpha\beta}^{\mathrm{D}}\Big\}\cdot(\mathbf{D}_\beta) \\
&+ \sum_{\beta=1}^{n}\Big\{\mathbf{C}_{\alpha\beta}^{\mathrm{W}}\Big\}\cdot(\mathbf{W}_\beta) \\
&+ c_\alpha^{\mathbf{N}}\,,
\end{aligned}
\tag{7.55}
$$

where the term $c_\alpha^{\mathbf{N}}$ accounts for the remaining non-linear contributions and depends on all constitutive variables. It must vanish in thermodynamic equilibrium, i. e.

$$
c_\alpha^{\mathbf{N}}\big|_{\mathrm{E}} = 0\,,
\tag{7.56}
$$

to be in accordance with $(7.12)_1$. The tensors $\mathbf{C}_{\alpha\beta}^{\mathrm{D}}$ and $\mathbf{C}_{\alpha\beta}^{\mathrm{W}}$ are defined as

$$
\begin{aligned}
\mathbf{C}_{\alpha\beta}^{\mathrm{D}} := \sum_{\delta=1}^{m}\Big\{\sum_{\gamma=1}^{m} c_{\alpha\beta\gamma\delta}^{\mathrm{D}\rho}\,\mathrm{sym}\left(\nabla\rho_\delta\otimes\nabla\rho_\gamma\right) &+ \sum_{\gamma=1}^{n-1} c_{\alpha\beta\gamma\delta}^{\mathrm{D}\nu\rho}\,\mathrm{sym}\left(\nabla\rho_\delta\otimes\nabla\nu_\gamma\right)\Big\} \\
+ \sum_{\delta=1}^{n-1}\Big\{\sum_{\gamma=1}^{n-1} c_{\alpha\beta\gamma\delta}^{\mathrm{D}\nu}\,\mathrm{sym}\left(\nabla\nu_\delta\otimes\nabla\nu_\gamma\right)\Big\}\,,
\end{aligned}
\tag{7.57}
$$

and

$$
\begin{aligned}
\mathbf{C}_{\alpha\beta}^{\mathrm{W}} := \sum_{\delta=1}^{m}\Big\{\sum_{\gamma=1}^{m} c_{\alpha\beta\gamma\delta}^{\mathrm{W}\rho}\,\mathrm{skw}\left(\nabla\rho_\delta\otimes\nabla\rho_\gamma\right) &+ \sum_{\gamma=1}^{n-1} c_{\alpha\beta\gamma\delta}^{\mathrm{W}\nu\rho}\,\mathrm{skw}\left(\nabla\rho_\delta\otimes\nabla\nu_\gamma\right)\Big\} \\
+ \sum_{\delta=1}^{n-1}\Big\{\sum_{\gamma=1}^{n-1} c_{\alpha\beta\gamma\delta}^{\mathrm{W}\nu}\,\mathrm{skw}\left(\nabla\nu_\delta\otimes\nabla\nu_\gamma\right)\Big\}\,.
\end{aligned}
\tag{7.58}
$$

The coefficients arising in (7.55), (7.57) and (7.58) are specified according to

$$
\begin{aligned}
c_\alpha^{\dot\theta} &:= (c_\alpha)_{,\dot\theta}\big|_{\mathsf{n}=0} = \hat c_\alpha^{\dot\theta}\Big(\theta,\ \vec\rho,\ \vec\nu,\ \nabla\vec\rho,\ \nabla\vec\nu,\ \vec{\mathbf{B}},\ \vec{\vec{\mathbf{Z}}}\Big)\,,\\
c_{\alpha\beta}^{\theta\rho}\,\mathbf{I} &:= (c_\alpha)_{,\nabla\theta\nabla\rho_\beta}\big|_{\boldsymbol{\mathcal{Y}}=0} = \hat c_{\alpha\beta}^{\theta\rho}\Big(\theta,\ \vec\rho,\ \vec\nu,\ \vec{\mathbf{B}},\ \vec{\vec{\mathbf{Z}}}\Big)\,\mathbf{I}\,,\\
c_{\alpha\beta}^{\theta\nu}\,\mathbf{I} &:= (c_\alpha)_{,\nabla\theta\nabla\nu_\beta}\big|_{\boldsymbol{\mathcal{Y}}=0} = \hat c_{\alpha\beta}^{\theta\nu}\Big(\theta,\ \vec\rho,\ \vec\nu,\ \vec{\mathbf{B}},\ \vec{\vec{\mathbf{Z}}}\Big)\,\mathbf{I}\,,\\
c_{\alpha\beta\gamma}^{v\rho}\,\mathbf{I} &:= (c_\alpha)_{,\mathbf{v}_\beta\nabla\rho_\gamma}\big|_{\boldsymbol{\mathcal{Y}}=0} = \hat c_{\alpha\beta\gamma}^{v\rho}\Big(\theta,\ \vec\rho,\ \vec\nu,\ \vec{\mathbf{B}},\ \vec{\vec{\mathbf{Z}}}\Big)\,\mathbf{I}\,,\\
c_{\alpha\beta\gamma}^{v\nu}\,\mathbf{I} &:= (c_\alpha)_{,\mathbf{v}_\beta\nabla\nu_\gamma}\big|_{\boldsymbol{\mathcal{Y}}=0} = \hat c_{\alpha\beta\gamma}^{v\nu}\Big(\theta,\ \vec\rho,\ \vec\nu,\ \vec{\mathbf{B}},\ \vec{\vec{\mathbf{Z}}}\Big)\,\mathbf{I}\,,\\
c_{\alpha\beta}^{\mathrm{D}}\,\boldsymbol{\mathcal{I}} &:= (c_\alpha)_{,\mathbf{D}_\beta}\big|_{\boldsymbol{\mathcal{Y}}=0} = \hat c_{\alpha\beta}^{\mathrm{D}}\Big(\theta,\ \vec\rho,\ \vec\nu,\ \vec{\mathbf{B}},\ \vec{\vec{\mathbf{Z}}}\Big)\,\boldsymbol{\mathcal{I}}\,,\\
c_{\alpha\beta\gamma\delta}^{\mathrm{D}\rho}\,\boldsymbol{\mathcal{I}} &:= (c_\alpha)_{,\mathbf{D}_\beta\nabla\rho_\gamma\nabla\rho_\delta}\big|_{\boldsymbol{\mathcal{Y}}=0} = \hat c_{\alpha\beta\gamma\delta}^{\mathrm{D}\rho}\Big(\theta,\ \vec\rho,\ \vec\nu,\ \vec{\mathbf{B}},\ \vec{\vec{\mathbf{Z}}}\Big)\,\boldsymbol{\mathcal{I}}\,,\\
c_{\alpha\beta\gamma\delta}^{\mathrm{D}\nu\rho}\,\boldsymbol{\mathcal{I}} &:= (c_\alpha)_{,\mathbf{D}_\beta\nabla\nu_\gamma\nabla\rho_\delta}\big|_{\boldsymbol{\mathcal{Y}}=0} = \hat c_{\alpha\beta\gamma\delta}^{\mathrm{D}\nu\rho}\Big(\theta,\ \vec\rho,\ \vec\nu,\ \vec{\mathbf{B}},\ \vec{\vec{\mathbf{Z}}}\Big)\,\boldsymbol{\mathcal{I}}\,,\\
c_{\alpha\beta\gamma\delta}^{\mathrm{D}\nu}\,\boldsymbol{\mathcal{I}} &:= (c_\alpha)_{,\mathbf{D}_\beta\nabla\nu_\gamma\nabla\nu_\delta}\big|_{\boldsymbol{\mathcal{Y}}=0} = \hat c_{\alpha\beta\gamma\delta}^{\mathrm{D}\nu}\Big(\theta,\ \vec\rho,\ \vec\nu,\ \vec{\mathbf{B}},\ \vec{\vec{\mathbf{Z}}}\Big)\,\boldsymbol{\mathcal{I}}\,,\\
c_{\alpha\beta\gamma\delta}^{\mathrm{W}\rho}\,\boldsymbol{\mathcal{I}} &:= (c_\alpha)_{,\mathbf{W}_\beta\nabla\rho_\gamma\nabla\rho_\delta}\big|_{\boldsymbol{\mathcal{Y}}=0} = \hat c_{\alpha\beta\gamma\delta}^{\mathrm{W}\rho}\Big(\theta,\ \vec\rho,\ \vec\nu,\ \vec{\mathbf{B}},\ \vec{\vec{\mathbf{Z}}}\Big)\,\boldsymbol{\mathcal{I}}\,,\\
c_{\alpha\beta\gamma\delta}^{\mathrm{W}\nu\rho}\,\boldsymbol{\mathcal{I}} &:= (c_\alpha)_{,\mathbf{W}_\beta\nabla\nu_\gamma\nabla\rho_\delta}\big|_{\boldsymbol{\mathcal{Y}}=0} = \hat c_{\alpha\beta\gamma\delta}^{\mathrm{W}\nu\rho}\Big(\theta,\ \vec\rho,\ \vec\nu,\ \vec{\mathbf{B}},\ \vec{\vec{\mathbf{Z}}}\Big)\,\boldsymbol{\mathcal{I}}\,,\\
c_{\alpha\beta\gamma\delta}^{\mathrm{W}\nu}\,\boldsymbol{\mathcal{I}} &:= (c_\alpha)_{,\mathbf{W}_\beta\nabla\nu_\gamma\nabla\nu_\delta}\big|_{\boldsymbol{\mathcal{Y}}=0} = \hat c_{\alpha\beta\gamma\delta}^{\mathrm{W}\nu}\Big(\theta,\ \vec\rho,\ \vec\nu,\ \vec{\mathbf{B}},\ \vec{\vec{\mathbf{Z}}}\Big)\,\boldsymbol{\mathcal{I}}\,,
\end{aligned}
\tag{7.59}
$$

where $\boldsymbol{\mathcal{I}}$ is the fourth order unit tensor and all coefficients are symmetric with respect to constituent indices that arise in the derivatives, e. g.

$$
c_{\alpha\beta\gamma}^{v\nu} = c_{\alpha\gamma\beta}^{v\nu}\,. \tag{7.60}
$$

The coefficients have to satisfy further restrictions, namely, first, according to $(4.7)_1$ the identity

$$
\sum_{\alpha=1}^{n} \bar\rho_\alpha c_\alpha = 0 \tag{7.61}
$$

must be assured and thus the conditions

$$
\begin{aligned}
&\sum_{\alpha=1}^{n} \bar{\rho}_\alpha c^\theta_\alpha = 0, \qquad \sum_{\alpha=1}^{n} \bar{\rho}_\alpha c^{\theta\rho}_{\alpha\beta} = 0, \qquad \sum_{\alpha=1}^{n} \bar{\rho}_\alpha c^{\theta\nu}_{\alpha\beta} = 0,\\
&\sum_{\alpha=1}^{n} \bar{\rho}_\alpha c^{\mathbf{v}\rho}_{\alpha\beta\gamma} = 0, \qquad \sum_{\alpha=1}^{n} \bar{\rho}_\alpha c^{\mathbf{v}\nu}_{\alpha\beta\gamma} = 0, \qquad \sum_{\alpha=1}^{n} \bar{\rho}_\alpha c^{\mathrm{D}}_{\alpha\beta} = 0,\\
&\sum_{\alpha=1}^{n} \bar{\rho}_\alpha \mathbf{C}^{\mathrm{D}}_{\alpha\beta} = 0, \qquad \sum_{\alpha=1}^{n} \bar{\rho}_\alpha \mathbf{C}^{\mathrm{W}}_{\alpha\beta} = 0, \qquad \sum_{\alpha=1}^{n} \bar{\rho}_\alpha c^{\mathbf{N}}_\alpha = 0
\end{aligned}
\tag{7.62}
$$

must hold. These conditions can, for instance, be met by expressing the nth coefficient in terms of the $(n-1)$ first ones as follows:

$$
c^{\theta\rho}_{n\beta} = \sum_{\alpha=1}^{n-1} \frac{\bar{\rho}_\alpha}{\bar{\rho}_n} c^{\theta\rho}_{\alpha\beta} = \sum_{\alpha=1}^{n-1} \left\{ \frac{\bar{\xi}_\alpha}{1 - \sum_{\gamma=1}^{n-1} \bar{\xi}_\gamma} \right\} c^{\theta\rho}_{\alpha\beta}, \qquad \text{etc,} \tag{7.63}
$$

as follows from (7.62).

Second, by invoking the principle of objectivity for the constitutive quantity c_α (see Section 4.6, eqs. (4.50) and (4.51)) i. e.,

$$
\sum_{\beta=1}^{n} (c_\alpha)_{,\mathbf{v}_\beta} = 0 \qquad \text{and} \qquad \sum_{\beta=1}^{n} (c_\alpha)_{,\mathbf{W}_\beta} = 0 \ , \tag{7.64}
$$

we require that

$$
\begin{aligned}
&\sum_{\beta=1}^{n} c^{v\rho}_{\alpha\beta\gamma} = 0 \ , \qquad \begin{cases} \alpha = 1,\dots,n \ , \\ \gamma = 1,\dots,m \ , \end{cases}\\
&\sum_{\beta=1}^{n} c^{v\nu}_{\alpha\beta\gamma} = 0 \ , \qquad \begin{cases} \alpha = 1,\dots,n \ , \\ \gamma = 1,\dots,n-1 \ , \end{cases}\\
&\sum_{\beta=1}^{n} (c^{\mathbf{N}}_\alpha)_{,\mathbf{v}_\beta} = 0 \ , \qquad \alpha = 1,\dots,n \ ,\\
&\sum_{\beta=1}^{n} \mathbf{C}^{\mathrm{W}}_{\alpha\beta} = 0 \ , \qquad \alpha = 1,\dots,n \ .
\end{aligned}
\tag{7.65}
$$

The ‘isotropic’ expansion of n_α is analogously expressed as follows:

$$
\begin{aligned}
n_\alpha = n_\alpha^{\dot\theta}(\dot\theta) &+ \Big\{\sum_{\beta=1}^{m} n_{\alpha\beta}^{\theta\rho}(\nabla\rho_\beta) + \sum_{\beta=1}^{n-1} n_{\alpha\beta}^{\theta\nu}(\nabla\nu_\beta)\Big\}\cdot(\nabla\theta)\\
&+ \sum_{\beta=1}^{n}\Big\{\sum_{\gamma=1}^{m} n_{\alpha\beta\gamma}^{\mathbf{v}\rho}(\nabla\rho_\gamma) + \sum_{\gamma=1}^{n-1} n_{\alpha\beta\gamma}^{\mathbf{v}\nu}(\nabla\nu_\gamma)\Big\}\cdot(\mathbf{v}_\beta)\\
&+ \sum_{\beta=1}^{n}\Big\{n_{\alpha\beta}^{\mathrm{D}}\mathbf{I} + \mathbf{N}_{\alpha\beta}^{\mathrm{D}}\Big\}\cdot(\mathbf{D}_\beta)\\
&+ \sum_{\beta=1}^{n}\Big\{\mathbf{N}_{\alpha\beta}^{\mathrm{W}}\Big\}\cdot(\mathbf{W}_\beta)\\
&+ n_\alpha^{\mathbf{N}}\ ,
\end{aligned}
\tag{7.66}
$$

where the term $n_\alpha^{\mathbf{N}}$ accounts for the remaining non-linear contributions and depends on all constitutive variables. It must vanish in thermodynamic equilibrium, i. e.

$$
n_\alpha^{\mathbf{N}}\big|_{\mathrm{E}} = 0\ , \tag{7.67}
$$

to be in accordance with $(7.12)_2$. The tensors $\mathbf{N}_{\alpha\beta}^{\mathrm{D}}$ and $\mathbf{N}_{\alpha\beta}^{\mathrm{W}}$ are defined as

$$
\begin{aligned}
\mathbf{N}_{\alpha\beta}^{\mathrm{D}} := \sum_{\delta=1}^{m}\Big\{\sum_{\gamma=1}^{m} n_{\alpha\beta\gamma\delta}^{\mathrm{D}\rho}\,\mathrm{sym}\big(\nabla\rho_\delta\otimes\nabla\rho_\gamma\big) &+ \sum_{\gamma=1}^{n-1} n_{\alpha\beta\gamma\delta}^{\mathrm{D}\nu\rho}\,\mathrm{sym}\big(\nabla\rho_\delta\otimes\nabla\nu_\gamma\big)\Big\}\\
+ \sum_{\delta=1}^{n-1}\Big\{\sum_{\gamma=1}^{n-1} n_{\alpha\beta\gamma\delta}^{\mathrm{D}\nu}\,\mathrm{sym}\big(\nabla\nu_\delta\otimes\nabla\nu_\gamma\big)\Big\}\ ,
\end{aligned}
\tag{7.68}
$$

and

$$
\begin{aligned}
\mathbf{N}_{\alpha\beta}^{\mathrm{W}} := \sum_{\delta=1}^{m}\Big\{\sum_{\gamma=1}^{m} n_{\alpha\beta\gamma\delta}^{\mathrm{W}\rho}\,\mathrm{skw}\big(\nabla\rho_\delta\otimes\nabla\rho_\gamma\big) &+ \sum_{\gamma=1}^{n-1} n_{\alpha\beta\gamma\delta}^{\mathrm{W}\nu\rho}\,\mathrm{skw}\big(\nabla\rho_\delta\otimes\nabla\nu_\gamma\big)\Big\}\\
+ \sum_{\delta=1}^{n-1}\Big\{\sum_{\gamma=1}^{n-1} n_{\alpha\beta\gamma\delta}^{\mathrm{W}\nu}\,\mathrm{skw}\big(\nabla\nu_\delta\otimes\nabla\nu_\gamma\big)\Big\}\ .
\end{aligned}
\tag{7.69}
$$

The coefficients arising in (7.66), (7.68) and (7.69) are specified according to

$$
\begin{aligned}
n_\alpha^{\dot\theta} &:= (n_\alpha)_{,\dot\theta}\big|_{\mathsf{n}=0} = \hat n_\alpha^{\dot\theta}\Big(\theta,\ \vec\rho,\ \vec\nu,\ \vec\nabla\rho,\ \vec\nabla\nu,\ \vec{\mathbf{B}},\ \vec{\vec{\mathbf{Z}}}\Big)\ ,\\
n_{\alpha\beta}^{\theta\rho}\,\mathbf{I} &:= (n_\alpha)_{,\nabla\theta\nabla\rho_\beta}\big|_{\boldsymbol{\mathcal{Y}}=0} = \hat n_{\alpha\beta}^{\theta\rho}\Big(\theta,\ \vec\rho,\ \vec\nu,\ \vec{\mathbf{B}},\ \vec{\vec{\mathbf{Z}}}\Big)\,\mathbf{I}\ ,\\
n_{\alpha\beta}^{\theta\nu}\,\mathbf{I} &:= (n_\alpha)_{,\nabla\theta\nabla\nu_\beta}\big|_{\boldsymbol{\mathcal{Y}}=0} = \hat n_{\alpha\beta}^{\theta\nu}\Big(\theta,\ \vec\rho,\ \vec\nu,\ \vec{\mathbf{B}},\ \vec{\vec{\mathbf{Z}}}\Big)\,\mathbf{I}\ ,\\
n_{\alpha\beta\gamma}^{v\rho}\,\mathbf{I} &:= (n_\alpha)_{,\mathbf{v}_\beta\nabla\rho_\gamma}\big|_{\boldsymbol{\mathcal{Y}}=0} = \hat n_{\alpha\beta\gamma}^{v\rho}\Big(\theta,\ \vec\rho,\ \vec\nu,\ \vec{\mathbf{B}},\ \vec{\vec{\mathbf{Z}}}\Big)\,\mathbf{I}\ ,\\
n_{\alpha\beta\gamma}^{v\nu}\,\mathbf{I} &:= (n_\alpha)_{,\mathbf{v}_\beta\nabla\nu_\gamma}\big|_{\boldsymbol{\mathcal{Y}}=0} = \hat n_{\alpha\beta\gamma}^{v\nu}\Big(\theta,\ \vec\rho,\ \vec\nu,\ \vec{\mathbf{B}},\ \vec{\vec{\mathbf{Z}}}\Big)\,\mathbf{I}\ ,\\
n_{\alpha\beta}^{\mathrm{D}}\,\boldsymbol{\mathcal{I}} &:= (n_\alpha)_{,\mathbf{D}_\beta}\big|_{\boldsymbol{\mathcal{Y}}=0} = \hat n_{\alpha\beta}^{\mathrm{D}}\Big(\theta,\ \vec\rho,\ \vec\nu,\ \vec{\mathbf{B}},\ \vec{\vec{\mathbf{Z}}}\Big)\,\boldsymbol{\mathcal{I}}\ ,\\
n_{\alpha\beta\gamma\delta}^{\mathrm{D}\rho}\,\boldsymbol{\mathcal{I}} &:= (n_\alpha)_{,\mathbf{D}_\beta\nabla\rho_\gamma\nabla\rho_\delta}\big|_{\boldsymbol{\mathcal{Y}}=0} = \hat n_{\alpha\beta\gamma\delta}^{\mathrm{D}\rho}\Big(\theta,\ \vec\rho,\ \vec\nu,\ \vec{\mathbf{B}},\ \vec{\vec{\mathbf{Z}}}\Big)\,\boldsymbol{\mathcal{I}}\ ,\\
n_{\alpha\beta\gamma\delta}^{\mathrm{D}\nu\rho}\,\boldsymbol{\mathcal{I}} &:= (n_\alpha)_{,\mathbf{D}_\beta\nabla\nu_\gamma\nabla\rho_\delta}\big|_{\boldsymbol{\mathcal{Y}}=0} = \hat n_{\alpha\beta\gamma\delta}^{\mathrm{D}\nu\rho}\Big(\theta,\ \vec\rho,\ \vec\nu,\ \vec{\mathbf{B}},\ \vec{\vec{\mathbf{Z}}}\Big)\,\boldsymbol{\mathcal{I}}\ ,\\
n_{\alpha\beta\gamma\delta}^{\mathrm{D}\nu}\,\boldsymbol{\mathcal{I}} &:= (n_\alpha)_{,\mathbf{D}_\beta\nabla\nu_\gamma\nabla\nu_\delta}\big|_{\boldsymbol{\mathcal{Y}}=0} = \hat n_{\alpha\beta\gamma\delta}^{\mathrm{D}\nu}\Big(\theta,\ \vec\rho,\ \vec\nu,\ \vec{\mathbf{B}},\ \vec{\vec{\mathbf{Z}}}\Big)\,\boldsymbol{\mathcal{I}}\ ,\\
n_{\alpha\beta\gamma\delta}^{\mathrm{W}\rho}\,\boldsymbol{\mathcal{I}} &:= (n_\alpha)_{,\mathbf{W}_\beta\nabla\rho_\gamma\nabla\rho_\delta}\big|_{\boldsymbol{\mathcal{Y}}=0} = \hat n_{\alpha\beta\gamma\delta}^{\mathrm{W}\rho}\Big(\theta,\ \vec\rho,\ \vec\nu,\ \vec{\mathbf{B}},\ \vec{\vec{\mathbf{Z}}}\Big)\,\boldsymbol{\mathcal{I}}\ ,\\
n_{\alpha\beta\gamma\delta}^{\mathrm{W}\nu\rho}\,\boldsymbol{\mathcal{I}} &:= (n_\alpha)_{,\mathbf{W}_\beta\nabla\nu_\gamma\nabla\rho_\delta}\big|_{\boldsymbol{\mathcal{Y}}=0} = \hat n_{\alpha\beta\gamma\delta}^{\mathrm{W}\nu\rho}\Big(\theta,\ \vec\rho,\ \vec\nu,\ \vec{\mathbf{B}},\ \vec{\vec{\mathbf{Z}}}\Big)\,\boldsymbol{\mathcal{I}}\ ,\\
n_{\alpha\beta\gamma\delta}^{\mathrm{W}\nu}\,\boldsymbol{\mathcal{I}} &:= (n_\alpha)_{,\mathbf{W}_\beta\nabla\nu_\gamma\nabla\nu_\delta}\big|_{\boldsymbol{\mathcal{Y}}=0} = \hat n_{\alpha\beta\gamma\delta}^{\mathrm{W}\nu}\Big(\theta,\ \vec\rho,\ \vec\nu,\ \vec{\mathbf{B}},\ \vec{\vec{\mathbf{Z}}}\Big)\,\boldsymbol{\mathcal{I}}\ ,
\end{aligned}
\tag{7.70}
$$

where all coefficients are symmetric with respect to constituent indices that arise in the derivatives, e. g.

$$
n_{\alpha\beta\gamma}^{v\nu} = n_{\alpha\gamma\beta}^{v\nu}\ . \tag{7.71}
$$

As a constitutive quantity, the volume fraction production rate density n_α has to satisfy the principle of objectivity which is formulated in (4.50) and (4.51). Applying these relations to n_α, i. e.

$$
\sum_{\beta=1}^{n}(n_\alpha)_{,\mathbf{v}_\beta} = 0 \qquad \text{and} \qquad \sum_{\beta=1}^{n}(n_\alpha)_{,\mathbf{W}_\beta} = 0\ , \tag{7.72}
$$

we require that

$$
\begin{aligned}
&\textstyle\sum_{\beta=1}^{n} n^{v\rho}_{\alpha\beta\gamma} = 0\ , && \begin{cases}\alpha = 1,\dots,n\ ,\\ \gamma = 1,\dots,m\ ,\end{cases}\\
&\textstyle\sum_{\beta=1}^{n} n^{v\nu}_{\alpha\beta\gamma} = 0\ , && \begin{cases}\alpha = 1,\dots,n\ ,\\ \gamma = 1,\dots,n-1\ ,\end{cases}\\
&\textstyle\sum_{\beta=1}^{n} (n^{\mathbf{N}}_\alpha)_{,\mathbf{v}_\beta} = 0\ , && \alpha = 1,\dots,n\ ,\\
&\textstyle\sum_{\beta=1}^{n} \mathbf{N}^{\mathrm{W}}_{\alpha\beta} = 0\ , && \alpha = 1,\dots,n\ .
\end{aligned}
\tag{7.73}
$$

We remark that the 'isotropic' expansion of n_α in (7.66) to (7.73) is analogous to that of c_α (see (7.55) to (7.65)) except that in general one has (compare with 7.61)

$$
\sum \bar{\rho}_\alpha n_\alpha \neq 0\ . \tag{7.74}
$$

Extra entropy flux k

If we apply the same expansion to the extra entropy flux, $\mathbf{k}$, we know from (6.58) that $\mathbf{k}$ is given by a vector combination of $\mathbf{u}_\alpha$ (or, equivalently of $\mathbf{v}_\alpha$) so that

$$
\mathbf{k} = \sum (k^v_\alpha)\mathbf{v}_\alpha + \mathbf{k}^{\mathbf{N}}\ , \tag{7.75}
$$

where the coefficients k^v_α are defined as

$$
k^v_\alpha \mathbf{I} := \mathbf{k}_{,\mathbf{v}_\alpha}\big|_{\mathcal{Y}=0} = \hat{k}^v_\alpha\Big(\theta,\ \vec{\rho},\ \vec{\nu},\ \vec{\mathbf{B}},\ \vec{\vec{\mathbf{Z}}}\Big)\ \mathbf{I}\ , \qquad \alpha = 1,\dots,n \tag{7.76}
$$

and $\mathbf{k}^{\mathbf{N}}$, again, subsumes all other contributions of the constitutive variables and, due to the principle of objectivity $\mathbf{k}$ satisfies the sum relations (see (4.50))

$$
\sum \mathbf{k}_{,\mathbf{v}_\beta} = \mathbf{0} \qquad \Rightarrow \qquad \sum k^v_\beta = 0, \quad \sum (\mathbf{k}^{\mathbf{N}})_{,\mathbf{v}_\beta} = \mathbf{0}\ . \tag{7.77}
$$

These restrict the independences of the parameters k^v_β, $(\mathbf{k}^{\mathbf{N}})_{,\mathbf{v}_\beta}$.

7.4 Final Representations for $\bar{\mathbf{m}}^i_\alpha\big|_{\mathrm{E}}$, $\mathbf{q}\big|_{\mathrm{E}}$, $\bar{\mathbf{T}}_\alpha\big|_{\mathrm{E}}$ and $\eta\big|_{\mathrm{E}}$

The purpose of the above 'isotropic' expansion is to simplify the derivatives of $\vec{c}$, $\vec{n}$ and $\mathbf{k}$ arising in the constitutive relations for the equilibrium quantities $\bar{\mathbf{m}}^i_\alpha\big|_{\mathrm{E}}$, $\mathbf{q}\big|_{\mathrm{E}}$, $\bar{\mathbf{T}}_\alpha\big|_{\mathrm{E}}$ and $\eta\big|_{\mathrm{E}}$ (see (7.48), (7.41), (7.44) and (7.38)). The derivatives of $\vec{c}$, $\vec{n}$ and $\mathbf{k}$ arising in those formulae have, in the representation of the 'isotropic' expansion, the following form

$$
\begin{aligned}
\mathbf{k}_{,\rho_\alpha \mathbf{v}_\beta}\big|_{\mathrm{E}} &= (k^v_\beta)_{,\rho_\alpha}\,\mathbf{I} + (\mathbf{k}^{\mathbf{N}})_{,\rho_\alpha \mathbf{v}_\beta}\big|_{\mathrm{E}},\\
\mathbf{k}_{,\nu_\alpha \mathbf{v}_\beta}\big|_{\mathrm{E}} &= (k^v_\beta)_{,\nu_\alpha}\,\mathbf{I} + (\mathbf{k}^{\mathbf{N}})_{,\nu_\alpha \mathbf{v}_\beta}\big|_{\mathrm{E}},\\
(c_\alpha)_{,\mathbf{v}_\beta}\big|_{\mathrm{E}} &= \sum_{\gamma=1}^{m} c^{\mathbf{v}\rho}_{\alpha\beta\gamma}(\nabla\rho_\gamma) + \sum_{\gamma=1}^{n-1} c^{\mathbf{v}\nu}_{\alpha\beta\gamma}(\nabla\nu_\gamma) + (c^{\mathbf{N}}_\alpha)_{,\mathbf{v}_\beta}\big|_{\mathrm{E}},\\
(n_\alpha)_{,\mathbf{v}_\beta}\big|_{\mathrm{E}} &= \sum_{\gamma=1}^{m} n^{\mathbf{v}\rho}_{\alpha\beta\gamma}(\nabla\nu_\gamma) + \sum_{\gamma=1}^{n-1} n^{\mathbf{v}\nu}_{\alpha\beta\gamma}(\nabla\nu_\gamma) + (n^{\mathbf{N}}_\alpha)_{,\mathbf{v}_\beta}\big|_{\mathrm{E}}
\end{aligned}
\tag{7.78}
$$

for $\bar{\mathbf{m}}^i_\alpha\big|_{\mathrm{E}}$. For the equilibrium energy flux, $\mathbf{q}\big|_{\mathrm{E}}$, the derivatives

$$
\begin{aligned}
(c_\alpha)_{,\nabla\theta}\big|_{\mathrm{E}} &= \sum_{\beta=1}^{m} c^{\theta\rho}_{\alpha\beta}(\nabla\rho_\beta) + \sum_{\beta=1}^{n-1} c^{\theta\nu}_{\alpha\beta}(\nabla\nu_\beta) + (c^{\mathbf{N}}_\alpha)_{,\nabla\theta}\big|_{\mathrm{E}},\\
(n_\alpha)_{,\nabla\theta}\big|_{\mathrm{E}} &= \sum_{\beta=1}^{m} n^{\theta\rho}_{\alpha\beta}(\nabla\rho_\beta) + \sum_{\beta=1}^{n-1} n^{\theta\nu}_{\alpha\beta}(\nabla\nu_\beta) + (n^{\mathbf{N}}_\alpha)_{,\nabla\theta}\big|_{\mathrm{E}}
\end{aligned}
\tag{7.79}
$$

are needed and for the constituent CAUCHY stress tensor $\bar{\mathbf{T}}_\alpha\big|_{\mathrm{E}}$, (7.44), we need the relations

$$
\begin{aligned}
\mathbf{k}_{,\mathbf{v}_\beta}\big|_{\mathrm{E}} &= k^v_\alpha \mathbf{I} + \mathbf{k}^{\mathbf{N}}_{,\mathbf{v}_\beta}\big|_{\mathrm{E}},\\
(c_\alpha)_{,\mathbf{D}_\beta}\big|_{\mathrm{E}} &= c^{\mathrm{D}}_{\alpha\beta}\mathbf{I} + \mathbf{C}^{\mathrm{D}}_{\alpha\beta} + (c^{\mathbf{N}}_\alpha)_{,\mathbf{D}_\beta}\big|_{\mathrm{E}},\\
(n_\alpha)_{,\mathbf{D}_\beta}\big|_{\mathrm{E}} &= n^{\mathrm{D}}_{\alpha\beta}\mathbf{I} + \mathbf{N}^{\mathrm{D}}_{\alpha\beta} + (n^{\mathbf{N}}_\alpha)_{,\mathbf{D}_\beta}\big|_{\mathrm{E}}\ ,
\end{aligned}
\tag{7.80}
$$

whilst for the entropy $\eta\big|_{\mathrm{E}}$ (7.38)

$$
\begin{aligned}
(c_\alpha)_{,\dot\theta}\big|_{\mathrm{E}} &= c^{\dot\theta}_\alpha + (c^{\mathbf{N}}_\alpha)_{\dot\theta}\big|_{\mathrm{E}}\ ,\\
(n_\alpha)_{,\dot\theta}\big|_{\mathrm{E}} &= n^{\dot\theta}_\alpha + (n^{\mathbf{N}}_\alpha)_{\dot\theta}\big|_{\mathrm{E}}
\end{aligned}
\tag{7.81}
$$

are needed.

If we now substitute the above representations into the equilibrium expressions $\bar{\mathbf{m}}^i_\alpha\big|_{\mathrm{E}}$, $\mathbf{q}\big|_{\mathrm{E}}$, $\bar{\mathbf{T}}_\alpha\big|_{\mathrm{E}}$ and $\eta_\alpha\big|_{\mathrm{E}}$, and if we further subsume the non-linear contributions $(\mathbf{k}^{\mathbf{N}})_{,\nu_\alpha \mathbf{v}_\beta}\big|_{\mathrm{E}}$, $(c^{\mathbf{N}}_\alpha)_{,\mathbf{v}_\beta}\big|_{\mathrm{E}}$, $(n^{\mathbf{N}}_\alpha)_{,\mathbf{v}_\beta}\big|_{\mathrm{E}}$, etc. into $\mathbf{m}^{\mathbf{N}}_\alpha$ for $\mathbf{m}^i_\alpha$, $\mathbf{q}^{\mathbf{N}}$ for $\mathbf{q}$, $\mathbf{T}^{\mathbf{N}}_\alpha$ for $\mathbf{T}_\alpha$ and $\eta^{\mathbf{N}}$ for η, respectively, we obtain from (7.48)

$$
\begin{aligned}
\bar{\mathbf{m}}^i_\beta\big|_\mathrm{E} = \sum_{\alpha=1}^{m} \Big\{ & \rho_\alpha^{-1}\bar{p}_\alpha\big|_\mathrm{E}(\delta_{\alpha\beta}-\bar{\xi}_\beta)\mathbf{I} - \theta(k^v_\beta)_{,\rho_\alpha}\mathbf{I} \\
& - \sum_{\gamma=1}^{n} \bar{\rho}_\gamma \, \mu^G_{\gamma \mathrm{I}}\big|_\mathrm{E} \, c^{\mathbf{v}\rho}_{\gamma\beta\alpha} + \sum_{\gamma=1}^{n} \bar{\iota}_\gamma\big|_\mathrm{E} n^{\mathbf{v}\rho}_{\gamma\beta\alpha} \Big\} \, (\nabla\rho_\alpha) \\
+ \sum_{\alpha=1}^{n-1} \Big\{ & \zeta_\alpha\big|_\mathrm{E}(\delta_{\alpha\beta}-\bar{\xi}_\beta)\mathbf{I} - \mathbf{c}_{,\mathbf{v}_\beta}\big|_\mathrm{E} - \theta(k^v_\beta)_{,\nu_\alpha}\mathbf{I} \\
& - \sum_{\gamma=1}^{n} \bar{\rho}_\gamma \, \mu^G_{\gamma \mathrm{I}}\big|_\mathrm{E} \, c^{\mathbf{v}\nu}_{\gamma\beta\alpha} + \sum_{\gamma=1}^{n} \bar{\iota}_\gamma\big|_\mathrm{E} n^{\mathbf{v}\nu}_{\gamma\beta\alpha} \Big\} \, (\nabla\nu_\alpha) \\
+ \sum_{\alpha=1}^{n} & \rho \left(\Psi^G_I\right)_{,\bar{\mathbf{Z}}_\alpha}\big|_\mathrm{E} \left(\bar{\mathbf{\Phi}}_\alpha\right)_{,\mathbf{v}_\beta}\big|_\mathrm{E} \\
+ \bar{\mathbf{m}}^{\mathbf{N}}_\beta & \big|_\mathrm{E} \, .
\end{aligned}
\tag{7.82}
$$

With the help of the definitions (7.8), (7.6) and the auxiliary result $(7.3)_3$, $\mathbf{c}_{,\mathbf{v}_\beta}\big|_\mathrm{E}$ takes the form

$$
\mathbf{c}_{,\mathbf{v}_\beta}\big|_\mathrm{E} = \Big\{ \left(\rho_n \boldsymbol{\Delta}^{*n}_D + \zeta_n \mathbf{I}\right)\mathbf{u}_n \Big\}_{,\mathbf{v}_\beta}\big|_\mathrm{E} = \mathfrak{c}_\beta \mathbf{I} \, . \tag{7.83}
$$

Since $\boldsymbol{\Delta}^{*n}_D\big|_\mathrm{E} = \mathbf{0}$, see $(7.27)_1$, the new abbreviation, $\mathfrak{c}_\beta$, is defined by

$$
\mathfrak{c}_\beta := \zeta_n\big|_\mathrm{E}(\delta_{n\beta}-\bar{\xi}_\beta) \, . \tag{7.84}
$$

With the results (7.79) and [**A16**] the equilibrium energy flux vector, $\mathbf{q}\big|_\mathrm{E}$, transforms into

$$
\begin{aligned}
\lambda^\varepsilon_{,\dot\theta}\big|_\mathrm{E}\mathbf{q}\big|_\mathrm{E} = {} & \theta^{-1}\sum_{\beta=1}^{m}\Big\{\sum_{\alpha=1}^{n}\bar{\rho}_\alpha \mu^G_{\alpha \mathrm{I}}\big|_\mathrm{E} c^{\theta\rho}_{\alpha\beta} - \sum_{\alpha=1}^{n}\iota_\alpha n^{\theta\rho}_{\alpha\beta}\Big\}(\nabla\rho_\beta) \\
& + \theta^{-1}\sum_{\beta=1}^{n-1}\Big\{\sum_{\alpha=1}^{n}\bar{\rho}_\alpha \mu^G_{\alpha \mathrm{I}}\big|_\mathrm{E} c^{\theta\nu}_{\alpha\beta} - \sum_{\alpha=1}^{n}\iota_\alpha n^{\theta\nu}_{\alpha\beta}\Big\}(\nabla\nu_\beta) \\
& + \theta^{-1}\sum_{\alpha=1}^{n}\rho\left(\Psi^G_I\right)_{,\bar{\mathbf{Z}}_\alpha}\big|_\mathrm{E}\left(\bar{\mathbf{\Phi}}_\alpha\right)_{,\nabla\theta}\big|_\mathrm{E} \\
& + \lambda^\varepsilon_{,\dot\theta}\big|_\mathrm{E}\mathbf{q}^\mathbf{N}\big|_\mathrm{E} \, .
\end{aligned}
\tag{7.85}
$$

Next, applying this expansion procedure to the entropy (7.38) and using $(7.59)_1$, i. e. $(c_\alpha)_{,\dot\theta}\big|_{\mathsf{n}=0} = c^{\dot\theta}_\alpha$ and $(7.70)_1$, i. e. $(n_\alpha)_{,\dot\theta}\big|_{\mathsf{n}=0} = n^{\dot\theta}_\alpha$, yields

$$
\begin{aligned}
\eta\big|_{\mathrm{E}} &= -\frac{1}{(1/\lambda^{\varepsilon})_{,\theta}\big|_{\mathrm{E}}}\left\{(\Psi^G_{I,\theta})\big|_{\mathrm{E}} + \frac{\mathcal{P}_\theta\big|_{\mathrm{E}}}{\rho\lambda^{\varepsilon}\big|_{\mathrm{E}}}\right\} \\
&= \frac{-1}{(1/\lambda^{\varepsilon})_{,\theta}\big|_{\mathrm{E}}}\left\{(\Psi^G_{I,\theta})\big|_{\mathrm{E}} + \sum_{\alpha=1}^{n}\left(\Psi^G_I\right)_{,\bar{\mathbf{Z}}_\alpha}\big|_{\mathrm{E}}\cdot(\bar{\boldsymbol{\Phi}}_\alpha)_{,\dot{\theta}}\big|_{\mathrm{E}}\right. \\
&\qquad \left. + \sum_{\alpha=1}^{n}\left\{\nu_\alpha(\underset{\alpha}{\mu}{}^G_{\mathrm{I}})\big|_{\mathrm{E}}\, c_\alpha^{\dot{\theta}} - \frac{\iota_\alpha\big|_{\mathrm{E}}}{\rho}\nu_\alpha n_\alpha^{\dot{\theta}}\right\}\right\}\,. \qquad (7.86)
\end{aligned}
$$

Here $(1/\lambda^{\varepsilon})_{,\theta}\big|_{\mathrm{E}} = 1$, $c_\alpha^{\dot{\theta}} = 0$ and $n_\alpha^{\dot{\theta}} = 0$ if $\dot{\theta} \notin \mathbb{S}$. In general, however, $(1/\lambda^{\varepsilon})_{,\theta}\big|_{\mathrm{E}} \neq 1$ even if $c_\alpha^{\dot{\theta}} = 0$ and $n_\alpha^{\dot{\theta}} = 0$.

Applying the 'isotropic' expansion, i. e. relations (7.80), to (7.44) and using again [**A16**] we deduce

$$
\begin{aligned}
\bar{\mathbf{T}}_\beta\big|_{\mathrm{E}} &= -\bar{\zeta}_\beta\big|_{\mathrm{E}}\mathbf{I} + 2\rho\ \mathrm{sym}\left(\Psi^G_{I,\mathbf{B}_\beta}\right)\big|_{\mathrm{E}}\mathbf{B}_\beta\big|_{\mathrm{E}} + \theta k^v_\beta\mathbf{I} \\
&\quad + \sum \rho\left(\Psi^G_I\right)_{,\bar{\mathbf{Z}}_\alpha}\big|_{\mathrm{E}}(\bar{\boldsymbol{\Phi}}_\alpha)_{,\mathbf{D}_\beta}\big|_{\mathrm{E}} \\
&\quad + \sum_{\alpha=1}^{n}\bar{\rho}_\alpha\ \underset{\alpha}{\mu}{}^G_{\mathrm{I}}\big|_{\mathrm{E}}\left\{c^{\mathrm{D}}_{\alpha\beta}\mathbf{I} + \mathbf{C}^{\mathrm{D}}_{\alpha\beta}\right\} \qquad (7.87) \\
&\quad - \sum_{\alpha=1}^{n}\bar{\iota}_\alpha\big|_{\mathrm{E}}\left\{n^{\mathrm{D}}_{\alpha\beta}\mathbf{I} + \mathbf{N}^{\mathrm{D}}_{\alpha\beta}\right\} + \bar{\mathbf{T}}^{\mathbf{N}}_\beta\big|_{\mathrm{E}}\,.
\end{aligned}
$$

It is convenient to collect all spherical contributions of $\bar{\mathbf{T}}_\beta\big|_{\mathrm{E}}$ to define a new pressure,

$$
\bar{\varpi}_\beta := \bar{\zeta}_\beta\big|_{\mathrm{E}} - \theta k^v_\beta - \sum_{\alpha=1}\bar{\rho}_\alpha\ \underset{\alpha}{\mu}{}^G_{\mathrm{I}}\big|_{\mathrm{E}}\ c^{\mathrm{D}}_{\alpha\beta} + \sum_{\alpha=1}^{n}\bar{\iota}_\alpha\big|_{\mathrm{E}}\ n^{\mathrm{D}}_{\alpha\beta}\,. \qquad (7.88)
$$

With this definition, (7.87) transforms into

$$
\begin{aligned}
\bar{\mathbf{T}}_\beta\big|_{\mathrm{E}} &= -\bar{\varpi}_\beta\mathbf{I} + 2\rho\ \mathrm{sym}\left(\Psi^G_{I,\mathbf{B}_\beta}\right)\big|_{\mathrm{E}}\mathbf{B}_\beta\big|_{\mathrm{E}} + \sum\rho\left(\Psi^G_I\right)_{,\bar{\mathbf{Z}}_\alpha}\big|_{\mathrm{E}}(\bar{\boldsymbol{\Phi}}_\alpha)_{,\mathbf{D}_\beta}\big|_{\mathrm{E}} \\
&\quad + \sum_{\alpha=1}^{n}\bar{\rho}_\alpha\ \underset{\alpha}{\mu}{}^G_{\mathrm{I}}\big|_{\mathrm{E}}\ \mathbf{C}^{\mathrm{D}}_{\alpha\beta} - \sum_{\alpha=1}^{n}\bar{\iota}_\alpha\big|_{\mathrm{E}}\ \mathbf{N}^{\mathrm{D}}_{\alpha\beta} + \bar{\mathbf{T}}^{\mathbf{N}}_\beta\big|_{\mathrm{E}}\,. \qquad (7.89)
\end{aligned}
$$

In this representation the explicit forms for Ψ^G_I and $\underset{\alpha}{\mu}{}^G_{\mathrm{I}}$ in terms of the invariants of vector and tensor valued variables may still yield additional isotropic contributions.

The sum of (7.89) over all constituents yields the equilibrium mixture CAUCHY stress tensor, $\mathbf{T}\big|_{\mathrm{E}}$, in the form

$$
\begin{aligned}
\mathbf{T}\big|_{\mathrm{E}} = \mathbf{T}_I\big|_{\mathrm{E}} &= \sum_{\beta=1}^{n} \bar{\mathbf{T}}_\beta\big|_{\mathrm{E}} \\
&= -\varpi\mathbf{I} + 2\rho \sum_{\beta=1}^{n} \left\{ \operatorname{sym}\left(\Psi^G_{I,\mathbf{B}_\beta}\right)\big|_{\mathrm{E}} \mathbf{B}_\beta\big|_{\mathrm{E}} \right\} \\
&\quad + \sum_{\beta=1}^{n}\sum_{\alpha=1}^{n} \rho \left(\Psi^G_I\right)_{,\bar{\mathbf{Z}}_\alpha}\big|_{\mathrm{E}} (\bar{\boldsymbol{\Phi}}_\alpha)_{,\mathbf{D}_\beta}\big|_{\mathrm{E}} \\
&\quad + \sum_{\beta=1}^{n}\sum_{\alpha=1}^{n} \bar{\rho}_\alpha\, \mu^G_{\alpha\mathrm{I}}\big|_{\mathrm{E}}\, \mathbf{C}^{\mathrm{D}}_{\alpha\beta} - \sum_{\beta=1}^{n}\sum_{\alpha=1}^{n} \bar{\iota}_\alpha\big|_{\mathrm{E}}\, \mathbf{N}^{\mathrm{D}}_{\alpha\beta} + \sum_{\beta=1}^{n} \bar{\mathbf{T}}^{\mathbf{N}}_\beta\big|_{\mathrm{E}},
\end{aligned}
\tag{7.90}
$$

where ϖ is defined as

$$
\varpi := \sum \bar{\varpi}_\beta = \sum_{\beta=1}^{n} \left\{ \bar{\zeta}_\beta\big|_{\mathrm{E}} - \sum_{\alpha=1}^{n} \bar{\rho}_\alpha\, \mu^G_{\alpha\mathrm{I}}\big|_{\mathrm{E}}\, c^{\mathrm{D}}_{\alpha\beta} + \sum_{\alpha=1}^{n} \bar{\iota}_\alpha\big|_{\mathrm{E}} n^{\mathrm{D}}_{\alpha\beta} \right\}. \tag{7.91}
$$

In (7.91) we have used the principle of objectivity to rule out $\sum k^v_\beta$, see (7.77).

Obviously, those terms in (7.90) and (7.91) that are related to the interaction supply rate density of mass are not allowed to arise in $\mathbf{T}\big|_{\mathrm{E}}$, as it is a mixture quantity. This drawback of our theory is exclusively due to the consideration of $\mathbf{W}_\alpha$ and $\mathbf{D}_\alpha$ as independent constitutive variables instead of $\mathbf{L}_\alpha$ ($\alpha = 1, \dots, n$). In the approach pursued here, we have two sets of coefficients, namely

$$
\left\{ c^{\mathrm{D}}_{\alpha\beta},\ \mathbf{C}^{\mathrm{D}}_{\alpha\beta} \right\} \quad \text{and} \quad \left\{ \mathbf{C}^{\mathrm{W}}_{\alpha\beta} \right\}, \tag{7.92}
$$

and when all $\mathbf{L}_\alpha$ are independent variables, only one such set of coefficients arises in the 'isotropic' expansion. The application of the principle of objectivity to the expansion of the interaction supply rate density of mass, when $\mathbf{L}_\alpha$ ($\alpha = 1, \dots, n$) are constitutive variables, would have allowed the restrictions

$$
\sum_{\beta=1}^{n} c^{\mathrm{L}}_{\alpha\beta} = \sum_{\beta=1}^{n} \left(c^{\mathrm{D}}_{\alpha\beta}\right) = 0 \quad \text{and} \quad \sum_{\beta=1}^{n} \mathbf{C}^{\mathrm{L}}_{\alpha\beta} = \sum_{\beta=1}^{n} \left(\mathbf{C}^{\mathrm{D}}_{\alpha\beta} + \mathbf{C}^{\mathrm{W}}_{\alpha\beta} \right) = \mathbf{0}, \tag{7.93}
$$

and thus the independence of the mixture CAUCHY stress tensor on the interaction supply rate densities of mass. As, in our approach, we are not dealing with the mixture CAUCHY stress tensor, but only with its constituent parts one is inclined to think that the above drawback does not come into play. This opinion is misleading, because in a mixture theory that correctly accounts for $\mathbf{L}_\alpha$ ($\alpha = 1, \dots, n$), the principle of objectivity affects the symmetric parts of the velocity gradients, too. Therefore, we are missing a restriction for the coefficients of the constituent stretching tensors arising in the constituent CAUCHY stress tensors. We therefore propose

$$\sum_{\beta=1}^{n} \mathbf{C}^{\mathrm{D}}_{\alpha\beta} = 0\ , \qquad \sum_{\beta=1}^{n} \mathbf{C}^{\mathrm{W}}_{\alpha\beta} = 0\ , \qquad (\alpha = 1, \dots, n) \tag{7.94}$$

as a sufficient condition satisfying $(7.93)_2$. This analysis implies that the constitutive coefficients $c^{\mathrm{D}}_{\alpha\beta}$ and $\mathbf{C}^{\mathrm{D,W}}_{\alpha\beta}$ must satisfy conditions $(7.93)_1$ and $(7.94)_{1,2}$ plus conditions (7.62). Therefore, since $c^{\mathrm{D}}_{\alpha\beta} = c^{\mathrm{D}}_{\beta\alpha}$, we have

$$\sum_{\alpha=1}^{n} \bar{\rho}_\alpha c^{\mathrm{D}}_{\alpha\beta} = 0 \qquad \text{and} \qquad \sum_{\alpha=1}^{n} c^{\mathrm{D}}_{\alpha\beta} = 0\ , \quad \beta \text{ fixed.} \tag{7.95}$$

These equations state that two of the n coefficients $c^{\mathrm{D}}_{\alpha\beta}$ (β fixed) are determined by the remaining ones. Solving (7.95) for $c^{\mathrm{D}}_{\alpha\beta}$ for $\alpha = n-1$, n yields

$$\begin{aligned} c^{\mathrm{D}}_{n-1\beta} &= \sum_{\alpha=1}^{n-2} \frac{\bar{\rho}_\alpha - \bar{\rho}_n}{\bar{\rho}_n - \bar{\rho}_{n-1}} c^{\mathrm{D}}_{\alpha\beta}\ , \\ c^{\mathrm{D}}_{n\beta} &= -\sum_{\alpha=1}^{n-2} \frac{\bar{\rho}_\alpha - \bar{\rho}_{n-1}}{\bar{\rho}_n - \bar{\rho}_{n-1}} c^{\mathrm{D}}_{\alpha\beta}\ . \end{aligned} \tag{7.96}$$

This, however requires that $\bar{\rho}_n \neq \bar{\rho}_{n-1}$. A similar calculation also applies to $\mathbf{C}^{\mathrm{D}}_{\alpha\beta}$, $\mathbf{C}^{\mathrm{W}}_{\alpha\beta}$.

The debris flow model presented in Chapter 8 does not suffer from this complexity, because, there, the interaction supply rate densities for mass will be set to zero ab initio.

For the next chapter, we also mention the results which emerge, when $\dot{\theta}$ is not considered an independent constitutive variable so that

[A17] $$\lambda^{\varepsilon}(\theta) = \theta^{-1}\ .$$

In this case Ψ^G reduces to the HELMHOLTZ free energy Ψ, and $\underset{\alpha}{\mu}{}^{G}_{\mathrm{I}}$ are the 'inner' parts of the usual GIBBS' free energies, $\underset{\alpha}{\mu}{}_{\mathrm{I}}$ ($\alpha = 1, \dots, n$). Furthermore, we obtain from [(5.30), line 1]

$$\mathcal{P}_\theta = 0, \qquad \mathcal{P}_{\nabla\theta} = \mathbf{0}\ , \tag{7.97}$$

as $\boldsymbol{\mathcal{F}}_{\dot{\theta}}$ vanishes identically. Thus, $(6.96)_1$ and $(6.96)_2$ reduce to

$$-(\Psi_I)_{,\theta} = \eta, \qquad (\Psi_I)_{,\nabla\theta} = \mathbf{0}\ . \tag{7.98}$$

Relations (7.98) are results known from many classical thermodynamic material theories. The constitutive law for $\mathbf{q}|_{\mathrm{E}}$ (see (7.85)) now takes the form

$$
\begin{aligned}
\lambda_{,\theta}^{\varepsilon}|_{\mathrm{E}}\mathbf{q}|_{\mathrm{E}} = & -\theta^{-1}\sum_{\beta=1}^{m}\Big\{\sum_{\alpha=1}^{n}\bar{\rho}_\alpha \mu^G_{\alpha\mathrm{I}}|_{\mathrm{E}} c^{\theta\rho}_{\alpha\beta} - \sum_{\alpha=1}^{n}\boldsymbol{\iota}_\alpha n^{\theta\rho}_{\alpha\beta}\Big\}(\nabla\rho_\beta) \\
& -\theta^{-1}\sum_{\beta=1}^{n-1}\Big\{\sum_{\alpha=1}^{n}\bar{\rho}_\alpha \mu^G_{\alpha\mathrm{I}}|_{\mathrm{E}} c^{\theta\nu}_{\alpha\beta} - \sum_{\alpha=1}^{n}\boldsymbol{\iota}_\alpha n^{\theta\nu}_{\alpha\beta}\Big\}(\nabla\nu_\beta) \\
& -\theta^{-1}\sum_{\alpha=1}^{n}\rho\left(\Psi_I^G\right)_{,\bar{\mathbf{Z}}_\alpha}|_{\mathrm{E}}\left(\bar{\boldsymbol{\Phi}}_\alpha\right)_{,\nabla\theta}|_{\mathrm{E}} \\
& +\mathbf{q}^{\mathbf{N}}|_{\mathrm{E}}\,.
\end{aligned}
\tag{7.99}
$$

Moreover, assumption [**A17**] implies no structural changes of the constitutive laws for $\bar{\mathbf{m}}^i_\beta$, η or $\bar{\mathbf{T}}_\beta$ ($\beta = 1,\dots,n$) (see (7.82), (7.88) and (7.89)). However, for the sake of completeness, we state these constitutive equations here:

$$
\begin{aligned}
\bar{\mathbf{T}}_\beta|_{\mathrm{E}} = & -\bar{\varpi}_\beta\mathbf{I} + 2\rho\ \mathrm{sym}\left(\Psi_{I,\mathbf{B}_\beta}\right)|_{\mathrm{E}}\mathbf{B}_\beta|_{\mathrm{E}} + \sum\rho\left(\Psi_I\right)_{,\bar{\mathbf{Z}}_\alpha}|_{\mathrm{E}}(\bar{\boldsymbol{\Phi}}_\alpha)_{,\mathbf{D}_\beta}|_{\mathrm{E}} \\
& + \sum_{\alpha=1}^{n}\bar{\rho}_\alpha\ \mu^G_{\alpha\mathrm{I}}|_{\mathrm{E}}\ \mathbf{C}^{\mathrm{D}}_{\alpha\beta} - \sum_{\alpha=1}^{n}\bar{\boldsymbol{\iota}}_\alpha|_{\mathrm{E}}\ \mathbf{N}^{\mathrm{D}}_{\alpha\beta} + \bar{\mathbf{T}}^{\mathbf{N}}_\beta|_{\mathrm{E}}\,,
\end{aligned}
\tag{7.100}
$$

$$
\begin{aligned}
\bar{\mathbf{m}}^i_\beta|_{\mathrm{E}} = & \textstyle\sum_{\alpha=1}^{m}\Big\{\rho_\alpha^{-1}\bar{p}_\alpha|_{\mathrm{E}}(\delta_{\alpha\beta}-\bar{\xi}_\beta)\mathbf{I} - \theta(k^v_\beta)_{,\rho_\alpha}\mathbf{I} \\
& \textstyle -\sum_{\gamma=1}^{n}\bar{\rho}_\gamma\ \mu^G_{\gamma\mathrm{I}}|_{\mathrm{E}}\ c^{\mathbf{v}\rho}_{\gamma\beta\alpha} + \sum_{\gamma=1}^{n}\bar{\boldsymbol{\iota}}_\gamma|_{\mathrm{E}} n^{\mathbf{v}\rho}_{\gamma\beta\alpha}\Big\}\ (\nabla\rho_\alpha) \\
& \textstyle +\sum_{\alpha=1}^{n-1}\Big\{\zeta_\alpha|_{\mathrm{E}}(\delta_{\alpha\beta}-\bar{\xi}_\beta)\mathbf{I} - \mathbf{c}_{,\mathbf{v}_\beta}|_{\mathrm{E}} - \theta(k^v_\beta)_{,\nu_\alpha}\mathbf{I} \\
& \textstyle -\sum_{\gamma=1}^{n}\bar{\rho}_\gamma\ \mu^G_{\gamma\mathrm{I}}|_{\mathrm{E}}\ c^{\mathbf{v}\nu}_{\gamma\beta\alpha} + \sum_{\gamma=1}^{n}\bar{\boldsymbol{\iota}}_\gamma|_{\mathrm{E}} n^{\mathbf{v}\nu}_{\gamma\beta\alpha}\Big\}\ (\nabla\nu_\alpha) \\
& \textstyle +\sum_{\alpha=1}^{n}\rho\left(\Psi_I^G\right)_{,\bar{\mathbf{Z}}_\alpha}|_{\mathrm{E}}\left(\bar{\boldsymbol{\Phi}}_\alpha\right)_{,\mathbf{v}_\beta}|_{\mathrm{E}} \\
& +\bar{\mathbf{m}}^{\mathbf{N}}_\beta|_{\mathrm{E}}\,,
\end{aligned}
\tag{7.101}
$$

with $\bar{\varpi}_\beta$ given in (7.88).

A complete exploitation of the equilibrium conditions also requires the satisfaction of the minimality requirements (7.19) for thermodynamic equilibrium. The exploitation of these is, however, better done with specialized constitutive relations.

It is worthwhile to discuss in some detail the results that have been reached up to now. We start with the properties of the constituent stresses and interaction forces in thermodynamic equilibrium. The main results are collected in the formulae (7.100), (7.101), (7.88), (7.85), (7.45), (7.7) and (7.9).

The equilibrium stress tensor (7.100) comprises six terms, a pressure, an elastic and a frictional contribution in the first line of (7.100), and two terms that trace back to non-vanishing interaction rate densities of mass and volume fractions as well as terms that are due to higher order non-linearities, the latter three in the second line of (7.100). We know of no formulation where this last term would have a non-zero value. The first term in the first line may be interpreted as 'pressure', but as shown by (7.88), it is itself composed of a number of terms of different origin,

- $\bar{\zeta}_\beta\big|_{\mathrm{E}}$ which may be called 'true constituent pressure',
- θk^v_β of purely thermodynamic nature. Its origin is the extra entropy flux $\mathbf{k}$ in (6.16) or (6.59), of which the isotropic representation is given in (7.75), (7.76). Nothing is known about this term and no obvious arguments can presently be given that θk^v_β should be different from zero. So, we shall later in [**A23**] set $k^v_\beta = 0$ for all β.
- The remaining members of (7.88) may only differ from zero, when mass and volume fraction interaction rate densities are different from zero.

Looking at the 'true constituent pressure' $\bar{\zeta}_\beta\big|_{\mathrm{E}}$, it follows from its definition, see (7.7), that

$$\zeta_\alpha := \begin{cases} \beta^G_\alpha - \rho_n \Psi^G_I + \varsigma, & \alpha = 1, \dots, n-1, \\ -\rho_n \Psi^G_I + \varsigma, & \alpha = n, \end{cases} \tag{7.102}$$

which, for the first $n-1$ components, is given by the configuration pressure and for all constituents by the inner free energy and the saturation pressure. It is evident from the definition above that $\bar{\zeta}_\alpha$ has the form

$$\bar{\zeta}_\alpha = \nu_\alpha \zeta_n + \nu_\alpha \beta^G_\alpha\ , \quad (\alpha = 1, \dots, n-1)\ , \qquad \zeta_n = -\rho_n \Psi^G_I + \varsigma\ . \tag{7.103}$$

Only the first term on the right-hand side has the structure 'volume fraction times an internal pressure' where this internal pressure is the saturation pressure plus a contribution due to the inner free energy. This part is reminiscent of what is known as 'pressure equilibrium', which, later, will be discussed in connection with the assumption of pressure equilibrium that applies to all pressure terms not just to some parts of the constituent pressure. The last two members of the pressure term (7.88) are the contributions due to the inter-constituent mass and volume fraction interaction rate densities; they obviously resemble the structure of interaction terms. Only if these productions are not present and the inner free energy does not depend on the constituent volume fractions, and when $k^v_\beta = 0$, the equilibrium pressure is given by

$$\bar{\zeta}_\alpha\big|_{\mathrm{E}} = \nu_\alpha \zeta_n \qquad (\alpha = 1, \dots, n)\ . \tag{7.104}$$

Let us look next at the remaining contributions to the constituent equilibrium stress tensors (7.100). They consist in the first line of elastic and frictional contributions and both will, below, further be analysed. Here, it may suffice to mention that the fluid and solid constituents need separate attention. Moreover, compressible and density preserving constituents need to be treated separately. We shall not go into any depth discussing the contributions on the second line of (7.100) due to constituent mass and volume fraction interaction rate densities except that the equilibrium stress of one constituent is affected by contributions from all other constituents with non-vanishing mass and volume fraction interaction rate densities. The coefficients are given by the inner parts of the constituent free enthalpies (chemical potentials, Gibbs free energies) defined in (6.109)-(6.111). Notice also that the coefficient $\boldsymbol{\iota}_\alpha$ defined in (6.110) and given by the LAGRANGE multipliers λ^ε and λ^ν_α only arises in connection with the constituent volume fraction production rate densities.

Consider next the constituent interaction forces. Their equilibrium values are given in (7.101). Apart from the very last non-linear term they consist of contributions that are proportional to the density gradients and similar ones that are proportional to the volume fraction gradients. Contributions of the former are only present for compressible constituents. The coefficients of the density gradient $\nabla\rho_\alpha$ represent (*i*) a contribution due to the thermodynamic pressure (6.101), (*ii*) due to the extra entropy flux (7.75), (later to be set to zero), (*iii*) due to the constituent free enthalpies (6.109) and (*iv*) the LAGRANGE multipliers in λ^ν_α, defined in (6.119). Alternatively, the coefficients of the volume fraction densities $\nabla\nu_\alpha$ are (*i*) due to the true pressure (7.7), (*ii*) due to the extra entropy flux (7.75), (7.76) (later to be set to zero), (*iii*) due to the free enthalpies (6.109), and (*iv*), due to the LAGRANGE multipliers in (6.119). These terms are very similar to one another, and they also have their correspondences in the equilibrium stresses.

We note that, modulo our assumptions, the formulae revealing the equilibrium properties of the constituent CAUCHY stress tensors and the constituent interaction forces are rational deductions from the Second Law of Thermodynamics. In fact, there is no flexibility in their choice. Thermodynamic arguments have resulted in their precise definition. We may now also give a partial answer to the question whether the formulae provide room for a different splitting of the divergence of the constituent stress tensor and the interaction force. In principle, the formulae (7.87) to (7.101) show how terms arising in the constituent stress tensors and interaction forces can be moved from one to the other. They must appear in the interaction forces as the divergence of a second order tensor or the gradient of a scalar. Or products of scalars with gradients can be complemented to such terms. For instance $a\nabla b$

is replaced by $\nabla(ab) - b\nabla a$. Of course, there is a multitude of such possibilities by way of the gradient operator of any pressure term or the divergence of any stress contribution. Such transformations may, however, destroy the property that the newly defined interaction force, summed over all constituents, must vanish. Moreover, the thermodynamic structure explained in connection with the properties (7.50) may be destroyed in this way. In addition, such transformations are always connected with differentiations which destroy the global structure and therefore possibly weak formulation of the balance laws. It is for these reasons that we do not recommend such transformations unless, of course, one wishes to search for the equivalence or non-equivalence of seemingly different formulations.

Chapter 8
Reduced Model

Abstract The intention of this chapter is to see whether (i) well known formulations of binary mixture models can be derived from the thermodynamic model, (ii) classical hypo-plasticity is deducible from the frictional evolution equation and (iii) the popular assumption of pressure equilibrium is justified. To this end, we ignore mass and volume fraction interaction rate densities, restrict considerations to isothermal processes, ignore higher order non-linearities in the constitutive relations and use the principle of phase separation. These assumptions transform the equilibrium stresses, heat flux and interaction forces to considerably simplified forms. Furthermore, the analysis shows that classical hypo-plasticity can be reconstructed with the introduction of a new objective time derivative for the stress-like variable. Non-equilibrium contributions to the stresses and interaction forces are also briefly discussed.
It is, finally, shown that the assumption of pressure equilibrium precludes the application of frictional stresses in equilibrium. This unphysical assumption is therefore replaced by a thermodynamic closure condition that is more flexible and less restrictive. It allows for frictional stresses in thermodynamic equilibrium and therefore is sufficiently general for applications to mixture theories.

In the previous chapters we developed a theory for an isotropic visco-elasto-plastic heat conducting mixture of n constituents, (i) in which mass interactions between the constituents may occur, (ii) some or all of the constituents are density preserving in the sense that they possess constant constituent mass densities, (iii) which is saturated in the sense that no void spaces are present in the mixture, (iv) that ignores constituent energy interactions, (v) which is capable of measuring the distribution and evolution of submacroscopic structures by means of new internal variables and corresponding balance laws and (vi) allows in the linearised case for a hyperbolic governing equation for the temperature distribution.

In the sequel, we aim to reduce the above theory to a model that is sufficiently simple to be numerically solvable but, equally, allows for the descrip-

L. Schneider, K. Hutter, *Solid-Fluid Mixtures of Frictional Materials in Geophysical and Geotechnical Context*, Advances in Geophysical and Environmental Mechanics and Mathematics, DOI 10.1007/978-3-642-02968-4_8,

tion of the main properties of debris flows, namely, (i) fluidisation in a thin shear band close to the bed, (ii) particle size segregation, (iii) shear stresses present in thermodynamic equilibrium and (iv) velocity differences of the fluid and the solid grains.

To this end, we commence this chapter with the basic physical assumptions (Section 8.1), e. g. a binary mixture postulate, no mass-interaction, etc. followed by 'artificial' assumptions (Section 8.2) on the free energy, Ψ^G, the constitutive quantities for hypo-plasticity, $\bar{\mathbf{\Phi}}_\alpha$, and on the non-equilibrium parts of the constitutive quantities. Along with the latter suppositions coefficients are introduced that allow the specification of the material. At the end of this chapter (Section 8.3) we inspect the reduced field equations under the strong assumption of '*pressure equilibrium*' (to be specified) and under another, new and presumably weaker, assumption introduced by HUTTER et al. [63].

8.1 Physical Assumptions

We model debris flows here as saturated mixtures of two constituents, where we interpret the first constituent as solid grains and the second as a fluid. Thus, the Greek indices take the identifiers s, for the solid and f, for the fluid. As a consequence of the saturation condition, [**A7**], the volume fraction for the fluid, ν_f, is replaced by $(1-\nu_s)$ and, as we have already seen, an independent constraint field s arises for which the field equations (to be specified) have to be solved. We also assume that both constituents are density-preserving and thus, in the context of assumption [**A6**], we prescribe

$$\rho_s = \text{const.}, \qquad \rho_f = \text{const.}\,, \qquad (m=0)\,. \tag{8.1}$$

The binary mixture concept, in which the solid constituent is not split into a number of separate components, implies that different characterizations of the solid component by the grain size or differences in resilience, etc., are not accounted for. Furthermore, we shall also exclude melting of the solid particles in the moving process. This would in most situations require a three constituent or even more detailed mixture concept. This excludes very large landslides in which the frictional heat will melt the rock and – after solidification of the molten rock – generate so-called frictionites. We formalised this in assumption [**A2**] which in this binary mixture model reduces to

$$c_s = c_f = 0\,. \tag{8.2}$$

As a consequence of (4.39) and (8.2) the volume fraction production rate densities, n_s and n_f, vanish and the mass- and volume fraction balance equations turn into

$$\begin{aligned} \partial \nu_s + \nabla \cdot (\nu_s \mathbf{v}_s) &= 0 \; , \\ \partial \nu_s - \nabla \cdot (\mathbf{v}_f) + \nabla \cdot (\nu_s \mathbf{v}_f) &= 0 \; . \end{aligned} \tag{8.3}$$

By subtracting these two equations one obtains

$$\nabla \cdot (\nu_s \mathbf{v}_s + (1 - \nu_s)\mathbf{v}_f) = 0 \; , \tag{8.4}$$

which may replace one of the equations (8.3). If we now define by

$$\mathbf{v}_{\text{vol}} = \nu_s \mathbf{v}_s + (1 - \nu_s)\mathbf{v}_f \tag{8.5}$$

the *volume-weighted mixture velocity*, (8.4) states that

$$\nabla \cdot (\mathbf{v}_{\text{vol}}) = 0 \; . \tag{8.6}$$

This result is sufficiently significant to state it in words: *The volume-weighted mixture velocity is solenoidal.* We emphasize this property, because in the literature (primarily of fluvial hydraulics) it is often used without explicitly mentioning that the mixture velocity is volume-weighted rather than mass weighted (=barycentric). Its simplicity also yields modelling and computational advantages. For a formal comparison of volume and mass weighted mixture concepts, see CHEN & TAI [27].

In the previous chapters we were, besides other things, concerned with thermal processes involving the temperature, θ, its gradient, $\nabla\theta$, and its material time derivative, $\dot{\theta}$, and obtained e. g. the results (6.69), $(6.96)_{1,2,3}$ and (7.36). However, for the sake of simplicity, from now on in this chapter

[A18] We regard debris flows as isothermal processes, i. e. each material element of the mixture is thought to exhibit the same temperature for all times.

Consequently, we can omit the temperature-related quantities, θ, $\nabla\theta$ and $\dot{\theta}$ in the constitutive law, [**A8**], (cf. HUTTER et al. [63]), so the mixture reduced energy balance (4.1) is no longer of interest. As a consequence, the problem of describing debris flows becomes purely mechanical.

We have also mentioned above that the concept of frictional, rate-independent behaviour does not make sense for (viscous) *fluids*. We, therefore, omit the internal variable, $\bar{\mathbf{Z}}_f$, in all constitutive laws and also disregard for the fluid constituent its evolution law.

The incorporation of the above simplifications into the constitutive law [**A8**] then yields

$$\mathsf{C} = \hat{\mathsf{C}}\Big(\nu_s,\ \nabla\nu_s,\ \mathbf{v}_{fs},\ \mathbf{B}_s,\ \mathbf{B}_f,\ \mathbf{D}_s,\ \mathbf{D}_f,\ \mathbf{W}_{fs},\ \bar{\mathbf{Z}}_s\Big), \qquad \text{for} \qquad \mathsf{C} := \Big\{\bar{\mathbf{T}}_s,\ \bar{\mathbf{T}}_f,\ \bar{\mathbf{m}}^i_s\Big\}\ . \tag{8.7}$$

As is proper in continuum theories of saturated mixtures the saturation pressure, ς, is not treated as an independent constitutive quantity in the constitutive law (8.7)[1]. We also recall the principle of objectivity, see Section 4.6, (4.45), in which we choose

$$\mathbf{a} = -\mathbf{v}_f\ , \qquad \boldsymbol{\Omega}^* = -\mathbf{W}_f \tag{8.8}$$

and define[2]

$$\mathbf{v}_{fs} := \mathbf{v}_f - \mathbf{v}_s, \qquad \mathbf{W}_{fs} := \mathbf{W}_f - \mathbf{W}_s \tag{8.9}$$

as an objective difference velocity of the solid and the fluid and the difference of the solid and fluid vorticity tensors, respectively. These variables are used in relations (8.7) which now obey the principle of material objectivity. Furthermore, we have omitted in $(8.7)_2$ the fluid interaction force $\bar{\mathbf{m}}^i_f$, because $\bar{\mathbf{m}}^i_f = -\bar{\mathbf{m}}^i_s$.

For the 'inner' part of the HELMHOLTZ free energy, we have the following dependencies

$$\Psi^G_I = \hat{\Psi}^G_I\Big(\nu_s,\ \mathbf{B}_s,\ \mathbf{B}_f,\ \bar{\mathbf{Z}}_s\Big)\ . \tag{8.10}$$

As Ψ^G is independent of ρ_α, see (8.1), the true thermodynamic pressures, p^G_s and p^G_f, are not present in this model. We further remark, that Ψ^G_I only depends on equilibrium quantities, and thus, the identifier $(\cdot)\big|_{\mathrm{E}}$ can be omitted for all quantities derived from Ψ^G_I, i. e. β^G_s, ζ_α $(\alpha = s, f)$ and $\Psi^G_{I,\mathbf{x}^J}$, where $\mathbf{x}^J \in \big\{\mathbf{B}_s,\ \mathbf{B}_f,\ \bar{\mathbf{Z}}_s\big\}$.

With all these simplifications the constitutive laws for the equilibrium quantities $\mathbf{q}\big|_{\mathrm{E}}$, $\bar{\mathbf{T}}_\beta\big|_{\mathrm{E}}$ $(\beta = s, f)$ and $\bar{\mathbf{m}}^i_s\big|_{\mathrm{E}}$ (see (7.99) to (7.101)) take the forms

$$\mathbf{q}\big|_{\mathrm{E}} = \mathbf{q}^{\mathbf{N}}\big|_{\mathrm{E}}\ , \tag{8.11}$$

$$\bar{\mathbf{T}}_s\big|_{\mathrm{E}} = -\bar{\varpi}_s\mathbf{I} + 2\rho\,\mathrm{sym}(\Psi^G_{I,\mathbf{B}_s})\mathbf{B}_s + \rho\Psi^G_{I,\bar{\mathbf{Z}}_s}(\bar{\boldsymbol{\Phi}}_s)_{,\mathbf{D}_s}\big|_{\mathrm{E}} + \bar{\mathbf{T}}^{\mathbf{N}}_s\big|_{\mathrm{E}}\ , \tag{8.12}$$

$$\bar{\mathbf{T}}_f\big|_{\mathrm{E}} = -\bar{\varpi}_f\mathbf{I} + 2\rho\,\mathrm{sym}(\Psi^G_{I,\mathbf{B}_f})\mathbf{B}_f + \bar{\mathbf{T}}^{\mathbf{N}}_f\big|_{\mathrm{E}}\ , \tag{8.13}$$

---

[1] To make a constraint variable an independent constitutive quantity as often done is actually rather controversial: The saturation pressure is in any boundary value problem uniquely defined up to an arbitrary constant ($\nabla\varsigma = \mathbf{0}$). So, a dependence of a constitutive quantity on ς is not unique!

[2] Analogously $\mathbf{v}_{sf} = \mathbf{v}_s - \mathbf{v}_f = -\mathbf{v}_{fs}$. So, either $\mathbf{v}_{fs}$ or $\mathbf{v}_{sf}$ is the generic variable and both are equivalent to one another.

$$
\begin{aligned}
\bar{\mathbf{m}}_s^i\big|_{\mathrm{E}} &= \Big\{\big(\zeta_s - \bar{\xi}_s\zeta_s\big) + \bar{\xi}_s\zeta_f - \theta(k_s^v)_{,\nu_s}\Big\}\nabla\nu_s \\
&\qquad +\rho\left(\Psi_I^G\right)_{,\bar{\mathbf{Z}}_s}\left(\bar{\mathbf{\Phi}}_s\right)_{,\mathbf{v}_s}\big|_{\mathrm{E}} + \bar{\mathbf{m}}_s^{\mathbf{N}}\big|_{\mathrm{E}} \\
&= \Big\{\zeta_s - \bar{\xi}_s\big(\zeta_s - \zeta_f\big) - \theta(k_s^v)_{,\nu_s}\Big\}\nabla\nu_s \\
&\qquad +\rho\left(\Psi_I^G\right)_{,\bar{\mathbf{Z}}_s}\left(\bar{\mathbf{\Phi}}_s\right)_{,\mathbf{v}_s}\big|_{\mathrm{E}} + \bar{\mathbf{m}}_s^{\mathbf{N}}\big|_{\mathrm{E}} \\
&\overset{(7.7)}{=} \Big\{\beta_s^G\big(1 - \bar{\xi}_s\big) - \rho_f\Psi_I^G + \varsigma - \theta(k_s^v)_{,\nu_s}\Big\}\nabla\nu_s \\
&\qquad +\rho\left(\Psi_I^G\right)_{,\bar{\mathbf{Z}}_s}\left(\bar{\mathbf{\Phi}}_s\right)_{,\mathbf{v}_s}\big|_{\mathrm{E}} + \bar{\mathbf{m}}_s^{\mathbf{N}}\big|_{\mathrm{E}} ,
\end{aligned}
\tag{8.14}
$$

in which

$$
\bar{\varpi}_s = \nu_s\big(\beta_s^G - \rho_f\Psi_I^G + \varsigma\big) - \theta k_s^v , \tag{8.15}
$$

$$
\bar{\varpi}_f = (1-\nu_s)\big(-\rho_f\Psi_I^G + \varsigma\big) - \theta k_f^v . \tag{8.16}
$$

The equilibrium momentum interaction force, $\bar{\mathbf{m}}_f^i\big|_{\mathrm{E}}$ follows from $\bar{\mathbf{m}}_s^i\big|_{\mathrm{E}}$ via

$$
\sum \bar{\mathbf{m}}_\alpha^i\big|_{\mathrm{E}} = \sum \bar{\mathbf{m}}_\alpha\big|_{\mathrm{E}} = \mathbf{0} \qquad \Rightarrow \qquad \bar{\mathbf{m}}_f^i\big|_{\mathrm{E}} = -\bar{\mathbf{m}}_s^i\big|_{\mathrm{E}} . \tag{8.17}
$$

We know from hydrostatics, that fluids in thermodynamic equilibrium can only sustain spherical stresses, i. e. pressures, and thus the second and third term in (8.13) can only have the form

$$
2\rho\ \mathrm{sym}(\Psi^G_{I,\mathbf{B}_f})\mathbf{B}_f + \bar{\mathbf{T}}_f^{\mathbf{N}}\big|_{\mathrm{E}} =: \ \pi_f\mathbf{I} , \tag{8.18}
$$

where π_f is a scalar which depends only on equilibrium variables. In addition, we know from the definition of $\bar{\varpi}_f$, (7.88), that it contains the independent saturation pressure, ς. Thus, and if one so desires, $\bar{\varpi}_f$ itself rather than ς or ζ_f could be regarded as an independent quantity which is not determined by constitutive relations but from the solution of the field equations. Without loss of generality, it is therefore permissible to incorporate π_f into $\bar{\varpi}_f$, and we are left with

$$
\bar{\mathbf{T}}_f\big|_{\mathrm{E}} = -\bar{\varpi}_f\mathbf{I} . \tag{8.19}
$$

[**A19**] In the sequel, for simplicity, we will omit the higher non-linear contributions to the equilibrium constitutive laws.

Although [**A19**] can be confirmed by convincing physical reasoning, i. e. (i) the absence of heat flux in thermodynamic equilibrium and (ii) the disregard of non-linear terms in the equilibrium fluid momentum interaction (cf. Svendsen & Hutter [115]) we will regard [**A19**] as an *ad hoc* assumption.

Considering the above arguments, we finally end up with

$$\mathbf{q}\big|_{\mathrm{E}} = \mathbf{0} \,, \tag{8.20}$$

$$\bar{\mathbf{T}}_s\big|_{\mathrm{E}} = -\bar{\varpi}_s\mathbf{I} + 2\rho\,\mathrm{sym}(\Psi^G_{I,\mathbf{B}_s})\mathbf{B}_s + \rho\Psi^G_{I,\bar{\mathbf{Z}}_s}(\bar{\mathbf{\Phi}}_s)_{,\mathbf{D}_s}\big|_{\mathrm{E}} \,, \qquad \bar{\varpi}_s = \nu_s\big(\beta^G_s - \rho_f\Psi^G_I + \varsigma\big) - \theta k^v_s \,, \tag{8.21}$$

$$\bar{\mathbf{T}}_f\big|_{\mathrm{E}} = -\bar{\varpi}_f\mathbf{I} \,, \qquad \bar{\varpi}_f = (1-\nu_s)\big(-\rho_f\Psi^G_I + \varsigma\big) - \theta k^v_f - \pi_f \,, \tag{8.22}$$

$$\begin{aligned} \bar{\mathbf{m}}^i_s\big|_{\mathrm{E}} &= \Big\{\beta^G_s(1-\bar{\xi}_s) - \rho_f\Psi^G_I + \varsigma - \theta(k^v_s)_{,\nu_s}\Big\}\nabla\nu_s \\ &\quad + \rho\left(\Psi^G_I\right)_{,\bar{\mathbf{Z}}_s}\left(\bar{\mathbf{\Phi}}_s\right)_{,\mathbf{v}_s}\big|_{\mathrm{E}} \,, \end{aligned} \tag{8.23}$$

where, in particular, the elastic and hypo-plastic parts of (8.21) still have to be discussed in greater detail.

If we combine $(8.21)_2$ and $(8.22)_2$ and use relation $(7.77)_2$, we obtain for the pressure of the mixture Cauchy stress tensor the expression

$$\varpi = \bar{\varpi}_s + \bar{\varpi}_f = \nu_s\beta^G_s + \big(-\rho_f\Psi^G_I + \varsigma\big) - \pi_f \,. \tag{8.24}$$

This pressure cannot be regarded as a very meaningful concept, because it contains solid and fluid properties as well as properties of saturation. We regard the partial stresses, $\mathbf{T}_s\big|_{\mathrm{E}}$ and $\mathbf{T}_f\big|_{\mathrm{E}}$, as the better entities characterising the state of normal stresses.

8.2 'Artificial' Assumptions

Constituent Cauchy stress tensors

We have already pointed out that all constitutive quantities can be decomposed into equilibrium and non-equilibrium parts, i. e.

$$\bar{\mathbf{T}}_s = \bar{\mathbf{T}}_s\big|_{\mathrm{E}} + \bar{\mathbf{T}}_s\big|_{\mathrm{N}} \,, \qquad \bar{\mathbf{T}}_f = \bar{\mathbf{T}}_f\big|_{\mathrm{E}} + \bar{\mathbf{T}}_f\big|_{\mathrm{N}} \,, \qquad \bar{\mathbf{m}}^i_s = \bar{\mathbf{m}}^i_s\big|_{\mathrm{E}} + \bar{\mathbf{m}}^i_s\big|_{\mathrm{N}} \,. \tag{8.25}$$

The heat flux vector, $\mathbf{q}$, is set aside, as it is unimportant for isothermal processes. In the last section we have found representations for the equilibrium parts of the constitutive quantities, but in particular the equilibrium solid stress tensor, $\bar{\mathbf{T}}_s\big|_{\mathrm{E}}$, which consists of a constraint (cs), an elastic (es) and a frictional (fric) (hypo-plastic) part, i. e.,

$$\bar{\mathbf{T}}_s\big|_{\mathrm{E}} = \underbrace{-\bar{\varpi}_s \mathbf{I}}_{\bar{\mathbf{T}}_{\mathrm{cs}}} + \underbrace{2\rho\,\mathrm{sym}(\Psi^G_{I,\mathbf{B}_s})\mathbf{B}_s}_{\bar{\mathbf{T}}_{\mathrm{es}}} + \underbrace{\rho\Psi^G_{I,\bar{\mathbf{Z}}_s}(\bar{\mathbf{\Phi}}_s)_{,\mathbf{D}_s}\big|_{\mathrm{E}}}_{\bar{\mathbf{T}}_{\mathrm{fric}}} \tag{8.26}$$

requires further modelling. Let us make the constitutive relations for $\bar{\mathbf{T}}_{\mathrm{es}}$ and $\bar{\mathbf{T}}_{\mathrm{fric}}$ more specific. To this end, we assume the 'inner' free energy to have the form

$$\Psi^G_I = \sum \Psi^G_\alpha = \tilde{\Psi}^G_s\Big(\nu_s,\ \mathbf{B}_s,\ \bar{\mathbf{Z}}_s\Big) + \tilde{\Psi}^G_f\Big((1-\nu_s),\ \mathbf{B}_f\Big). \tag{8.27}$$

By prescribing this representation for Ψ^G_s and Ψ^G_f we have used the[3]

[**A20**] '*Principle of phase separation*' introduced by PASSMAN et al. [103], which requires the '*material-specific*' constitutive quantities for constituent K_α, to depend only on those constitutive variables that belong to the same constituent. This principle does not apply to the remaining quantities, e. g. those for the whole mixture or those describing interactions between the constituents.

We remark that for single-material bodies the 'principle of phase separation' reduces to the well known *principle of equipresence*, TRUESDELL & NOLL [121] . We further notice that the principle must likely be wrong when exchange processes between the constituents take place.

In order to specify the elastic parts, of the constituent CAUCHY stress tensors we isolate the elastic and frictional effects in Ψ^G_I separately by assuming (cf. HUTTER et al. [63])

[**A21**] $$\Psi^G_I = \hat{\Psi}^G_{fric}(\nu_s,\ \bar{\mathbf{Z}}_s) + \hat{\Psi}^G_{es}(\mathbf{B}_s) + \hat{\Psi}^G_{ef}(\mathbf{B}_f)\ . \tag{8.28}$$

Here the indices '*fric*', '*es*' and '*ef*' stand for 'friction', 'elastic-solid' and 'elastic-fluid', respectively. The last two terms in (8.28) are thought to account for the elastic contributions of the solid and fluid, respectively. In $\hat{\Psi}^G_{fric}$, on

[3] [**A20**] was introduced into the literature much earlier by MORLAND [89], however by not declaring it a 'principle'. In MORLAND [90] and subsequent papers [91], [92] it was re-iterated on and the poor terminology 'effective' was changed to 'intrinsic' which we call 'true'.

the other hand, we have subsumed all other dependencies of Ψ_I^G. It is believed that the representation of $\hat{\Psi}_{fric}^G$ in [**A21**] is able to describe all effects of the visco-elasto-plastic binary mixture, except those of elasticity.

We know from the representation theory of isotropic functions, that isotropic scalar-valued functions of a single symmetric tensor, such as $\hat{\Psi}_{es}^G$ and $\hat{\Psi}_{ef}^G$, can only depend on the invariants of this tensor (cf. OGDEN [101]). Consequently, those two functions exhibit the following dependencies

$$\Psi_{es}^G = \check{\Psi}_{es}^G\big(\mathrm{I}_{\mathbf{B}_s},\ \mathrm{II}_{\mathbf{B}_s},\ \mathrm{III}_{\mathbf{B}_s}\big), \qquad \Psi_{ef}^G = \check{\Psi}_{ef}^G\big(\mathrm{I}_{\mathbf{B}_f},\ \mathrm{II}_{\mathbf{B}_f},\ \mathrm{III}_{\mathbf{B}_f}\big), \tag{8.29}$$

where the invariants for a general symmetric second-order tensor, $\mathbf{A}$, are defined according to

$$\mathrm{I}_{\mathbf{A}} = \mathrm{tr}(\mathbf{A}), \qquad \mathrm{II}_{\mathbf{A}} = \tfrac{1}{2}\Big(\left(\mathrm{I}_{\mathbf{A}}\right)^2 - \mathrm{I}_{\mathbf{A}^2}\Big), \qquad \mathrm{III}_{\mathbf{A}} = \det(\mathbf{A})\ . \tag{8.30}$$

With these results in mind, we can now turn the attention to the contributions of Ψ_I^G in the elastic parts of the constituent CAUCHY stress tensors. If we, first, ignore the arguments for π_f (see Section 8.1, (8.18)) for a moment, the elastic part of the fluid CAUCHY stress tensor, $\big(\bar{\mathbf{T}}_f\big)_{\mathrm{ef}}$, can be written in the forms

$$\begin{aligned}
\big(\bar{\mathbf{T}}_f\big)_{\mathrm{ef}} &= 2\rho\ \mathrm{sym}\left(\ (\Psi_I^G)_{,\mathbf{B}_f}\right)\mathbf{B}_f \overset{[\mathbf{A21}]}{=} 2\rho\ \mathrm{sym}\left(\ (\hat{\Psi}_{ef}^G)_{,\mathbf{B}_f}\right)\mathbf{B}_f \\
&= 2\rho\ \mathrm{sym}\left(\frac{\partial\check{\Psi}_{ef}^G}{\partial\mathrm{I}_{\mathbf{B}_f}}\frac{\partial\mathrm{I}_{\mathbf{B}_f}}{\partial\mathbf{B}_f} + \frac{\partial\check{\Psi}_{ef}^G}{\partial\mathrm{II}_{\mathbf{B}_f}}\frac{\partial\mathrm{II}_{\mathbf{B}_f}}{\partial\mathbf{B}_f} + \frac{\partial\check{\Psi}_{ef}^G}{\partial\mathrm{III}_{\mathbf{B}_f}}\frac{\partial\mathrm{III}_{\mathbf{B}_f}}{\partial\mathbf{B}_f}\right)\mathbf{B}_f\ , \\
&= 2\rho\left(\frac{\partial\check{\Psi}_{ef}^G}{\partial\mathrm{I}_{\mathbf{B}_f}}\mathbf{I} + \frac{\partial\check{\Psi}_{ef}^G}{\partial\mathrm{II}_{\mathbf{B}_f}}\left(\mathrm{I}_{\mathbf{B}_f}\mathbf{I} - \mathbf{B}_f\right) + \frac{\partial\check{\Psi}_{ef}^G}{\partial\mathrm{III}_{\mathbf{B}_f}}\mathrm{III}_{\mathbf{B}_f}\mathbf{B}_f^{-1}\right)\mathbf{B}_f\ ,
\end{aligned} \tag{8.31}$$

where the chain rule of differentiation has been used. For a general second-rank tensor $\mathbf{A}$ the above derivatives take the forms (cf. HUTTER & JÖHNK [62])

$$\frac{\partial\mathrm{I}_{\mathbf{A}}}{\partial\mathbf{A}} = \mathbf{I}, \qquad \frac{\partial\mathrm{II}_{\mathbf{A}}}{\partial\mathbf{A}} = \left(\mathrm{I}_{\mathbf{A}}\mathbf{I} - \mathbf{A}\right), \qquad \frac{\partial\mathrm{III}_{\mathbf{A}}}{\partial\mathbf{A}} = \mathrm{III}_{\mathbf{A}}\mathbf{A}^{-1}\ . \tag{8.32}$$

If we apply these results with $\mathbf{A} = \mathbf{B}_f$ in (8.31), we see that the term in parentheses is already symmetric, which justifies the last line in (8.31).

In Section 8.1 we mentioned that in thermodynamic equilibrium, fluids can only sustain spherical stresses. Consequently, only those derivatives of the invariants are to be considered which allow $(\bar{\mathbf{T}}_f)_{\mathrm{ef}}$ to become proportional to the unit tensor, $\mathbf{I}$. This situation can only be reached if we require

$$\frac{\partial\check{\Psi}_{ef}^G}{\partial\mathrm{I}_{\mathbf{B}_f}} = \frac{\partial\check{\Psi}_{ef}^G}{\partial\mathrm{II}_{\mathbf{B}_f}} = 0\ . \tag{8.33}$$

It follows that $\check{\Psi}^G_{ef}$ cannot depend on $\mathrm{I}_{\mathbf{B}_f}$ and $\mathrm{II}_{\mathbf{B}_f}$. Furthermore, we observed in Section 4.3 that only for mixtures with non-vanishing mass interactions, i. e. $c_\alpha \neq 0$, the variables $\mathbf{B}_\alpha$ and ρ_α ($\alpha = 1, \ldots, m$) are independent of one another. Thus, in the present model for which $c_\alpha = 0$ the assumption of constant true mass densities and [**A2**] allow the conclusion[4]

$$\det(\mathbf{B}_f) = \mathrm{III}_{\mathbf{B}_f} = \text{const.}\,, \qquad \det(\mathbf{B}_s) = \mathrm{III}_{\mathbf{B}_s} = \text{const.} \tag{8.34}$$

and therefore, $\check{\Psi}^G_{ef}$ cannot depend on $\mathrm{III}_{\mathbf{B}_f}$ either, i. e.,

$$\frac{\partial \check{\Psi}^G_{ef}}{\partial \mathrm{III}_{\mathbf{B}_f}} = 0\,. \tag{8.35}$$

We obtain from (8.33) and (8.35) that

$$\frac{\partial \hat{\Psi}^G_{ef}}{\partial \mathbf{B}_f} \overset{[\mathbf{A21}]}{=} \frac{\partial \Psi^G_I}{\partial \mathbf{B}_f} = \mathbf{0}\,. \tag{8.36}$$

Thus, Ψ^G_I cannot be a function of $\mathbf{B}_f$. This, together with [**A19**], (8.18) and (8.19) implies that π_f can be neglected.

The elastic part of the solid CAUCHY stress tensor, on the other hand, becomes[4]

$$\bar{\mathbf{T}}_{\text{es}} = 2\rho \left(\frac{\partial \check{\Psi}^G_{es}}{\partial \mathrm{I}_{\mathbf{B}_s}} + \mathrm{I}_{\mathbf{B}_s} \frac{\partial \check{\Psi}^G_{es}}{\partial \mathrm{II}_{\mathbf{B}_s}} \right) \mathbf{B}_s - 2\rho\, \frac{\partial \check{\Psi}^G_{es}}{\partial \mathrm{II}_{\mathbf{B}_s}}\, \mathbf{B}_s^2\,. \tag{8.37}$$

With this relation we have reached the point, where, except for the postulate of an explicit representation for the elastic part of the solid free energy, Ψ^G_{es}, no other simplification can be performed. The simple choice

$$\begin{aligned} \rho \check{\Psi}^G_{es} &= C_1 \left(\mathrm{I}_{\mathbf{B}_s} - 3\right) + C_2 \left(\mathrm{II}_{\mathbf{B}_s} - 3\right), \\ C_1 &= \tfrac{1}{2}\mu \left(\tfrac{1}{2} + \beta\right) = \text{const.}, \quad C_2 = \tfrac{1}{2}\mu \left(\tfrac{1}{2} - \beta\right) = \text{const.}\,, \end{aligned} \tag{8.38}$$

---

[4] The density-preserving assumption for a constituent K_α whose mass production rate is not present, $c_\alpha = 0$, implies according to (4.20) that it also preserves its volume along its own trajectory. Hence $c_\alpha = 0$ also means $\det \mathbf{F}_\alpha$ = constant. Otherwise stated, the constituent motion is isochoric, and $\mathbf{F}_\alpha$ and $\mathbf{B}_\alpha$ are unimodular. So, the elastic stress $\bar{\mathbf{T}}_{es}$ cannot depend on $\mathrm{III}_{\mathbf{B}_s}$.

If c_α were not zero, then $\mathbf{B}_\alpha$ would not be unimodular and density-preserving could not imply volume-preserving of constituent K_α. Insensitivity of $\bar{\mathbf{T}}_{es}$ to solid volume changes would then require that (8.37) holds true if $\mathbf{B}_s$ is replaced by

$$\mathbf{B}_s^{\text{unimod}} := (\det \mathbf{B}_s)^{-1/3} \mathbf{B}_s\,.$$

This then simply would mean that there is no bulk elastic response.

which is attributed to MOONEY & RIVLIN (cf. RIVLIN & SAUNDERS [109]), leads to

$$\bar{\mathbf{T}}_{\text{es}} = 2(C_1 + C_2 \mathrm{I}_{\mathbf{B}_s})\mathbf{B}_s - 2C_2\mathbf{B}_s^2 \ . \tag{8.39}$$

In (8.38), μ can be interpreted as the *shear modulus of the solid grains* and β as a modelling parameter. For the special case of $\beta = \frac{1}{2}$, we attain a fairly simple representation, namely

$$\bar{\mathbf{T}}_{\text{es}} = \mu \mathbf{B}_s \ , \tag{8.40}$$

which is denoted *Neo-Hookean* behaviour. Despite the simple structure, $\bar{\mathbf{T}}_{es}$ in (8.40) still allows finite deformations, but, of course, its accuracy reduces quickly with the extent of the deformation. For the *natural* configuration, i. e. the situation of an undistorted mixture, which ought to be stress-free, the solid left CAUCHY-GREEN tensor, $\mathbf{B}_s$, reduces to $\mathbf{I}$. Therefore, the independent pressure field (here $\bar{\varpi}_f$) has to be chosen in such a way that in the undistorted mixture no stresses are present. A simple alternative would be to replace $\mathbf{B}_s$ in (8.39) and (8.40) by $\mathbf{E}_s := \frac{1}{2}(\mathbf{B}_s - \mathbf{I})$. With the elastic free energy given as $\hat{\Psi}_{es}^G(\mathbf{E}_s)$ we then obtain instead of (8.37),

$$\bar{\mathbf{T}}_{\text{es}} = 2\rho \left(\frac{\partial \check{\hat{\Psi}}_{es}^G}{\partial \mathrm{I}_{\mathbf{E}_s}} + \mathrm{I}_{\mathbf{E}_s} \frac{\partial \check{\hat{\Psi}}_{es}^G}{\partial \mathrm{II}_{\mathbf{E}_s}} \right) \mathbf{E}_s - 2\rho \, \frac{\partial \check{\hat{\Psi}}_{es}^G}{\partial \mathrm{II}_{\mathbf{E}_s}} \, \mathbf{E}_s^2 \ , \tag{8.41}$$

and (8.38) changes to

$$\begin{aligned} \rho \check{\hat{\Psi}}_{es}^G &= \check{C}_1 \left(\mathrm{I}_{\mathbf{E}_s} - 3 \right) + \check{C}_2 \left(\mathrm{II}_{\mathbf{E}_s} - 3 \right) , \\ \check{C}_1 &= \tfrac{1}{2}\check{\mu} \left(\tfrac{1}{2} + \check{\beta} \right) = \text{const.}, \quad C_2 = \tfrac{1}{2}\check{\mu} \left(\tfrac{1}{2} - \check{\beta} \right) = \text{const.} \ , \end{aligned} \tag{8.42}$$

so that

$$\bar{\mathbf{T}}_{\text{es}} = 2(\check{C}_1 + \check{C}_2 \mathrm{I}_{\mathbf{E}_s})\mathbf{E}_s - 2\check{C}_2\mathbf{E}_s^2 \ , \tag{8.43}$$

from which, with $\check{\beta} = \frac{1}{2}$, we get

$$\bar{\mathbf{T}}_{\text{es}} = \check{\mu} \mathbf{E}_s \ . \tag{8.44}$$

Even better, however, is to choose the elastic stress contribution as used by geotechnical engineers.

The frictional part of the solid stress (8.26),

$$\bar{\mathbf{T}}_{\text{fric}} = \rho \Psi_{I,\bar{\mathbf{Z}}_s}^G(\bar{\mathbf{\Phi}}_s)_{,\mathbf{D}_s} \big|_{\mathrm{E}} = \rho \Psi_{\text{fric},\bar{\mathbf{Z}}_s}^G(\bar{\mathbf{\Phi}}_s)_{,\mathbf{D}_s} \big|_{\mathrm{E}}, \tag{8.45}$$

is still an unknown function of the equilibrium quantities, since so far no representation has been given for Ψ_{fric}^G and $\bar{\mathbf{\Phi}}_s$. We follow the argumentation of TEUFEL [117] who formulated the following postulate:

[A22] $\bar{\mathbf{T}}_{\text{fric}}$ is collinear to $\bar{\mathbf{Z}}_s$, i. e.

$$\bar{\mathbf{T}}_{\text{fric}} = \rho\delta\bar{\mathbf{Z}}_s, \qquad \delta = \text{constant} . \tag{8.46}$$

With the choice [**A22**] a special functional relation has been chosen for the solid frictional stress. Substituting (8.46) into (8.45) allows by way of integration an explicit determination of Ψ^G_{fric}; so, [**A22**] is not a genuine assumption but rather a convenient choice by which hypo-plastic behaviour can be demonstrated. Using assumption (8.46), i. e. substituting $\bar{\mathbf{Z}}_s = \bar{\mathbf{T}}_{\text{fric}}/(\rho\delta)$ into the evolution equation for $\bar{\mathbf{Z}}_s$, (4.36), yields

$$\begin{aligned}
&\frac{1}{\rho\delta}\overset{\circ}{\bar{\mathbf{T}}}_{\text{fric}} - \frac{1}{\rho^2\delta}\frac{\mathrm{d}^s\rho}{\mathrm{d}t}\bar{\mathbf{T}}_{\text{fric}} \\
&\quad = \frac{1}{\rho\delta}\left\{\frac{\mathrm{d}^s\,\bar{\mathbf{T}}_{\text{fric}}}{\mathrm{d}t} - [\mathbf{\Omega}_s,\ \bar{\mathbf{T}}_{\text{fric}}] - \frac{1}{\rho}\frac{\mathrm{d}^s\rho}{\mathrm{d}t}\bar{\mathbf{T}}_{\text{fric}}\right\} \\
&\quad = \bar{\mathbf{\Phi}}_s\Big(\frac{1}{\rho\delta}\bar{\mathbf{T}}_{\text{fric}},\cdot\Big) ,
\end{aligned} \tag{8.47}$$

where the dot indicates additional dependencies, say on ν_s and $\mathbf{D}_s$. If we use in addition

$$\begin{aligned}
\frac{\mathrm{d}^s\rho}{\mathrm{d}t} &= \rho_s\frac{\mathrm{d}^s\nu_s}{\mathrm{d}t} + \rho_f\frac{\mathrm{d}^s(1-\nu_s)}{\mathrm{d}t} \\
&= (\rho_s - \rho_f)\left(\frac{\partial\nu_s}{\partial t} + \nabla(\nu_s)\mathbf{v}_s\right) \\
&= -\nu_s(\rho_s - \rho_f)\nabla\cdot\mathbf{v}_s ,
\end{aligned} \tag{8.48}$$

which is obtained from (8.3) and the saturation condition [**A7**], (8.47) reduces to

$$\frac{1}{\rho\delta}\left\{\overset{\circ}{\bar{\mathbf{T}}}_{\text{fric}} + \nu_s\,\frac{\rho_s-\rho_f}{\rho}\,(\nabla\cdot\mathbf{v}_s)\,\bar{\mathbf{T}}_{\text{fric}}\right\} = \bar{\mathbf{\Phi}}_s\Big(\frac{1}{\rho\delta}\bar{\mathbf{T}}_{\text{fric}},\cdot\Big) . \tag{8.49}$$

So far, we are dealing with a general constitutive quantity, $\bar{\mathbf{\Phi}}_s$, for which the hypo-plastic behaviour has not been explicitly described, but could be. We note that to model hypo-elastic behaviour, $\bar{\mathbf{\Phi}}_s$ must be linear in $\mathbf{D}_s$; however, hypo-elasticity cannot capture the fact that the material behaviour of debris in slow or rapid flows is, in general, different in extension from compression (cf. KOLYMBAS [76]). To incorporate this property, we prescribe

$\bar{\mathbf{\Phi}}_s$ to have a hypo-plastic structure. Similarly to the approach outlined in Section 4.4, we require

$$\bar{\mathbf{\Phi}}_s = \hat{\mathbf{\Phi}}_s\big(\cdot, \bar{\mathbf{Z}}_s, \mathbf{D}_s\big) \tag{8.50}$$

to be positively homogenous of the first degree in $\bar{\mathbf{Z}}_s$ and $\mathbf{D}_s$. Analogously to (4.31), we also decompose $\bar{\mathbf{\Phi}}_s$ into an operator which is linear in $\mathbf{D}_s$ and another one which is non-linear in $\mathbf{D}_s$, i. e.

$$\bar{\mathbf{\Phi}}_s = \boldsymbol{\mathcal{L}}\big(\cdot,\, \bar{\mathbf{Z}}_s, \mathbf{D}_s\big) + \boldsymbol{\mathcal{N}}\big(\cdot,\, \bar{\mathbf{Z}}_s, \mathbf{D}_s\big)\ . \tag{8.51}$$

Following the proposal (4.32) we now assume the representation

$$\bar{\mathbf{\Phi}}_s = \mathrm{f}_1(\cdot)\,\big(\mathsf{L}\big(\bar{\mathbf{Z}}_s\big)\mathbf{D}_s + \mathrm{f}_2(\cdot)\mathbf{N}\big(\bar{\mathbf{Z}}_s\big)\,|\mathbf{D}_s|\big)\ , \tag{8.52}$$

where the norm of $\mathbf{D}_s$ is defined as

$$|\mathbf{D}_s| := \sqrt{\mathrm{tr}(\mathbf{D}_s^2)}\ , \tag{8.53}$$

as in (4.33) and f_1 and f_2 are the coefficients of *barotropy* and *pyknotropy*, which may depend on the variables $\mathbb{S}$. The tensors L and $\mathbf{N}$ are of fourth and second order, respectively. Representation (8.52) satisfies automatically the requirement of positive homogeneity in $\mathbf{D}_s$. If we require homogeneity of $\bar{\mathbf{\Phi}}_s$ with respect to $\bar{\mathbf{Z}}_s$, (8.49) can be reduced to the form

$$\begin{aligned} \overset{\circ}{\bar{\mathbf{T}}}_{\text{fric}} + \ \nu_s\,\frac{\rho_s - \rho_f}{\rho}\,\big(\nabla\cdot\mathbf{v}_s\big)\,\bar{\mathbf{T}}_{\text{fric}} \\ = \ \mathrm{f}_1(\cdot)\,\big(\mathsf{L}\big(\bar{\mathbf{T}}_{\text{fric}}\big)\mathbf{D}_s + \mathrm{f}_2(\cdot)\mathbf{N}\big(\bar{\mathbf{T}}_{\text{fric}}\big)\,|\mathbf{D}_s|\big)\ . \end{aligned} \tag{8.54}$$

This representation of the hypo-plastic evolution law is close to that postulated by WU & KOLYMBAS [129]. The differences are those due to the binary mixture, and an additional term bilinear in $\nabla\cdot\mathbf{v}_s$ and $\bar{\mathbf{T}}_{\text{fric}}$. However for $\rho_f = 0$, agreeing with the dry granular case, (8.54) reduces to the form previously derived by SVENDSEN et al. [116]. By using the idea of TEUFEL [117], we define the new objective time derivative[5]

$$\overset{\diamond}{\bar{\mathbf{Z}}}_s := \frac{\mathrm{d}^s\,\bar{\mathbf{Z}}_s}{\mathrm{d}t} - \big[\mathbf{\Omega}\ ,\ \bar{\mathbf{Z}}_s\big] - \nu_s\,\frac{(\rho_s - \rho_f)}{\rho}\,(\nabla\cdot\mathbf{v}_s)\bar{\mathbf{Z}}_s\ , \tag{8.55}$$

which in view of [**A22**] immediately leads to a form of the hypo-plastic stress evolution equation agreeing with that of WU & KOLYMBAS [129], i. e.

$$\overset{\circ}{\bar{\mathbf{T}}}_{\text{fric}} = \ \mathrm{f}_1(\cdot)\,\big(\mathsf{L}\big(\bar{\mathbf{T}}_{\text{fric}}\big)\mathbf{D}_s + \mathrm{f}_2(\cdot)\mathbf{N}\big(\bar{\mathbf{T}}_{\text{fric}}\big)\,|\mathbf{D}_s|\big)\ . \tag{8.56}$$

---

[5] On the basis that $\bar{\mathbf{Z}}_s$ and $\overset{\circ}{\bar{\mathbf{Z}}}$ are objective symmetric tensors, it is trivial to show that $\overset{\diamond}{\bar{\mathbf{Z}}}$ is objective. Indeed, ρ_s, ρ_f, ρ and ν_s are objective scalars, as is $\nabla\cdot\mathbf{v}_s$.

Obviously, and importantly to recognise, the new objective time derivative does change the above thermodynamic analysis, but only the result for the solid equilibrium CAUCHY stress tensor is affected by these changes. The incorporation of

$$\overset{\diamond}{\bar{\mathbf{Z}}}_s = \bar{\mathbf{\Phi}}_s \ , \tag{8.57}$$

instead of (4.36), into (5.10) leads to

$$\begin{aligned}\bar{\mathbf{T}}_s\big|_{\mathrm{E}} &= -\bar{\varpi}_s \mathbf{I} + 2\rho\,\mathrm{sym}(\Psi^G_{I,\mathbf{B}_s})\mathbf{B}_s \\ &\quad + \rho\Psi^G_{I,\bar{\mathbf{Z}}_s}(\bar{\mathbf{\Phi}}_s)_{,\mathbf{D}_s}\big|_{\mathrm{E}} + \nu_s(\rho_s-\rho_f)\big(\Psi^G_{I,\bar{\mathbf{Z}}_s}\cdot\bar{\mathbf{Z}}_s\big)\,\mathbf{I}\end{aligned} \tag{8.58}$$

rather than (8.21). The fact that the last term in (8.58) is spherical allows its incorporation into $\bar{\varpi}_s$, which therefore has the form

$$\bar{\varpi}_s = \nu_s\big(\beta_s - \rho_f\Psi^G_I + \varsigma\big) - \nu_s(\rho_s-\rho_f)\big(\Psi^G_{I,\bar{\mathbf{Z}}_s}\cdot\bar{\mathbf{Z}}_s\big) - \theta k^v_s \ . \tag{8.59}$$

Thus, by changing $\overset{\circ}{\bar{\mathbf{Z}}}_s$ to $\overset{\diamond}{\bar{\mathbf{Z}}}_s$ an additional contribution to the solid pressure arises. This pressure contains contributions from the configration pressure, saturation pressure, the free energy Ψ^G_I and extra entropy flux $\mathbf{k}$.

Assumption [**A22**] also implies that

$$\delta\bar{\mathbf{Z}}_s = \Psi^G_{\mathrm{fric},\bar{\mathbf{Z}}_s}(\bar{\mathbf{\Phi}}_s)_{,\mathbf{D}_s}\big|_{\mathrm{E}} \ , \tag{8.60}$$

which follows from (8.45). When f_1, f_2, $\mathsf{L}(\bar{\mathbf{Z}}_s)$ and $\mathbf{N}(\bar{\mathbf{Z}}_s)$ in $\bar{\mathbf{\Phi}}_s$ are specified, the integration of (8.60) with respect to $\bar{\mathbf{Z}}_s$ leads to a representation for Ψ^G_{fric}. Consequently, the assumption [**A22**] and the choice of $\bar{\mathbf{\Phi}}_s$ determine the form of Ψ^G_{fric}.

The obvious drawback of this hypo-plastic approach is the lack of differentiability of $\bar{\mathbf{\Phi}}_s$ at $\mathbf{D}_s = \mathbf{0}$ and therefore the singularity of $\bar{\mathbf{T}}_{\mathrm{fric}}$ in thermodynamic equilibrium (see (8.26)). To circumvent this situation, SVENDSEN et al. [116] proposed a so-called *non-standard analysis* which for the purpose here is too complicated. We may try to regularize the problem by replacing $(\mathbf{D}_s/\,|\mathbf{D}_s|)$ which arises in $(\bar{\mathbf{\Phi}}_s)_{,\mathbf{D}_s}$ by[6]

$$\frac{\mathbf{D}_s}{\varepsilon + |\mathbf{D}_s|}, \qquad 0 < \varepsilon \ll 1 \ . \tag{8.61}$$

Regularizing the problem in such a way has the advantage that the limit

$$\lim_{\mathbf{D}_s\to 0}\frac{\mathbf{D}_s}{\varepsilon + |\mathbf{D}_s|} = \mathbf{0} \tag{8.62}$$

[6] Cf. FANG, WANG, HUTTER [43, 40], however, the regularization proposal (8.61) is well known in the rheological literature.

is finite, in fact zero, but, on the other hand, this procedure contradicts the requirement that $\bar{\mathbf{\Phi}}_s$ is positively homogenous of first order in $\mathbf{D}_s$ and consequently, $\bar{\mathbf{T}}_{\text{fric}}$ does not have a rate-independent part. Nevertheless, we are convinced that for very small values of ε the term $\left\{\mathbf{D}_s/(\varepsilon+|\mathbf{D}_s|)\right\}$ is only affected by ε in the vicinity of $\mathbf{D}_s = \mathbf{0}$. For rapid motions, i. e. steep velocity gradients and thus large values of $|\mathbf{D}_s|$, ε is negligibly small. However, with the introduction of (8.61) the equilibrium stress (8.58) of the solid no longer contains the frictional contribution, because this term now vanishes in equilibrium. This means that the equilibrium stresses will now have to be carried by the pressure like contributions and, above these, the elastic stresses. This may be somewhat unrealistic, but it is so only in a very small regime.

As an alternative method of regularization, we may apply the following approach: When starting from a state of rest, at which the strains and stresses must first be determined, an initial value problem of the quasi-static equations using the stress representation (8.58) without the original frictional term replaced by the frictional term with the regularization (8.61) is integrated in time. As soon as $|\mathbf{D}_s|$ has reached the value $10^n \times \varepsilon$, where n can be selected $(1 < n \leqslant 2)$, the actual value of $\left(\mathbf{D}_s/\,|\mathbf{D}_s|\right)$ is assigned to the equilibrium frictional stress in (8.58). Computations are then continued with the classical hypo-plastic equations. On the other hand, for a decelerating phase of the motion, the regularization (8.61) does not need to be introduced at all. If $|\mathbf{D}_s|$ reaches the value ε from above, we then may simply maintain this limiting value $\left(\mathbf{D}_s/\,|\mathbf{D}_s|\right)$ also for smaller values of $|\mathbf{D}_s|$ down to $|\mathbf{D}_s| = 0$ (essentially locking it to the equilibrium). This then defines the equilibrium value for the stress according to (8.58).[7] Reloading phases of a dynamical process can then be started from this 'frozen' equilibrium state. This procedure corresponds to the approach of non-standard analysis.

When reviewing an earlier version of this manuscript Bob Svendsen noted:

> 'Yet another possibility to analyse thermodynamic equilibrium for hypoplastic materials may be offered by non-convex analysis. If we consider the 'hypo-elastic' form
>
> $$\bar{\mathbf{\Phi}}_s = \mathsf{L}(\mathbf{B}_s, \mathbf{Z}_s)\mathbf{D}_s \tag{8.63}$$
>
> for $\bar{\mathbf{\Phi}}_s$, then (8.63) fulfills the differentiability requirement since it is linear in $\mathbf{D}_s$. Here, $\mathsf{L}(\mathbf{B}_s, \mathbf{Z}_s)$ represents a fourth-order-tensor valued isotropic function, and as mentioned already, it is common in the soil mechanics context to model the void ratio e as a function of $\mathbf{B}_s$, which is reflected in the dependence of $\bar{\mathbf{\Phi}}_s$ on $\mathbf{B}_s$. From (8.63) and (8.58) above follows, in particular, the form
>
> $$\left(\bar{\mathbf{T}}_s\big|_{\mathrm{E}}\right)^{\text{hypoelastic}} = 2\rho\,\mathrm{sym}(\Psi^G_{I,\mathbf{B}_s})\mathbf{B}_s + \rho\mathbf{L}^{\mathrm{T}}\Psi^G_{I,\bar{\mathbf{Z}}_s} \tag{8.64}$$

[7] This procedure is obviously 'mesh' dependent, the mesh being given by ε.

for the equilibrium CAUCHY stress. Models for granular materials based on (8.63) in the realm of soil mechanics have been considered by, e.g., STUTZ [114], ROMANO [111], as well as DAVIS & MULLENGER [31]; they were criticized by GUDEHUS [47]. The basic problem of (8.63) is that it cannot capture the fact that the material behaviour of granular materials is in general different in extension than in compression. In contrast to (8.63), the *hypoplastic* form (see (8.52))

$$\bar{\mathbf{\Phi}}_s = \mathbf{L}\left(\mathbf{B}_s, \bar{\mathbf{Z}}_s\right)\mathbf{D}_s + \mathbf{N}\left(\mathbf{B}_s, \bar{\mathbf{Z}}_s\right)|\mathbf{D}_s| \tag{8.65}$$

for $\bar{\mathbf{\Phi}}_s$ does account for the fact that the material behaviour of granular materials is in general different in extension than in compression, i. e., via the second term non-linear in $\mathbf{D}_s$. Note that, in contrast to (8.63), (8.65) is not (FRÉCHET) differentiable in $\mathbf{D}_s$ at $\mathbf{D}_s = \mathbf{0}$ since the Euclidean norm is not. Consequently, standard concepts of thermodynamic equilibrium which presume such differentiability are not applicable to the hyperplastic case.

One suggestion for further work on this issue was made by SVENDSEN et al. [116] in the form of non-standard analysis. An alternative possibility not suggested by them is convex analysis (e. g., ROCKAFELLAR [110]) and the calculus of variations (i. e., for rate problems). To look into this briefly, note that the hypoplastic form (8.65) of $\bar{\mathbf{\Phi}}_s$ results in the quasi-bilinear form[8]

$$\Gamma = \boldsymbol{\Sigma}_s \cdot \mathbf{D}_s + \sigma_s|\mathbf{D}_s| \tag{8.66}$$

of the dissipation-rate density Γ, with the stress-like quantities

$$\begin{aligned} \boldsymbol{\Sigma}_s &:= \mathbf{T}_s - 2\rho\,\mathrm{sym}(\Psi^G_{I,\mathbf{B}_s})\mathbf{B}_s + \rho\mathbf{L}^{\mathrm{T}}\Psi^G_{I,\bar{\mathbf{Z}}_s}\ , \\ \sigma_s &:= -\Psi^G_{I,\bar{\mathbf{Z}}_s} \cdot \mathbf{N}\ , \end{aligned} \tag{8.67}$$

independent of $\mathbf{D}_s$. As a function of $\mathbf{D}_s$, note that Γ is closed, convex, and positive-homogeneous of order one. Now, as discussed, $\Gamma_{,\mathbf{D}_s}$ does not exist at $\mathbf{D}_s = \mathbf{0}$. On the other hand, the subdifferential

$$\partial\Gamma(\mathbf{D}) := \{(\boldsymbol{\Gamma}_*, \sigma_*)\mid \boldsymbol{\Gamma}_*\cdot(\mathbf{D}_* - \mathbf{D}) + \sigma_*(|\mathbf{D}_*| - |\mathbf{D}|) \le \Gamma(\mathbf{D}_*) - \Gamma(\mathbf{D}),\ \forall \mathbf{D}_*\} \tag{8.68}$$

of Γ in the context of (8.66) does exist there. Since thermodynamic equilibrium represents a minimum of Γ with respect to $\mathbf{D}_s$ at $\mathbf{0}$, a necessary and sufficient condition for such equilibrium is

$$(\mathbf{0}, 0) \in \partial\Gamma(\mathbf{0})\ . \tag{8.69}$$

In particular, (8.67) implies that this can only be the case if $\mathbf{N}$ is perpendicular to $\Psi^G_{I,\bar{\mathbf{Z}}_s}$. Since Ψ^G_I is an isotropic function of $\mathbf{Z}_s$, this could be

[8] The equilibrium variables $\mathbf{B}_s$ and $\mathbf{Z}_s$ are left out of the notation for simplicity here.

the case in general iff N is (i), deviatoric, (ii), perpendicular to $\mathbf{Z}_s$, and, (iii), perpendicular to the cofactor of $\mathbf{Z}_s$. In the simplest case, i. e., if Ψ_I^G were a function of the first invariant of $\mathbf{Z}_s$ alone, then only (i) would have to hold. The forms for N found in the literature (e. g., Wu et al. [128]) used to model the failure of various types of soils, however, satisfy none of these conditions. Consequently, such models for granular materials would appear to possess no state of thermodynamic equilibrium in the standard sense. More generally, note that the assumption of an additional principle such as maximum dissipation would imply the generalized normal from $(\boldsymbol{\Sigma}, \sigma) \in \partial\Gamma(\mathbf{D}_s)$ in this context.'

It is to be seen how this approach will open new avenues in elasto-visco-plasticity of granular mixtures.

For the modelling of $\bar{\mathbf{T}}_{\text{fric}}$ there still remains the specification of L and $\mathbf{N}$. In general, both tensors are allowed to depend on the following set of constitutive variables

$$\left\{\nu_s,\ \nabla\nu_s,\ \mathbf{v}_{fs},\ \mathbf{B}_s,\ \mathbf{B}_f,\ \mathbf{W}_{fs},\ \mathbf{D}_f,\ \bar{\mathbf{Z}}_s\right\}, \tag{8.70}$$

but considering all these variables leads to very complex isotropic representations of L and $\mathbf{N}$. Therefore, we here adopt the 'principle of phase separation', [**A20**], and abandon those quantities which are related, (i) to the interaction of the constituents, (ii) to the mixture and (iii) to the fluid constituent, i. e. $\mathbf{v}_{fs}$, $\mathbf{W}_{fs}$, $\mathbf{B}_f$ and $\mathbf{D}_f$. In the hypo-plastic single-material theory of Svendsen et al. [116], $\boldsymbol{\Phi}$ is assumed to depend only on the set $\{\mathbf{B},\ \mathbf{Z},\ \mathbf{D}\}$. In the present model we are left with the equivalent quantities $\mathbf{B}_s$, $\bar{\mathbf{Z}}_s$ and $\mathbf{D}_s$, but owing to the mixture character of the model the quantities ν_s and $\nabla\nu_s$ should also arise. To disregard the latter contributions is hardly feasible and therefore, if we want to use the representations for L and $\mathbf{N}$ proposed in the literature for single-body hypo-plasticity,[9] their adaption is necessary. The existing recent literature on hypo-plastic constitutive modelling and parameter identification for special choices of the operators L and $\mathbf{N}$ clearly point at a dominant role played by the void ratio $e = (1-\nu_s)/\nu_s$.

As we are presently not dealing with specific problems we leave the choice of L and $\mathbf{N}$ open, but draw the reader's attention to the footnote below.

Final equilibrium constitutive laws

One of the major achievements of the present work is the prescription of the following constitutive laws for $\bar{\mathbf{T}}_s\big|_{\mathrm{E}}$, $\bar{\mathbf{T}}_f\big|_{\mathrm{E}}$, $\bar{\mathbf{m}}_s^i\big|_{\mathrm{E}}$ and $\mathbf{q}\big|_{\mathrm{E}}$. They were found, (i) by using the well-known principles and rules of material modelling (see

---

[9] Cf. Svendsen et al. [116], Kolymbas [74]-[76], Niemunis [99], von Wolffersdorff [126], Bauer [8, 9], Masin [84], Wu [127], Wu & Kolymbas [129], Chambon [26]-[25], Darve [29, 30].

Chapter 4), (ii) by taking into account the MÜLLER-LIU entropy principle and its exploitation, (iii) by performing an 'isotropic' expansion about the point $\mathcal{Y} = \mathbf{0}$ (see (7.53)) and (iv) using the assumptions [**A1**] to [**A22**]. The results read as follows

$$
\begin{aligned}
\bar{\mathbf{T}}_s\big|_{\mathrm{E}} &= -\,\bar{\varpi}_s \mathbf{I} + \mu \mathbf{B}_s + \rho\delta\bar{\mathbf{Z}}_s\ , \\
\bar{\mathbf{T}}_f\big|_{\mathrm{E}} &= -\,\bar{\varpi}_f \mathbf{I}\ , \\
\bar{\mathbf{m}}_s^i\big|_{\mathrm{E}} &= \Big\{\beta_s^G(1-\bar{\xi}_s) - \rho_f \Psi_I^G + \varsigma - \theta\big(k_s^v\big)_{,\nu_s}\Big\}\nabla\nu_s \\
&\qquad + \rho\left(\Psi_I^G\right)_{,\bar{\mathbf{Z}}_s}\left(\bar{\mathbf{\Phi}}_s\right)_{,\mathbf{v}_s}\big|_{\mathrm{E}} \\
&= -\,\bar{\mathbf{m}}_f^i\big|_{\mathrm{E}}\ , \\
&\left(\mathbf{q}\big|_{\mathrm{E}} = \mathbf{0}\right)\ ,
\end{aligned}
\tag{8.71}
$$

with

$$
\begin{aligned}
\bar{\varpi}_s &= \nu_s\big(\beta_s^G - \rho_f\Psi_I^G + \varsigma\big) - \theta k_s^v - \nu_s(\rho_s - \rho_f)\big(\Psi_{I,\bar{\mathbf{Z}}_s}^G \cdot \bar{\mathbf{Z}}_s\big)\ , \\
\bar{\varpi}_f &= (1-\nu_s)\big(-\rho_f\Psi_I^G + \varsigma\big) - \theta k_f^v\ .
\end{aligned}
\tag{8.72}
$$

The total pressure of the mixture CAUCHY stress tensor becomes

$$
\varpi = \bar{\varpi}_s + \bar{\varpi}_f = \nu_s\beta_s^G + \big(-\rho_f\Psi_I^G + \varsigma\big) - \nu_s(\rho_s - \rho_f)\big(\Psi_{I,\bar{\mathbf{Z}}_s}^G \cdot \bar{\mathbf{Z}}_s\big)\ . \tag{8.73}
$$

The formulae (8.72), (8.73) are interesting by the fact how friction contributes to the total pressure. If the solid and fluid densities are the same ($\rho_f = \rho_s$), then the last terms of $(8.72)_1$ and (8.73) obviously vanish. In this case the solid is completely buoyant in the fluid and friction is expected to be minimal – in the equations (8.72) and (8.73) zero. In a dry granular material ($\rho_f = 0$) rubbing friction operates and the frictional pressure contribution is proportional to the partial density $\bar{\rho}_s = \nu_s\rho_s$. This form is adequate for a solid body with voids, ($\nu_s < 1$), or without voids ($\nu_s = 1$). These results appear to be reasonable.

Non-equilibrium contributions

So far, the findings were based on rather strong assumptions. Unfortunately, only a few rational arguments exist that allow the construction of reasonable constitutive laws for the quantities[10] $\bar{\mathbf{T}}_s\big|_{\mathrm{N}}$, $\bar{\mathbf{T}}_f\big|_{\mathrm{N}}$ and $\bar{\mathbf{m}}_s^i\big|_{\mathrm{N}}$. However,

[10] Note the subscript $(\cdot)|_{\mathrm{N}}$ denotes non-equilibrium contributions, whilst the superscript $(\cdot)^{\mathrm{N}}$ indicates a general non-linear expression.

since we are assuming isothermal conditions, the heat flux vector, $\mathbf{q}$, is of no interest. In the literature, often, physically rather dubious assumptions are made which facilitate the treatment and construction of solutions of certain applied problems. Sometimes they contradict the physics of the respective application. This is in particular the case for the assumption of 'pressure equilibrium' (see Section 8.4), but also for the aforementioned 'principle of phase separation', [**A20**]. KIRCHNER [71] pointed out that particle size segregation, which is present in every debris flow, can only be modelled with continuum mechanical methods if the latter principle is rejected. However, in most of the existing models both assumptions are made. If we want to restate in the subsequent treatment the non-equilibrium constitutive laws arising in the prominent literature, we must, however, adhere to the 'principle of phase separation'.

We recall the dependencies of the constitutive quantities $\mathsf{C} = \{\bar{\mathbf{T}}_s\ ,\ \bar{\mathbf{T}}_f, \bar{\mathbf{m}}_s^i\}$ as listed in (8.7). Applying, now, the 'principle of phase separation', [**A20**], to $\bar{\mathbf{T}}_s\big|_{\mathrm{N}}$ and $\bar{\mathbf{T}}_f\big|_{\mathrm{N}}$, these dependencies reduce to

$$\bar{\mathbf{T}}_s\big|_{\mathrm{N}} = \hat{\bar{\mathbf{T}}}_s\ \big|_{\mathrm{N}}\Big(\nu_s,\ \nabla\nu_s,\ \mathbf{B}_s,\ \mathbf{D}_s,\ \bar{\mathbf{Z}}_s\Big), \tag{8.74}$$

$$\bar{\mathbf{T}}_f\big|_{\mathrm{N}} = \hat{\bar{\mathbf{T}}}_f\ \big|_{\mathrm{N}}\Big(\nu_f,\ \mathbf{B}_f,\ \mathbf{D}_f\Big)\ . \tag{8.75}$$

The isotropic representations for $\bar{\mathbf{T}}_s\big|_{\mathrm{N}}$ and $\bar{\mathbf{T}}_f\big|_{\mathrm{N}}$ are still very complex and therefore, we shall neglect dependencies on $\mathbf{B}_s$, $\bar{\mathbf{Z}}_s$ and $\mathbf{B}_f$. Doing so, we are indeed loosing information, but as $\mathbf{B}_s$ and $\bar{\mathbf{Z}}_s$ affect the 'equilibrium' constitutive laws, their information is automatically carried over to non-equilibrium processes. The 'principle of phase separation' makes only sense in connection with constituent-specific constitutive quantities. Interaction supply rate densities, such as that for the solid momentum $\bar{\mathbf{m}}_s^i$, are by definition excluded from the application of this principle. Under these restrictive assumptions the isotropic representations of the two non-equilibrium CAUCHY stress tensors (8.74) and (8.75) take the forms

$$\begin{aligned}\bar{\mathbf{T}}_s\big|_{\mathrm{N}} = {}& \kappa_1^s \mathrm{I}_{\mathbf{D}_S}\ \mathbf{I} + \kappa_2^s\mathbf{D}_s + \kappa_3^s\mathbf{D}_s^2 \\ & + \kappa_4^s\mathbf{M}_s + \kappa_5^s\,\mathrm{sym}\,\big(\mathbf{M}_s\mathbf{D}_s\big) + \kappa_6^s\,\mathrm{sym}\,\big(\mathbf{M}_s\mathbf{D}_s^2\big),\end{aligned} \tag{8.76}$$

$$\bar{\mathbf{T}}_f\big|_{\mathrm{N}} = \kappa_1^f \mathrm{I}_{\mathbf{D}_f}\ \mathbf{I} + \kappa_2^f\mathbf{D}_f + \kappa_3^f\mathbf{D}_f^2\ , \tag{8.77}$$

where

$$\mathbf{M}_s := \nabla\nu_s \otimes \nabla\nu_s\ . \tag{8.78}$$

The following dependences are explicitly not included in the isotropic representations (8.76), (8.77):

- $\mathbf{M}_s^2$, since $\mathbf{M}_s^2 = \mathrm{I}_{\mathbf{M}_s}\mathbf{M}_s$,
- $\mathbf{M}_s^3$, since $\mathbf{M}_s^3 = (\mathrm{I}_{\mathbf{M}_s})^2\mathbf{M}_s$,
- $\mathbf{M}_s^2\mathbf{D}_s$, since $\mathbf{M}_s^2\mathbf{D}_s = (\mathrm{I}_{\mathbf{M}_s})\mathbf{M}_s\mathbf{D}_s$,
- $\mathbf{M}_s^2\mathbf{D}_s^2$, since $\mathbf{M}_s^2\mathbf{D}_s^2 = (\mathrm{I}_{\mathbf{M}_s})\mathbf{M}_s\mathbf{D}_s^2$.

The coefficients κ_{1-6}^s are functions of ν_s, $\mathrm{I}_{\mathbf{D}_s}$, $\mathrm{II}_{\mathbf{D}_s}$, $\mathrm{III}_{\mathbf{D}_s}$, $\mathrm{I}_{\mathbf{M}_s}$, $\mathrm{I}_{\mathbf{M}_s\mathbf{D}_s}$ and $\mathrm{I}_{\mathbf{M}_s\mathbf{D}_s^2}$. It can be shown with the help of definition (8.78) for $\mathbf{M}_s$ and the CAYLEY-HAMILTON theorem that $\mathrm{II}_{\mathbf{M}_s}$ and $\mathrm{III}_{\mathbf{M}_s}$ are identically zero. Therefore, in view of the properties of the above list, we also have $\mathrm{III}_{\mathbf{M}_s\mathbf{D}_s} = 0$ and $\mathrm{III}_{\mathbf{M}_s\mathbf{D}_s^2} = 0$. The invariants $\mathrm{II}_{\mathbf{M}_s\mathbf{D}_s}$ and $\mathrm{II}_{\mathbf{M}_s\mathbf{D}_s^2}$ vanish as well, because of the outlined proof provided in the footnote below[11]. Now, from the definition of thermodynamic equilibrium (see Section 7.2), i. e.,

$$\lim_{\mathrm{n}\to 0} \bar{\mathbf{T}}_s\big|_{\mathrm{N}} = \mathbf{0} \,, \tag{8.79}$$

and the fact, that $\nabla\nu_s$ is an equilibrium quantity, it immediately follows that $\kappa_4^s\big|_{\mathrm{E}}$ must vanish, i. e.

$$\kappa_4^s\big|_{\mathrm{E}} = 0 \,. \tag{8.80}$$

This implies that $\kappa_4^s = \kappa_4^s\big|_{\mathrm{N}}$, in general. A first guess may well simply be $\kappa_4^s\big|_{\mathrm{N}} = 0$.

---

[11] Let $\mathbf{a}$, $\mathbf{b}$ be vectors and $\mathbf{A}$ a second order tensor defined in three-dimensional space. Then

$$(\mathbf{a}\otimes\mathbf{b})\,\mathbf{A} = \mathbf{a}\otimes\mathbf{A}^{\mathrm{T}}\mathbf{b}, \quad \mathbf{A}\mathbf{a}\cdot\mathbf{b} = \mathbf{a}\cdot\mathbf{A}^{\mathrm{T}}\mathbf{b} \,.$$

So, with $\mathbf{A} = \mathbf{D} = \mathbf{D}^{\mathrm{T}}$ and $\mathbf{M} = \mathbf{a}\otimes\mathbf{a}$ we have

$$\begin{aligned}
\mathbf{MD} &= (\mathbf{a}\otimes\mathbf{a})\,\mathbf{D} = \mathbf{a}\otimes\mathbf{Da} \,,\\
(\mathbf{MD})^2 &= (\mathbf{a}\otimes\mathbf{Da})\,(\mathbf{a}\otimes\mathbf{Da}) = (\mathbf{a}\cdot\mathbf{Da})\,\mathbf{a}\otimes\mathbf{Da} \,,\\
\mathbf{MD}^2 &= (\mathbf{a}\otimes\mathbf{a})\,\mathbf{D}^2 = \mathbf{a}\otimes\mathbf{D}^2\mathbf{a} \,,\\
\left(\mathbf{MD}^2\right)^2 &= (\mathbf{a}\cdot\mathbf{D}^2\mathbf{a})\,\mathbf{a}\otimes\mathbf{D}^2\mathbf{a} = (\mathbf{Da}\cdot\mathbf{Da})\,\mathbf{a}\otimes\mathbf{D}^2\mathbf{a} \,,
\end{aligned}$$

as well as

$$\begin{aligned}
\mathrm{tr}\,(\mathbf{MD}) &= (\mathbf{a}\cdot\mathbf{Da}) \,,\\
\mathrm{tr}\,((\mathbf{MD})^2) &= (\mathbf{a}\cdot\mathbf{Da})^2 \,,\\
\mathrm{tr}\,(\mathbf{MD}^2) &= \mathbf{a}\cdot\mathbf{D}^2\mathbf{a} = \mathbf{Da}\cdot\mathbf{Da} \,,\\
\mathrm{tr}\,(\mathbf{MD}^2)^2 &= (\mathbf{Da}\cdot\mathbf{Da})^2 \,.
\end{aligned}$$

This implies

$$\begin{aligned}
\mathrm{II}_{\mathbf{MD}} &= \tfrac{1}{2}\left[(\mathrm{I}_{\mathbf{MD}})^2 - \mathrm{I}_{(\mathbf{MD})^2}\right] = 0 \,,\\
\mathrm{II}_{\mathbf{MD}^2} &= \tfrac{1}{2}\left[(\mathrm{I}_{\mathbf{MD}^2})^2 - \mathrm{I}_{(\mathbf{MD}^2)^2}\right] = 0 \,,
\end{aligned}$$

The representations (8.76), (8.77) are fairly complicated, and identification of the parameters by experiment or other means must be difficult. For this reason simplification of (8.76), (8.77) is desired simply for technical reasons. Such simplifications are the assumptions of *quasi-linearity* and *strict linearity*. The first of these simplifications allows the vector or tensor valued dependent constitutive quantities to depend explicitly and linearly on the vector- and tensor-valued independent constitutive variables, respectively. In contrast to the strict linearity, the coefficients introduced along with such a representation are functions not only of the scalar-valued constitutive variables (as for strict linearity), but also of invariants of the independent vector- and tensor-valued variables themselves.

Following HUTTER et al. [63] we assume, for simplicity, that $\bar{\mathbf{m}}^i_s\big|_{\mathrm{N}}$ and $\bar{\mathbf{T}}_f\big|_{\mathrm{N}}$ can be adequately modelled by their strict linear forms and $\bar{\mathbf{T}}_s\big|_{\mathrm{N}}$ by its quasi-linear form. In addition, we adopt the assumption that $\bar{\mathbf{T}}_s\big|_{\mathrm{N}}$ and $\bar{\mathbf{m}}^i_s\big|_{\mathrm{N}}$ are independent of the variable $\nabla\nu_s$. From a mathematical point of view there is no obvious reason for this assumption but at least for $\bar{\mathbf{m}}^i_s$ the information contained in $\nabla\nu_s$ is not entirely lost because its equilibrium part depends linearly on $\nabla\nu_s$. The equations that evolve from these considerations read as follows

$$\bar{\mathbf{T}}_s\big|_{\mathrm{N}} = \kappa^s_1 \mathrm{I}_{\mathbf{D}_s}\mathbf{I} + \kappa^s_2\mathbf{D}_s\ , \quad \bar{\mathbf{T}}_f\big|_{\mathrm{N}} = \kappa^f_1 \mathrm{I}_{\mathbf{D}_f}\mathbf{I} + \kappa^f_2\mathbf{D}_f\ , \quad \bar{\mathbf{m}}^i_s\big|_{\mathrm{N}} = m_{\mathbf{D}}\mathbf{v}_{fs}\ , \tag{8.81}$$

where κ^s_1 and κ^s_2 are, in general, functions of ν_s, $\mathrm{I}_{\mathbf{D}_s}$, $\mathrm{II}_{\mathbf{D}_s}$ and $\mathrm{III}_{\mathbf{D}_s}$, whilst κ^f_1 and κ^f_2 depend on $\nu_f = (1-\nu_s)$, and $m_{\mathbf{D}}$ is a function of ν_s. An explicit dependence on $\mathbf{M}_s$ has dropped out entirely from (8.81).

The quasi-linearity of the solid CAUCHY stress tensor which is expressed through the non-linear dependence of κ^s_1 and κ^s_2 on $\mathbf{D}_s$ reflects the strong non-linear stress-stretching behaviour that arises during creep or rapid shear of the granular part of the debris flow (cf. HUTTER et al. [63]). By excluding the dependence upon $\mathbf{D}^2$, the proposal (8.81) is not capable of modelling normal stress effects. These effects are not likely important in rapid granular flows and only come to bear when strong decelerations in the approach to the deposition are active. However, the above model properties for $\bar{\mathbf{T}}_s\big|_{\mathrm{N}}$ are in agreement with BAGNOLD's experiments (cf. HUTTER & RAJAGOPAL [64]).

The interpretation of parameters κ^s_1, κ^s_2, κ^f_1, κ^f_2 and $m_{\mathbf{D}}$ depends on the perspective which is taken when looking at relations (8.81). The parameter $\hat{m}_{\mathbf{D}}(\nu_s)$ is commonly known as a *drag coefficient* and can be prescribed in terms of the DARCY *permeability*. Considering the well-known representation

$$\begin{aligned} \bar{\mathbf{T}}_f\big|_{\mathrm{N}} &= \kappa_f \mathrm{I}_{\mathbf{D}_f}\mathbf{I} + 2\mu_f\mathbf{D}'_f\ , \\ &\kappa_f := \kappa^f_1 + \tfrac{1}{3}\kappa^f_2\ , \quad \mu_f := \tfrac{1}{2}\kappa^f_2\ , \end{aligned} \tag{8.82}$$

where $\mathbf{D}'_f = \{\mathbf{D}_f - \frac{1}{3}\mathrm{I}_{\mathbf{D}_f}\mathbf{I}\}$ is the deviatoric part of $\mathbf{D}_f$, one is inclined to regard $\kappa_f(\nu_s)$ and $\mu_f(\nu_s)$ as *bulk* and *shear viscosities* of the fluid constituent. For the model we could conclude from this interpretation that $\kappa_f = 0$ which would be due to the assumed resistance of the fluid to volume changes. Although this interpretation is convenient, it is only well founded for single-material theories and in a mixture concept must be treated with care. A similar conclusion also holds for $\bar{\mathbf{T}}_s\big|_{\mathrm{N}}$ which is either written as

$$\begin{aligned} \bar{\mathbf{T}}_s\big|_{\mathrm{N}} &= \kappa_s \mathrm{I}_{\mathbf{D}_s}\mathbf{I} + 2\mu_s \mathbf{D}'_s \ , \\ \kappa_s &:= \kappa_1^s + \tfrac{1}{3}\kappa_2^s \ , \quad \mu_s := \tfrac{1}{2}\kappa_2^s \ , \end{aligned} \tag{8.83}$$

where $\mathbf{D}'_s = \{\mathbf{D}_s - \frac{1}{3}\mathrm{I}_{\mathbf{D}_s}\mathbf{I}\}$ is the *deviatoric part* of $\mathbf{D}_s$, or in inversed form

$$\begin{aligned} &\mathbf{D}_s = \mathrm{K}_s \mathrm{I}_{\mathbf{T}_s|_{\mathrm{N}}}\mathbf{I} + \mathrm{B}_s\big(\bar{\mathbf{T}}_s\big|_{\mathrm{N}}\big)' \ , \\ &\text{with} \quad \big(\bar{\mathbf{T}}_s\big|_{\mathrm{N}}\big)' = \{\bar{\mathbf{T}}_s - \tfrac{1}{3}\mathrm{I}_{\bar{\mathbf{T}}_s}\mathbf{I}\}\big|_{\mathrm{N}} \ . \end{aligned} \tag{8.84}$$

The remaining task of finding appropriate constitutive laws for $\bar{\mathbf{T}}_s\big|_{\mathrm{N}}$, $\bar{\mathbf{T}}_f\big|_{\mathrm{N}}$ and $\bar{\mathbf{m}}_s^i\big|_{\mathrm{N}}$, is the specification of the explicit functional forms of the coefficients $m_{\mathbf{D}}$, κ_f, μ_f, κ_s and μ_s (or K_s and B_s, respectively). These functions shall be determined in such a way that (i) a large range of different debris flow experiments can be correctly reproduced by the model, (ii) the arising material parameters are only to be calibrated for the material, but not for special deformations or special geometries, (iii) only a small number of material parameters arises.

Parameterization of the non-equilibrium stresses (8.82), (8.83) leans on viscometric experiments, which are routinely performed in applied rheology. The focus in such studies is the determination of the parameters $\kappa_{f,s}$, $\mu_{f,s}$, K_s, B_s, as functions of their variables, which are temperature (here kept constant and not explicitly written), the volume fraction and the invariants of $\mathbf{D}_{s,f}$ or $\bar{\mathbf{T}}_s\big|_{\mathrm{N}}$. We propose the following restricted dependencies:

$$\begin{aligned} \kappa_{f,s} &= \hat{\kappa}_{f,s}(\nu_s, \mathrm{I}_{\mathbf{D}_{f,s}}) \ , \qquad & \mu_{f,s} &= \hat{\mu}_{f,s}(\nu_s, \mathrm{II}_{\mathbf{D}'_{f,s}}, \mathrm{III}_{\mathbf{D}'_{f,s}}) \ , \\ \mathrm{K}_s &= \hat{\mathrm{K}}_s(\nu_s, \mathrm{I}_{\mathbf{T}_s|_{\mathrm{N}}}) \ , & \mathrm{B}_s &= \hat{\mathrm{B}}_s(\nu_s, \mathrm{II}_{\bar{\mathbf{T}}'_s|_{\mathrm{N}}}, \mathrm{III}_{\bar{\mathbf{T}}'_s|_{\mathrm{N}}}) \ , \end{aligned} \tag{8.85}$$

in which the 'principle of phase separation' is assumed to hold. Relations (8.85) make the bulk viscosities to depend on ν_s and the first invariants of $\mathbf{D}_{s,f}$ or $\bar{\mathbf{T}}_s\big|_{\mathrm{N}}$, respectively. The shear viscosities $\mu_{s,f}$ or the shear fluidity B_s will depend on ν_s and the second and third invariants of $\mathbf{D}_s$ or $\bar{\mathbf{T}}_s\big|_{\mathrm{N}}$. These assumptions imply that volume changes affect only the bulk parameters, whilst the shear parameters do not. Conversely, the shear viscosities and shear fluidity are assumed to depend only on the second and third invariants of the stretching deviators $\mathbf{D}'_{s,f}$ and the stress deviator $\bar{\mathbf{T}}'_s\big|_{\mathrm{N}}$, respectively, and not on the first invariants. This implies that they respond primarily to shearing. In rheology of 'viscoplastic' liquids, as rheologists say, the above dependence of the shear parameters $\mu_{f,s}, \mathrm{B}_s$ on the third deviator invariants $\mathrm{III}_{\mathbf{D}'_s}$, $\mathrm{III}_{\bar{\mathbf{T}}'_s|_\mathrm{N}}$ is generally omitted. The likely reason is that in plate-cone viscometers the underlying deformation is simple shearing for which $\mathrm{III}_{\mathbf{D}'_s} = 0$, $\mathrm{III}_{\bar{\mathbf{T}}'_s|_\mathrm{N}} = 0$.[12]

Let us now focus on a number of idealised experiments:

(i) Isotropic extension-compression

In the laboratory such an experiment is not difficult to perform, but it may be very hard to deduce inferences for the identification of the bulk viscosities $\kappa_{s,f}$. So, this case is rather treated as a Gedanken experiment. We shall treat the fluid as volume (and density) preserving and set $\kappa_f \equiv 0$. For the solid a drained compression experiment is thought to be conducted. With

$$\mathbf{D}_s = \dot{\varepsilon}\mathbf{1}, \quad \mathbf{D}'_s = \mathbf{0}, \quad \mathrm{I}_{\mathbf{D}_s} = 3\dot{\varepsilon}, \quad \mathrm{II}_{\mathbf{D}'_s} = 0, \quad \mathrm{III}_{\mathbf{D}'_s} = 0\ , \tag{8.86}$$

and $\bar{\mathbf{T}}_s\big|_\mathrm{N} = \sigma\mathbf{1}$, one deduces from (8.83) that

$$\sigma = \kappa_s\left(\nu_s,\, 3\dot{\varepsilon}\right) 3\dot{\varepsilon}\ , \tag{8.87}$$

or with $\dot{\varepsilon}_\mathrm{vol} := 3\dot{\varepsilon}$,

$$\kappa_s\left(\nu_s,\, \dot{\varepsilon}_\mathrm{vol}\right) = \frac{\sigma}{\dot{\varepsilon}_\mathrm{vol}}\ . \tag{8.88}$$

It should be clear that in the performance of this isotropic compression experiment ν_s cannot be assumed to remain constant. As $(-\dot{\varepsilon}_\mathrm{vol})$ increases in a compression experiment, the compaction of the grains will also increase. It follows that equation (8.88) is only meaningful as long as volumetric strains remain small. We now introduce the [13]

Postulate:

(i) κ_s does not depend on $\dot{\varepsilon}_\mathrm{vol}$,
(ii) at densest packing, $\nu_s = \nu_{s\,\max}$,
(iii) when $\nu_s \leqslant \nu_{s\,\mathrm{crit}}$, $\kappa_s = \boldsymbol{\kappa}_s$.

[12] It is not difficult to show that triaxial experiments are needed in order that a dependence of the parameters (8.85) on the second and third invariants can be experimentally identified.

[13] $\nu_{s\,\mathrm{crit}}$ is the solid volume fraction at which the nominal particle distance is larger than, or equal to, the distance at which the particle contact ceases to exist. $\nu_s = \nu_{s\,\max}$ is the maximum solid volume fraction

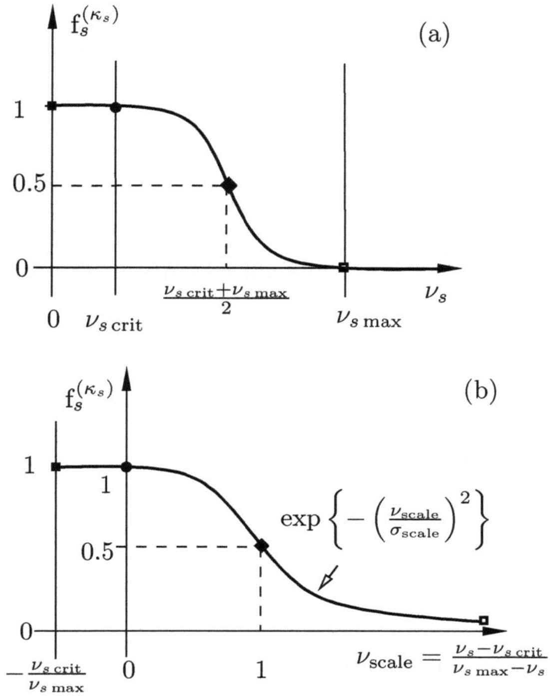

Fig. 8.1 (a) Dependence of the scaled solid bulk viscosity as a function of the solid volume fraction and (b) as parameterizated in (8.90) with $\sigma_{\text{scale}} = (1/\ln 2)^{(1/2)} = 1.2011$

With this postulate and the assumption that κ is monotonically decreasing with growing ν_s reading the zero value at $\nu_{s\,\max}$, we may parameterize κ_s as follows:

$$\kappa_s = \boldsymbol{\kappa}_s \mathrm{f}_s^{(\kappa_s)}(\nu_{\text{scale}}), \quad \nu_{\text{scale}} = \left(\frac{\nu_s - \nu_{s\,\text{crit}}}{\nu_{s\,\max} - \nu_s}\right), \tag{8.89}$$

where the shape of the function $\mathrm{f}_s^{(\kappa_s)}$ is given in Fig. 8.1. According to this graph, the bulk viscosity vanishes at densest packing and assumes the value $\boldsymbol{\kappa}_s$ at the critical packing and beyond, when $\nu_s \leqslant \nu_{s\,\text{crit}}$. It is further assumed that the value of κ_s stays constant for dilute concentrations. Apart from these assumptions the graph in Fig. 8.1(a) simply connects these limiting stages with a smooth curve. The function $\mathrm{f}_s^{(\kappa_s)}$ is of sigmoidal type, and may, for instance, be written as

$$\mathrm{f}_s^{(\kappa_s)} = \begin{cases} 1\,, & -\frac{\nu_{s\,\text{crit}}}{\nu_{s\,\max}} \leqslant \nu_{\text{scale}} \leqslant 0\,, \\ \exp\left[-\left(\frac{\nu_{\text{scale}}}{\sigma_{\text{scale}}}\right)^2\right], & 0 \leqslant \nu_{\text{scale}} \leqslant \infty. \end{cases} \tag{8.90}$$

If the value of $f_s^{(\kappa_s)}(\nu_{\text{scale}} = 1)$ is given by f_{mean}, then

$$\sigma_{\text{scale}} = \left(\frac{-1}{\ln(f_{\text{mean}})} \right)^{1/2} . \tag{8.91}$$

A concrete identification would consist in matching the graph of Fig. 8.1 with experimental results. As far as relations (8.89) and (8.90) are concerned, $\nu_{s\,\text{max}}$ and $\nu_{s\,\text{crit}}$ must be identified, which is not difficult, and $\boldsymbol{\kappa}_s$ must be determined, which may be more difficult and may require (semi) inverse modelling. The following first estimates are suggested.

$$\left.\begin{aligned} \nu_{s\,\text{max}} &= 0.75 \\ \nu_{s\,\text{crit}} &= 0.20 \\ \boldsymbol{\kappa}_s &= 10^{-3}\ \text{Pa s} \end{aligned}\right\} \text{only first estimates.} \qquad \sigma_{\text{scale}} = 1.20 \tag{8.92}$$

(ii) Simple shearing

The rheologically most popular and probably simplest experiment is viscometric shearing e. g. in an axi-symmetric cone-plate viscometer. We consider an experiment being conducted for the fluid[14] and solid in isolation. With

$$\mathbf{D}_{f,s} = \mathbf{D}'_{f,s} = \begin{pmatrix} 0 & \frac{1}{2}\dot\gamma & 0 \\ \frac{1}{2}\dot\gamma & 0 & 0 \\ 0 & 0 & 0 \end{pmatrix} , \tag{8.93}$$

for the fluid and the solid, one deduces

$$\mathrm{I}_{\mathbf{D}_{f,s}} = 0, \quad \mathrm{II}_{\mathbf{D}'_{f,s}} = \frac{\dot\gamma^2}{4}, \quad \mathrm{III}_{\mathbf{D}'_{f,s}} \equiv 0 , \tag{8.94}$$

and then obtains with

$$\bar{\mathbf{T}}_{f,s}\big|_{\mathrm{N}} = \begin{pmatrix} 0 & \tau & 0 \\ \tau & 0 & 0 \\ 0 & 0 & 0 \end{pmatrix} , \tag{8.95}$$

and (8.82), (8.83)

$$\mu_{f,s}(\nu_s, \frac{\dot\gamma^2}{4}, 0) = \frac{\tau}{\dot\gamma} . \tag{8.96}$$

---

[14] Most likely, the fluid in a debris flow will be loaded with silt to clay components of the debris that extends over a large range of particle diameters. Therefore, the fluid is not pure water, but a slurry with a certain concentration of the fine particles.

Monitoring τ and measuring $\dot{\gamma}$ allows identification of the functions $\mu_{s,f}$ in (8.96). It is obvious from the above formulae $(8.94)_3$ and (8.96) that simple shearing experiments cannot identify a $\mathrm{III}_{\mathbf{D}_{s,f}}$-dependence of the viscosity functions $\mu_{s,f}(\cdot)$. Applied rheologists, therefore, generally omit the third variable, $\mathrm{III}_{\mathbf{D}'_{s,f}}$, and also do not make the ν_s dependence explicit. Instead, we shall use the following

Postulate:

(i) The shear viscosity functions $\mu_{f,s}(\nu_s, \mathrm{II}_{\mathbf{D}'_{f,s}}, \mathrm{III}_{\mathbf{D}'_{f,s}})$ allow the product decomposition

$$\mu_{f,s} = \mathrm{M}_{f,s}(\mathrm{III}_{\mathbf{D}'_{f,s}})\bar{\eta}_{f,s}(\nu_s, \mathrm{II}_{\mathbf{D}'_{f,s}}) \,, \tag{8.97}$$

with a first estimate $\mathrm{M}_{f,s}(\mathrm{III}_{\mathbf{D}'_{f,s}}) = 1$.

(ii) The solid volume fraction as a variable enters only the functions $\bar{\eta}_{f,s}(\cdot)$. This dependence may again be separated from that of $\mathrm{II}_{\mathbf{D}'_{f,s}}$ as follows

$$\bar{\eta}_{f,s} = g_{f,s}(\nu_s)\bar{\bar{\eta}}_{f,s}(\mathrm{II}_{\mathbf{D}'_{f,s}}) \,, \tag{8.98}$$

or the coefficients in the parameterizations may be assumed to be ν_s-dependent.

With this postulate, we may identify the functions $\eta_{s,f}(\cdot)$ for a fixed solid volume fraction, formally treated to have a reference value.

The literature dealing with shear viscosity functions (they will be called here simply 'viscosity functions') is abundant; justification for all the proposals is not possible. Therefore, we restrict here considerations to what is referred to as *viscometry of fluids with yield stress.* These have been in the past few years the concern of many rheologists. Our attention here is to propose formulae, which embrace possibly all cases that may occur in debris flow modelling, so that only the identification of the parameters in specific situations is left to the user. In the literature, the behaviour is often called visco-plastic, but it is emphasised here that rate independence of the stress in terms of the stretching is not a necessity of this characterization. The subsequent analysis has been particularly influenced by contributions of ZHU et al. [130], MENDES & DUTRA [87] and the review article by ANCEY [5]. In the following, results by LUCA et al. [82] will be reported, and LUCA's text will closely be followed. To be in conformity with the rheological literature, we shall write

$$\tau^*_{f,s} = \bar{\bar{\eta}}_{f,s}\left(\frac{\dot{\gamma}^2}{4}\right)\dot{\gamma} = \eta_{f,s}\left(\dot{\gamma}\right)\dot{\gamma} \,, \tag{8.99}$$

where,

$$\tau^*_{f,s} = \frac{\tau}{\mathrm{M}_{f,s}\left(\mathrm{III}_{\mathbf{D}'_{f,s}}\right) g_{f,s}(\nu_s)} \tag{8.100}$$

is an appropriately scaled stress, i. e., $\mathrm{M}_{f,s}$ and $g_{f,s}$ are dimensionless, so that $\bar{\bar{\eta}}_{f,s}$ or $\eta_{f,s}$ has dimension [Pa s]. Clearly, $\bar{\bar{\eta}}_{f,s} = \bar{\bar{\eta}}_{f,s}(\dot{\gamma}^2/4) := \eta_{f,s}(\dot{\gamma})$. In what follows we shall drop the indices (s, f) and simply write η. The large variety of grain size distributions in soils from clay to gabbro, and the tremendous range of mixing of the soil with water in common debris flows, gives rise to a wide diversity of non-Newtonian viscous behaviour. The compound materials can arise as shear thinning[15] fluids for which the viscosity at $\dot{\gamma} = 0$ is very small, and shear thickening fluids for which the viscosity is bounded away from zero for $\dot{\gamma} = 0$. Many materials such as paints, slurries, pastes, but also debris are described by either the BINGHAM [12] or the HERSCHEL-BULKLEY [55] model, whose viscous constitutive behaviour is described by the constitutive relations

$$\begin{cases} \dot{\gamma} = 0, & \text{if} \quad \tau \leqslant \tau_0 \,, \\ \tau = \left(\mathrm{k} + \frac{\tau_0}{\dot{\gamma}}\right)\dot{\gamma}, & \text{if} \quad \tau \geqslant \tau_0 \,, \end{cases} \qquad \text{BINGHAM} \tag{8.101a}$$

$$\begin{cases} \dot{\gamma} = 0, & \text{if} \quad \tau \leqslant \tau_0 \,, \\ \tau = \left(\mathrm{k}\,\dot{\gamma}^{\lambda-1} + \frac{\tau_0}{\dot{\gamma}}\right)\dot{\gamma}, & \text{if} \quad \tau \geqslant \tau_0 \,, \end{cases} \qquad \text{HERSCHEL-BULKLEY} \tag{8.101b}$$

in which k, λ and τ_0 are model parameters. τ_0 is the yield stress below which no deformation occurs, and k and λ are viscosity parameters of Newtonian $(\lambda = 1)$ and power law $(\lambda \neq 1)$ behaviour, respectively, for excess stresses $(\tau - \tau_0) > 0$. Equations (8.101) can also be written in the form

$$\eta = \mathrm{k}\dot{\gamma}^{\lambda-1} + \frac{\tau_0}{\dot{\gamma}}, \quad \lambda > 0, \quad \mathrm{k} > 0, \quad \tau_0 \geqslant 0 \,, \tag{8.102}$$

and it is understood that, depending on the values of λ and τ_0, the viscosity may become infinitely large, when $\dot{\gamma} = 0$ is approached. When (i) $\lambda \in (0, 1)$ and (ii) $\lambda = 1$ and $\tau_0 \neq 0$, relation (8.102) describes shear thinning behaviour, whilst for $\lambda > 1$ shear thickening is described. Physically, arbitrarily large viscosity at arbitrarily small stretching is unrealistic, and so, the BINGHAM and HERSCHEL-BULKLEY models need to be amended by regularizing them.

Such attempts of regularizations have been proposed in the literature. PAPANASTASIOU [102] proposed such an amendment, involving a parameter m of dimension 'time',

[15] A fluid shows *shear thinning* or (*pseudo-plastic*)/*shear thickening* (or *dilatant*) behaviour, if its viscosity decreases/increases with increasing shear rate. Accordingly, a fluid is called shear thinning/thickening if it has only shear thinning/thickening behaviour for all $\dot{\gamma} \in [0, \infty)$.

$$\eta(\dot{\gamma}) = \mathrm{k} + \frac{\tau_0}{\dot{\gamma}}\left(1 - \exp(-m\dot{\gamma})\right), \quad m > 0 , \tag{8.103}$$

which for the BINGHAM model gives

$$\lim_{\dot{\gamma}\to 0} \eta(\dot{\gamma}) = \mathrm{k} + \tau_0 m \neq \infty . \tag{8.104}$$

For $\lambda \in (0,1)$ with or without yield stress, LUCA et al. [82] proposed

$$\eta(\dot{\gamma}) = \mathrm{k}\dot{\gamma}^{\lambda-1}\left(1 - \exp\left(-l\dot{\gamma}^{1-\lambda}\right)\right) + \frac{\tau_0}{\dot{\gamma}}\left(1 - \exp(-m\dot{\gamma})\right), \tag{8.105}$$
$$l > 0, \quad m > 0 ,$$

which implies the following limiting values:

$$\begin{aligned}
&\lim_{\dot{\gamma}\to 0} \eta(\dot{\gamma}) = \mathrm{k}l + \tau_0 m \neq 0 , \\
&\lim_{\dot{\gamma}\to \infty} \eta(\dot{\gamma}) \approx \mathrm{k}\dot{\gamma}^{\lambda-1} , \\
&\lim_{\dot{\gamma}\to 0} \eta'(\dot{\gamma}) = -\infty , \\
&\eta'(\dot{\gamma}) < 0 .
\end{aligned} \tag{8.106}$$

This shows that the singularity for η at zero stretching is removed, that the viscosity at large stretching shows power law behaviour with an exponent $(\lambda - 1) \in (-1, 0)$ and a slope, reaching $-\infty$ which is rather strange. It also maintains a persistently negative slope, $\eta'(\dot{\gamma}) < 0$. Thus, the essential features of the HERSCHEL-BULKLEY model are preserved as is the shear thinning property in the range $\lambda \in (0,1)$. In passing, we may also mention that MENDES & DUTRA [87] introduced instead of (8.105) the following regularization of the HERSCHEL-BULKLEY model

$$\eta(\dot{\gamma}) = \left(\mathrm{k}\dot{\gamma}^{\lambda-1} + \frac{\tau_0}{\dot{\gamma}}\right)\left(1 - \exp(-m\dot{\gamma})\right), \quad m > 0 . \tag{8.107}$$

Unlike (8.105), this viscosity formula shows shear thickening behaviour at zero shear rate, since $\lim_{\dot{\gamma}\to 0} \eta'(\dot{\gamma}) = +\infty$; moreover, it introduces a slope singularity of the viscosity at zero shear rate.

To also remove the slope singularity of the viscosity function (8.105) [see formula $(8.106)_3$], LUCA et al. [82] proposed the more suitable viscosity function

$$\eta(\dot{\gamma}) = \eta_1 \exp\left(-t_1\dot{\gamma}\right) + \frac{2}{\pi}\eta_2 \left(\frac{\dot{\gamma}}{\dot{\gamma}_r}\right)^{\lambda-1} \arctan\left(t_2\left(\frac{\dot{\gamma}}{\dot{\gamma}_r}\right)^{\beta}\right) \frac{\tau_0}{\dot{\gamma}}\left(1-\exp(-m\dot{\gamma})\right) , \qquad (8.108)$$

in which η_1 and η_2 are constant reference viscosities [Pa s], $\dot{\gamma}_r$ is a constant stretching [s^{-1}], τ_0 is the yield stress [Pa], t_1 and m are reference times [s] and λ, β and t_2 are dimensionless constants, for which numbers must be given subject to the following constraints:

$$\begin{aligned} &\eta_1 > 0, \quad \eta_2 > 0, \quad \tau_0 \geqslant 0, \quad \lambda \in (0,\, 1)\,, \\ &t_1 > 0, \quad t_2,\, m > 0, \quad \beta + \lambda - 2 > 0. \end{aligned} \qquad (8.109)$$

These guarantee the limits

$$\begin{aligned} &\lim_{\dot{\gamma}\to 0} \eta(\dot{\gamma}) = \eta_1 + \tau_0 m \neq \infty\,, \\ &\lim_{\dot{\gamma}\to 0} \eta\prime(\dot{\gamma}) = -\eta_1 t_1 - \tfrac{1}{2}\tau_0 m^2 = \text{finite}\,, \\ &\lim_{\dot{\gamma}\to \infty} \eta(\dot{\gamma}) \approx \eta_2 \left(\frac{\dot{\gamma}}{\dot{\gamma}_r}\right)^{\lambda-1}\,, \end{aligned} \qquad (8.110)$$

and the shear thinning properties prevail for at least large stretchings. Model (8.108) includes the BINGHAM - PAPANASTASIOU fluid (8.103), if $t_0 = 0$, $\eta_2 = 0$, $\tau_0 \neq 0$ and the model introduced by ZHU et al. [130] ($\eta_2 = 0$) as an extension of the DE KEE & TURCOTTE [32] proposal.

In the above, we started from the BINGHAM and HERSCHEL-BULKLEY models as two popular models describing the stress-deformation response of a large class of visco-plastic fluids. The behaviour in these formulae is described by the yield stress, but this feature led to stress-stretching relations with slope discontinuities at zero stretching. They become manifest in the formulae as an infinite viscosity at zero stretching. A similar singularity also arises for the power law fluid ($\tau_0 = 0$) when $\lambda \in (0, 1)$. For a fluid with yield stress this singularity becomes physically apparent as an abrupt transition from the viscous fluid to rigid solid behaviour, which generates mathematical-numerical complexities which one wishes to avoid. The intention in all improved viscosity proposals is to smooth-out these singularities. However, in so doing, the plastic, rate-independent response is formally replaced by a viscous, rate-dependent response. In the context of the model equations in this book, such a 'viscofaction' is even a mandatory smoothing operation, since the non-equilibrium stresses for which the above parameterizations are presented must necessarily vanish in thermodynamic equilibrium. As we know from the modelling of the hypo-plastic stress parameterization, the rate independent parts do not vanish in thermodynamic equilibrium.

From a purely practical point of view, it could also be criticized that the regularized final formulae (8.108)-(8.110) are overly complicated, and identification of the many parameters by experiments must be very difficult, if not impossible. However, since many parameters in the model (8.108) are introduced for regularization purposes, they need not be 'accurately' determined. Values can be estimated such that regularity is established and the graphs of the functions $\eta(\dot{\gamma})$ still mimic the experiments, which anyhow never allows inferences without errors, reasonably well.

In the above formulae, a dependence of the viscosity parameters on solid volume fraction has not been made explicit, but there are indications that such dependencies exist. For instance, ANCEY [5] reports work of HUSBAND et al. [57] and others, who identified clear yielding behaviour in suspensions with solid-volume fractions $\nu_s \leqslant 0.47$. They observed that this yield stress increased dramatically when the solid concentration approached its densest packing. ANCEY [5] mentions other supporting evidences for such a yield stress and quotes WILDEMUTH & WILLIAMS' [122] yield stress formula

$$\tau_0 = \left[A \frac{\nu_s/\nu_0 - 1}{1 - \nu_s/\nu_\infty} \right]^{1/m} , \tag{8.111}$$

in which A, ν_0, ν_∞ and m are parameters fitting their data. This demonstrates that yielding is associated with a size range $\nu_s \in (\nu_0, \nu_\infty)$ of solid volume fraction.

In slurries at moderate to small solid volume fraction, theoretical models predict Newtonian behaviour with viscosities whose value depends on the viscosity of the pure fluid, η_f^{pure}, and the solid volume fraction ν_s; this function is increasing with growing ν_s. Two famous formulae are

$$\begin{cases} \eta_f = \eta_f^{\text{pure}}(1 - 2.5\nu_s), & \text{EINSTEIN [36, 37]}, \\ \eta_f = \eta_f^{\text{pure}}(1 - 2.5\nu_s - 7.6\nu_s^2), & \text{BATCHELOR \& GREEN [7]} . \end{cases} \tag{8.112}$$

ANCEY [5] quotes a general formula supposed to be adequate beyond small ν_s^2-terms,

$$\eta_f = \eta_f^{\text{pure}} \left(1 - \frac{\nu_s}{\nu_{s\,\max}} \right)^{-2.5\nu_{s\,\max}} . \tag{8.113}$$

Relation (8.113) matches the EINSTEIN [36, 37] relation at small ν_s. Formula (8.113) becomes singular when $\nu_s \to \nu_{s\,\max}$ which is certainly unphysical. So, it can be only valid for ν_s sufficiently below $\nu_{s\,\max}$. For values of ν_s close to $\nu_{s\,\max}$, the yield stress will become important and the parameterization (8.108) should be used. For the latter, however, dependencies on the solid volume fraction have to our knowledge not been suggested.

Finally, it is emphasised that the parameterizations as suggested by rheometry are based on the relatively simple formulae (8.82), (8.83) and the application of these formulae to only two very special processes of isotropic extension/compression and simple shear. This does not permit identification of a third invariant dependence, as we have seen. To identify such possible dependencies compound deformations consisting of shear and normal strains are needed. To conduct such experiments for dynamic situations must be very difficult. Moreover, it is also clear that (8.82), (8.83) do not include dynamic normal stress effects for which the quadratic $\mathbf{D}_{f,s}$-dependences in (8.76), (8.77) must not be dropped. Such arguments explain that non-equilibrium stress parameterizations of the class (8.76), (8.77) will keep debris flow modellers busy for a long time until a complete satisfactory parameterization is known.

8.3 Final Constitutive Laws

The reduced forms of the constitutive laws for the solid and fluid CAUCHY stress tensors and the interaction supply rate density for the solid momentum now read

$$
\begin{aligned}
\bar{\mathbf{T}}_s =& \; -\bar{\varpi}_s\mathbf{I} + \bar{\mathbf{T}}_{\rm es}(\mathbf{B}_s) + \rho\delta\bar{\mathbf{Z}}_s + \lambda_s \mathrm{I}_{\mathbf{D}_s}\mathbf{I} + 2\mu_s\mathbf{D}_s' \,, \\
&\bar{\mathbf{\Phi}}_s = \mathrm{f}_1\Big(\mathbf{L}\left(\bar{\mathbf{Z}}_s\right)\mathbf{D}_s + \mathrm{f}_2\mathbf{N}\left(\bar{\mathbf{Z}}_s\right)|\mathbf{D}_s|\Big)\,, \\
\bar{\mathbf{T}}_f =& \; -\bar{\varpi}_f\mathbf{I} \quad + \kappa_f\mathrm{I}_{\mathbf{D}_f}\mathbf{I} + 2\mu_f\mathbf{D}_f' \,, \\
\bar{\mathbf{m}}_s^i =& \; \Big\{\beta_s^G(1-\bar{\xi}_s) - \rho_f\Psi_I^G + \varsigma - \theta\big(k_s^v\big)_{,\nu_s}\Big\}\nabla\nu_s \\
&+ \rho\left(\Psi_I^G\right)_{,\bar{\mathbf{Z}}_s}\left(\bar{\mathbf{\Phi}}_s\right)_{,\mathbf{v}_s}\Big|_{\rm E} + m_{\mathbf{D}}\mathbf{v}_{fs} \\
=& -\bar{\mathbf{m}}_f^i \,,
\end{aligned}
\tag{8.114}
$$

with

$$
\begin{aligned}
\bar{\varpi}_s &= \nu_s\big(\beta_s^G - \rho_f\Psi_I^G + \varsigma\big) - \theta k_s^v - \nu_s(\rho_s - \rho_f)\big(\Psi_{I,\bar{\mathbf{Z}}_s}^G \cdot \bar{\mathbf{Z}}_s\big)\,, \\
\bar{\varpi}_f &= (1-\nu_s)\big(-\rho_f\Psi_I^G + \varsigma\big) - \theta k_f^v \,,
\end{aligned}
\tag{8.115}
$$

and

$$
\varpi = \bar{\varpi}_s + \bar{\varpi}_f = \nu_s\beta_s^G + \big(-\rho_f\Psi_I^G + \varsigma\big) - \nu_s(\rho_s - \rho_f)\big(\Psi_{I,\bar{\mathbf{Z}}_s}^G \cdot \bar{\mathbf{Z}}_s\big) \tag{8.116}
$$

for the spherical contribution to the mixture CAUCHY stress tensor. $\bar{\mathbf{T}}_{\rm es}$ follows from any elastic potential $\bar{\Psi}_{es}^G$, see (8.38), ff.

8.4 An Alternative to the Assumption of 'Pressure Equilibrium'

To analyse the assumption of 'pressure equilibrium' we start with the collection of balance equations for the reduced model. Under the suppositions of isothermal conditions, [**A18**], density-preserving constituents, (8.1), saturation, [**A7**], and the disregard of interaction supply rate densities for mass, (8.2), the evolution law for the frictional behaviour, (4.36), the mass-, volume-fraction- and momentum balance equations, $(4.5)_{1,2}$ and (8.3) turn into

$$\begin{aligned}
\partial\nu_s + \nabla\cdot(\nu_s\mathbf{v}_s) &= 0\ ,\\
\partial\nu_s - \nabla\cdot\mathbf{v}_f + \nabla\cdot(\nu_s\mathbf{v}_f) &= 0\ ,\\
\nu_s\rho_s\left(\partial\mathbf{v}_s + (\nabla\mathbf{v}_s)\mathbf{v}_s\right) &= \nabla\cdot\bar{\mathbf{T}}_s + \bar{\mathbf{b}}_s + \bar{\mathbf{m}}^i_s\ ,\\
(1-\nu_s)\rho_f\left(\partial\mathbf{v}_f + (\nabla\mathbf{v}_f)\mathbf{v}_f\right) &= \nabla\cdot\bar{\mathbf{T}}_f + \bar{\mathbf{b}}_f + \bar{\mathbf{m}}^i_f\ ,\\
\overset{\diamond}{\bar{\mathbf{Z}}}_s &= \bar{\mathbf{\Phi}}_s\ .
\end{aligned}\tag{8.117}$$

Together with the constitutive laws for $\bar{\mathbf{T}}_s$, $\bar{\mathbf{T}}_f$, $\bar{\mathbf{m}}^i_s$, $\bar{\mathbf{m}}^i_f$ and $\bar{\mathbf{\Phi}}_s$, given in (8.114) and (8.115), a set of field equations can be constructed. For the solvability of this set we have to ensure that the number of equations is in conformity with the number of unknown variables arising in the field equations.

'Pressure equilibrium'

From (8.114) and (8.115) we observe that, besides $m_{\mathbf{D}}$, κ_f, μ_f, κ_s, μ_s, L and $\mathbf{N}$, we are still missing explicit expressions for Ψ^G_I and k^v_α ($\alpha = s,\ f$) which are necessary for the description of the evolution of β^G_s, $\Psi^G_{I,\bar{\mathbf{Z}}_s}$, $\bar{\varpi}_s$ and $\bar{\varpi}_f$, respectively. To avoid postulating representations for Ψ^G_I and k^v_α ($\alpha = s,\ f$) and to facilitate the construction of solutions of the field equations, in the literature on multiphase mixtures (cf. Pitman & Le [104]), the assumption of 'pressure equilibrium' is usually made. This assumption is not based on any physical principle, but rather on surmised 'feelings of adequacy'. For the model it reads

$$\varpi_s = \varpi_f = \pi. \tag{8.118}$$

In view of (8.115) this assumption can only hold if we

(i) abandon any hypo-plastic effects from the model, i. e. require that

$$\Psi^G_{I,\bar{\mathbf{Z}}_s} = \mathbf{0}\ , \tag{8.119}$$

(ii) require that

$$k_\alpha^v = 0 \qquad (\alpha = s, f), \tag{8.120}$$

which follows from (8.118) and automatically satisfies the principle of objectivity for $\mathbf{k}$ (see $(7.77)_2$) and

(iii) prescribe the HELMHOLTZ free energy to be in conformity with

$$\beta_s^G = \rho(\Psi_I^G)_{,\nu_s} = 0 \ , \tag{8.121}$$

implying that Ψ_I^G cannot be a function of ν_s.

The assumption states 'physically' that the spherical contribution to the mixture CAUCHY stress is distributed among the constituents according to their volume fractions. With (8.118) to (8.121) the constitutive laws (8.114) reduce to

$$\begin{aligned}
\bar{\mathbf{T}}_s &= -\nu_s\pi\mathbf{I} + \bar{\mathbf{T}}_{\text{es}}(\mathbf{B}_s) + \bar{\mathbf{T}}_s\big|_{\text{N}}, \\
\bar{\mathbf{T}}_f &= -(1-\nu_s)\pi\mathbf{I} + \bar{\mathbf{T}}_f\big|_{\text{N}}, \\
\bar{\mathbf{m}}_s^i &= \pi\nabla\nu_s + m_{\mathbf{D}}\mathbf{v}_{fs} \\
&= -\bar{\mathbf{m}}_f^i,
\end{aligned} \tag{8.122}$$

where, obviously, no frictional stress, $\bar{\mathbf{T}}_{\text{fric}}$ and no implicit volume fraction dependence arise, because

$$\Psi_I^G \neq \hat{\Psi}^G(\nu_s, \bar{\mathbf{Z}}_s) \ . \tag{8.123}$$

Therefore, the hypo-plastic balance law $(8.117)_5$ is no longer of interest, and the dependence of Ψ_I^G reduces to

$$\Psi_I^G = \hat{\Psi}_I^G(\mathbf{B}_s) \ . \tag{8.124}$$

For 'pressure equilibrium' the field equations read

$$\begin{aligned}
\partial\nu_s + \nabla\cdot(\nu_s\mathbf{v}_s) &= 0 \ , \\
\partial\nu_s - \nabla\cdot\mathbf{v}_f + \nabla\cdot(\nu_s\mathbf{v}_f) &= 0 \ , \\
\nu_s\rho_s\left(\partial\mathbf{v}_s + (\nabla\mathbf{v}_s)\mathbf{v}_s\right) &= -\nabla(\nu_s\pi) + \nabla\cdot\left(\bar{\mathbf{T}}_{\text{es}}(\mathbf{B}_s) + \bar{\mathbf{T}}_s\big|_{\text{N}}\right) \\
&\quad + \pi\nabla\nu_s + \bar{\mathbf{b}}_s + m_{\mathbf{D}}\mathbf{v}_{fs} \ , \\
(1-\nu_s)\rho_f\left(\partial\mathbf{v}_f + (\nabla\mathbf{v}_f)\mathbf{v}_f\right) &= -\nabla((1-\nu_s)\pi) + \nabla\cdot\left(\bar{\mathbf{T}}_f\big|_{\text{N}}\right) \\
&\quad - \pi\nabla\nu_s + \bar{\mathbf{b}}_f - m_{\mathbf{D}}\mathbf{v}_{fs} \ .
\end{aligned} \tag{8.125a}$$

To solve these equations in a well posed initial boundary value problem they must be complemented by an evolution equation for $\mathbf{B}_s$. This equation follows from the definition of $\mathbf{B}_s = \mathbf{F}_s \mathbf{F}_s^{\mathrm{T}}$ as[16]

$$\mathbf{B}_s' := \partial \mathbf{B}_s + (\nabla \mathbf{B}_s)\mathbf{v}_s = \mathbf{L}_s \mathbf{B}_s + \mathbf{B}_s \mathbf{L}_s^{\mathrm{T}} \ . \tag{8.125b}$$

If adequate initial values for ν_s, $\mathbf{v}_{f,s}$, $\mathbf{B}_s$, π are prescribed and there exist boundary conditions such that the resulting initial boundary value problem (IBVP) is not ill-posed,[17] we can, in principle, solve (8.125) for ν_s, $\mathbf{v}_s$, $\mathbf{v}_f$, $\mathbf{B}_s$ and π. Although this procedure seems convenient, we observe (see items (i) to (iii)) that 'pressure equilibrium' is a rough *ad hoc* assumption that destroys the structure of hypo-plasticity and that of configuration pressures; moreover, it rules out the linear dependence of $\mathbf{k}$ on $\mathbf{v}_s$ and $\mathbf{v}_f$ (see (7.75)). We conclude that the assumption of 'pressure equilibrium' is based on unnecessary restrictions and thus in general not appropriate for the modelling of debris flows. These unnecessary restrictions prevent first, the description of frictional stresses in thermodynamic equilibrium by means of a hypo-plastic stress contribution and second, eliminate consideration of the configuration pressure, β_s^G, which represents the driving force between the grains and between the fluid and the grains (cf. PASSMAN et al. [103]) and thus might be necessary for the description of particle size segregation in debris flows. Thus, we reject the pressure equilibrium assumption as a physical acceptable assumption; this is also confirmed by PASSMAN et al. [103]. In spite of this, the assumption is still popular and often used, see e. g. PITMAN & LE [104], IVERSON & DENLINGER [69]. It is pleasing, however, that the thermodynamic approach has proved the assumption to be superfluous, or replaceable by a more useful alternative.

Thermodynamic closure assumption

A less restrictive assumption which replaces 'pressure equilibrium' was proposed by HUTTER et al. [63] who simply suppose that

[**A23**] $$k_\alpha^v = 0 \qquad (\alpha = s, f) \ .$$

This assumption fixes the extra entropy flux without making it collinear with the mixture heat flux. Obviously, [**A23**] is an *ad hoc* assumption, too, but does not lose the possibility of modelling frictional effects by hypo-plasticity

[16] If the elastic strain, used in $\bar{\mathbf{T}}_{es}$, is not $\mathbf{B}_s$ but another strain measure, then the evolution equation for that variable must be used, e. g. for $\mathbf{E}_s = \frac{1}{2}(\mathbf{B}_s - \mathbf{I})$, $\mathbf{E}_s' := \partial \mathbf{E}_s + (\nabla \mathbf{E}_s)\mathbf{v}_s = \mathbf{L}_s \mathbf{E}_s + \mathbf{E}_s \mathbf{L}_s^{\mathrm{T}} + \mathbf{D}_s$.

[17] As pointed out by PASSMAN et al. [103] well-posedness is not always the case.

and maintains a possible dependence of Ψ_I^G on ν_s. With the above assumption we obtain

$$\varpi_f = -\rho_f \Psi_I^G + \varsigma \ , \qquad \varpi_s = \beta_s^G + \varpi_f - (\rho_s - \rho_f)\big(\Psi_{I,\bar{\mathbf{Z}}_s}^G \cdot \bar{\mathbf{Z}}_s\big) \ . \tag{8.126}$$

If we, therefore, regard ϖ_f as an independent field, rather than ς, we obtain the following set of field equations: The constitutive laws are expressed as

$$\begin{aligned}
\bar{\mathbf{T}}_s &= -\,\nu_s\Big\{\beta_s^G + \varpi_f - (\rho_s - \rho_f)\big(\Psi_{I,\bar{\mathbf{Z}}_s}^G \cdot \bar{\mathbf{Z}}_s\big)\Big\}\mathbf{I} + \bar{\mathbf{T}}_{\mathrm{es}}(\mathbf{B}_s) + \rho\delta\bar{\mathbf{Z}}_s + \bar{\mathbf{T}}_s\big|_{\mathrm{N}} \ , \\
\bar{\mathbf{T}}_f &= -\,(1-\nu_s)\varpi_f\mathbf{I} + \bar{\mathbf{T}}_f\big|_{\mathrm{N}} \ , \\
\bar{\mathbf{m}}_s^i &= \big\{\beta_s^G(1-\bar{\xi}_s) + \varpi_f\big\}\nabla\nu_s + \rho\left(\Psi_I^G\right)_{,\bar{\mathbf{Z}}_s}\big|_{\mathrm{E}}\left(\bar{\mathbf{\Phi}}_s\right)_{,\mathbf{v}_s}\big|_{\mathrm{E}} + m_{\mathbf{D}}\mathbf{v}_{fs} \\
&= -\,\bar{\mathbf{m}}_f^i \ , \\
\bar{\mathbf{\Phi}}_s &= \mathrm{f}_1\left(\mathbf{L}(\bar{\mathbf{Z}}_s)\mathbf{D}_s + \mathrm{f}_2\mathbf{N}(\bar{\mathbf{Z}}_s)\,|\mathbf{D}_s|\right) \ ,
\end{aligned} \tag{8.127}$$

and the corresponding balance laws (see(8.117)) have the form

$$\begin{aligned}
\partial\nu_s + \nabla\cdot(\nu_s\mathbf{v}_s) &= 0 \ , \\
\partial\nu_s - \nabla\cdot\mathbf{v}_f + \nabla\cdot(\nu_s\mathbf{v}_f) &= 0 \ , \\
\nu_s\rho_s\left(\partial\mathbf{v}_s + (\nabla\mathbf{v}_s)\mathbf{v}_s\right) &= \nabla\cdot\bar{\mathbf{T}}_s + \bar{\mathbf{b}}_s + \bar{\mathbf{m}}_s^i \ , \\
(1-\nu_s)\rho_f\left(\partial\mathbf{v}_f + (\nabla\mathbf{v}_f)\mathbf{v}_f\right) &= \nabla\cdot\bar{\mathbf{T}}_f + \bar{\mathbf{b}}_f + \bar{\mathbf{m}}_f^i \ , \\
\frac{\mathrm{d}^s\,\bar{\mathbf{Z}}_s}{\mathrm{d}t} - [\mathbf{\Omega}_s,\ \bar{\mathbf{Z}}_s] - \nu_s\frac{(\rho_s-\rho_f)}{\rho}\bar{\mathbf{Z}}_s(\nabla\cdot\mathbf{v}_s) &= \bar{\mathbf{\Phi}}_s \ , \\
\mathbf{B}_s' := \partial\mathbf{B}_s + (\nabla\mathbf{B}_s)\mathbf{v}_s &= \mathbf{L}_s\mathbf{B}_s + \mathbf{B}_s\mathbf{L}_s^{\mathrm{T}} \ .
\end{aligned} \tag{8.128}$$

If $\bar{\mathbf{T}}_{es}$ is given as $\bar{\mathbf{T}}_{es}(\mathbf{E}_s)$, where $\mathbf{E}_s = \frac{1}{2}\left(\mathbf{B}_s - \mathbf{I}\right)$, then the evolution equation for $\mathbf{E}_s$ is given by

$$\mathbf{E}_s' := \partial\mathbf{E}_s + (\nabla\mathbf{E}_s)\mathbf{v}_s = \mathbf{L}_s\mathbf{E}_s + \mathbf{E}_s\mathbf{L}_s^{\mathrm{T}} + \mathbf{D}_s \ . \tag{8.129}$$

Remarks:

- One of the equations $(8.128)_{1,2}$ could be replaced by the mixture volume balance
$$\nabla\cdot\mathbf{v}_{\mathrm{vol}} = \nabla\cdot\left(\nu_s\mathbf{v}_s + (1-\nu_s)\mathbf{v}_f\right) = 0 \ . \tag{8.130}$$
  Similarly, $(8.128)_3$ or $(8.128)_4$ could be replaced by the momentum balance relation for the mixture as a whole, but this equation does not offer computational advantages.

- There is still no restriction for the choice of the skew-symmetric tensor $\boldsymbol{\Omega}_s$, and, thus, we choose that form of $\boldsymbol{\Omega}_s$ which is the most convenient one for the investigations at hand. Furthermore, the above field equations have to be complemented by appropriate functions for the non-elastic part of the 'inner' free energy, $\hat{\Psi}^G_{fric}(\nu_s, \bar{\mathbf{Z}}_s)$, for the tensors $\hat{\mathbf{L}}(\nu_s, \nabla\nu_s, \mathbf{B}_s, \bar{\mathbf{Z}}_s)$ and $\hat{\mathbf{N}}(\nu_s, \nabla\nu_s, \mathbf{B}_s, \bar{\mathbf{Z}}_s)$ and for the coefficients $m_{\mathbf{D}}(\nu_s)$, $\lambda_s(\nu_s)$ and $\mu_s(\nu_s)$, $\kappa_f(\nu_s)$, $\mu_f(\nu_s)$, f_1 and f_2. Suppose that these functions are known and initial and boundary conditions are proposed such that the resulting IBVP is well-posed; then we are, in principle, in the position to solve the field equations for the variables $\{\nu_s, \mathbf{v}_s, \mathbf{v}_f\}$.
  Obviously, [**A23**] is a weaker assumption than that of 'pressure equilibrium', because it does not rule out hypo-plasticity and the configuration pressure β^G_s. The price we have to pay for [**A23**], i. e. for the gain of sensitivity of the model, is the need of an additional postulate for the Ψ^G_I and, presumably, the increased complexity of the resulting IBVP.
- According to $(8.127)_{1,2}$, 'pressure equilibrium' is recovered in this formulation only when

  (i) $\rho_s = \rho_f$ and $\beta^G_s = 0$,
  (ii) for $\rho_s \neq \rho_f$ when $\Psi^G_I \neq \hat{\Psi}^G(\nu_s, \bar{\mathbf{Z}}_s, \cdot)$.

  The second case is equivalent to (8.122), the first is not realistic for soil.

Chapter 9
Discussions and Conclusions

Abstract We review the essentials of the theory, discuss the highlights of the achievements and the limitations of the model equations, and give an outlook of some still unsolved problems.

Review

The thermodynamic analysis has been performed for a mixture of which the constitutive relations are of the class (4.37) or [**A7**], viz.,

$$\mathsf{C} = \hat{\mathsf{C}}(\underbrace{\theta,\ \dot{\theta},\ \nabla\theta}_{(i)},\ \underbrace{\vec{\rho},\ \nabla\vec{\rho}}_{(ii)},\ \underbrace{\vec{\nu},\ \nabla\vec{\nu}}_{(iii)},\ \vec{\mathbf{v}},\ \vec{\mathbf{B}},\ \vec{\mathbf{D}},\ \vec{\mathbf{W}},\ \vec{\mathbf{Z}})\ . \tag{9.1}$$

Consideration of the thermal variables (i) accounts for effects of heat conduction and thermo-elasto-visco-plastic behaviour (when $\dot{\theta}$ is incorporated we are also accounting for a finite speed of thermal pulses). Even though purely mechanical processes were in focus, this makes the general model equations also applicable for mixture flows in which heat transport may be important. In geological applications such as pyroclastic gravity currents – a mixture of hot debris and air with heat exchange to the atmosphere – the processes may, perhaps, have to be described by a model of this complexity. In this model, the air and the hot debris, however, have the same temperature. A formulation in which the various constituents are given by different temperatures is not contained in the results but may, perhaps, be necessary when describing e. g. lahars.

Consideration of the true densities (given by (ii) in (9.1)) as independent constitutive variables accounts for compressibility of the constituents. The thermodynamic pressures (6.101) are the corresponding stress variables. Incorporation of the compressibility is, however, hardly necessary for flows envisaged in the context of geophysical applications. A possible application may be a lahar in air which suffers during its motion such a large temperature

L. Schneider, K. Hutter, *Solid-Fluid Mixtures of Frictional Materials in Geophysical and Geotechnical Context*, Advances in Geophysical and Environmental Mechanics and Mathematics, DOI 10.1007/978-3-642-02968-4_9,

drop that the thermal volume expansion of the air can no longer be ignored. In snow avalanches a variable true density of the snow as an independent constitutive variable may open the possibility to account for compaction of the snow balls in motion. In a particle-laden turbulent boundary layer flow such as a powder avalanche or a pyroclastic flow from a volcano, supersonic conditions may develop in the gas of the flow, so that its compressibility can not be ignored. These effects can be accounted for by the dependencies (*ii*) in (9.1). Incorporation of the constituent volume fraction dependence (given by (*iii*) in (9.1)) allows a dynamic description of the internal redistribution of the constituents. The configurational pressures (6.102) that are due to this dependence are the driving forces to achieve such redistributions. Omission of both, (*ii*) and (*iii*) in (9.1), as independent variables eliminates, among other things, these pressures, so that in this case the saturation pressure (6.103) remains the only true pressure variable in the model. The dependence on $\vec{\mathbf{v}}$, $\vec{\mathbf{D}}$, $\vec{\mathbf{W}}$, and $\vec{\mathbf{Z}}$ allows that viscous and plastic effects are modelled; indeed, it was demonstrated that a dependence on $\vec{\mathbf{v}}$ resulted in non-equilibrium interaction forces that depend e. g. on the constituent difference velocities, and the incorporation of $\vec{\mathbf{D}}$ and $\vec{\mathbf{Z}}$ together with evolution equations for $\vec{\mathbf{Z}}$ led to constitutive relations for the CAUCHY stress tensor, which, by selecting the tensorial production rate density $\mathbf{\Phi}_\alpha$ in the stress like evolution equation (4.36), [**A4**], yields theories of which hypo-plasticity is one of many. It should further be noted that both dependencies on $\vec{\mathbf{B}}$ and $\vec{\mathbf{Z}}$ also give rise to equilibrium stress contributions, provided the internal free energy Ψ_I^G depends on $\vec{\mathbf{B}}$ and $\vec{\mathbf{Z}}$, see (7.100). The dependence on $\vec{\mathbf{B}}$ allows elastic effects to be modelled, but their incorporation needs to be carefully done for density-preserving constituents, since ρ_α and $\mathbf{B}_\alpha$ are related to one another when the constituent mass production rate density c_α of that constituent vanishes, see (4.21). In that case the free internal energy should depend on $\mathrm{I}_{\mathbf{B}_\alpha}$, $\mathrm{II}_{\mathbf{B}_\alpha}$, but not on $\mathrm{III}_{\mathbf{B}_\alpha}$, (see 8.37). Note further, that equation (9.1), and therefore our theory and the results it generates covers a large class of elasto-visco-plastic mixtures, but not all which have been proposed in the debris flow literature of single constituent continua. There are granular flow models, in which also objective time derivatives of $\vec{\mathbf{D}}$, the so-called RIVLIN-ERICKSEN tensors, have been introduced as additional independent constitutive variables. A thermodynamic theory for such mixtures has, however, not yet been developed.

The focus in the above applications has been on binary mixtures and 'reduced' theories, in which the mass and volume fraction production rate densities are set to zero. In mixtures for which the mass production rate density for constituent K_α does not vanish, $c_\alpha \neq 0$, formula (4.20) shows that density-preserving does not imply volume-preserving, because (4.20) reduces in this case to

$$(\det\mathbf{F}_\alpha \nu_\alpha)' = (\det\mathbf{F}_\alpha \nu_\alpha)\, c_\alpha \ , \tag{9.2}$$

which integrates to

$$(\det\mathbf{F}_\alpha\nu_\alpha) = (\det\mathbf{F}_{\alpha\,0}\nu_{\alpha\,0})\exp\left\{\int_{t_0}^{t} c_\alpha(\mathbf{x}(t'),t')dt'\right\} \tag{9.3}$$

with $\mathbf{F}_{\alpha\,0} = \mathbf{1}$. Here, t_0 is the initial time and $\nu_{\alpha\,0}$, $\mathbf{F}_{\alpha\,0}$ are the initial volume fraction and deformation gradient. This computation shows that in mixtures with non-vanishing mass production rate densities, 'density-preserving' and 'volume-preserving' of a particular constituent are distinct concepts, leading likely to distinct inferences. In particular, when $c_\alpha \neq 0$, the assumption of density-preserving still eliminates the density variables (*ii*) in (9.1), but parameterization of elastic effects does not constrain the form of the HELMHOLTZ-like free energy as a function of the invariants $\mathrm{I}_{\mathbf{B}_\alpha}$, $\mathrm{II}_{\mathbf{B}_\alpha}$, $\mathrm{III}_{\mathbf{B}_\alpha}$. Of course, the modeller is free to use a free energy parameterization, which agrees with that valid for $c_\alpha = 0$, but such a choice is not necessary. However, it may be advantageous, because it automatically merges into the required form when $c_\alpha = 0$. In either case, the volume fraction changes may be obtained from (9.3) by setting $\det\mathbf{F}_{\alpha\,0} = 1$.

Achievements

In this work we derived a general, thermodynamic consistent, multiphase model that accounts for saturation, mass-interaction and the resistance to volume changes of some or all constituents comprising the mixture. In addition, we extended the classical mixture theory by a new system of evolution equations and constitutive laws that are thought to predict, first, the evolution of internal structures present in immiscible mixtures and second, the plastic behaviour of solid constituents. The latter extension follows the hypoplastic single-material theory proposed by SVENDSEN et al. [116] and is, as far as we know, one of the first attempts to incorporate hypo-plasticity in a thermodynamical consistent way into mixture theories. This is obviously a rewarding result; however, the thermodynamic model presented here delivers also a straightforward method to generalize the modelling concept of hypo-plasticity for fluid-solid mixtures with more than one solid constituent. Such extensions are necessary if one intends to describe dynamic processes in which particle size separation ought to be described. When starting this research on dynamics of granular materials, it was not clear to us, how a hypo-plastic constitutive model of a mixture with more than one fluid and one solid component should be described. This is now absolutely clear, at least within the context of the constitutive class (9.1). In fact, the thermodynamic analysis performed in Chaps. 3-7 has been done for such a general case – except for the explicit constitutive parameterization. The remaining step as such is now straightforward, even though it is by no means easy.

As a geologically significant problem, consider a dynamical theory for a landslide, which allows determination of dynamical particle size separation. A mixture model for such a situation could consist of N_s solid constituents, each representing disjoint ranges of particle diameters. For these components, the formulations of Chaps. 3-7 yield well defined field equations, yet the 'crux', left for future research, will be to specify the parameterization for the free energy and the frictional stresses accounting for interaction forces, which favour the separation of the particle sizes. This will most likely mean that the 'principle of phase separation' is not valid. When a fluid component is present, such a model will also allow modelling abrasion and/or fragmentation as a dynamical process of the mixture.

All these results were achieved with the aid of a form of the Second Law of Thermodynamics, which was 'flexible' in its underlying basic postulates, but still sufficiently restricted to allow useful practical inferences for the emerging constitutive statements. More precisely, we formulated the Second Law of Thermodynamics as an entropy principle, and we requested as expression of the irreversibility of the Second Law of Thermodynamics that the entropy production rate density be non-negative for all processes which obey the field equations for the postulated constitutive class (here (9.1)) plus constraint conditions of saturation, density-preserving and field equations of internal variables. Following MÜLLER, the entropy flux vector and the coldness function (as a generalization of the absolute temperature) were not treated as primitive concepts, but rather as quantities to be delivered as inferences from the exploitation of the entropy principle. This concept stands in contrast to the approach of the Second Law of Thermodynamics via the CLAUSIUS-DUHEM inequality, in which entropy flux and absolute temperature are given by *a priori estimates.*The results obtained with the more flexible version of MÜLLER's entropy principle prove that this enhanced flexibility is compelling, because the exploitation of the entropy principle via the so-called LIU identities shows (with the aid of a number of simplifying ad-hoc assumptions) expressions for the mixture entropy flux vectors $\mathbf{k}$ and $\boldsymbol{\phi}^{\rho\eta}$, respectively, which are in conflict with results that could be obtained with the CLAUSIUS-DUHEM inequality. The results show that in contrast to the CLAUSIUS-DUHEM inequality postulate, $\boldsymbol{\phi}^{\rho\eta}$ does not only depend on the heat flux vector, but also on a linear combination of the constituent diffusion velocities. Furthermore, with the exploitation of the LIU identities and the assumption that the LAGRANGE parameter λ^{ε} would depend only on the temperature and its material time derivative, $\lambda^{\varepsilon} = \lambda^{\varepsilon}(\theta, \dot{\theta})$, all LAGRANGE parameters except one could be explicitly expressed in terms of constitutive quantities. The exception is the nth solid volume fraction which is related to the saturation pressure, the constraint variable maintaining the saturation condition.

One further revealing result, obtained by the exploitation of the entropy inequality, in particular the identical satisfaction of the LIU identities, is, how the equilibrium values of the interaction forces, entropy, heat flux vector and CAUCHY stress tensor depend on the production rate densities of constituent mass, volume fraction and frictional stress variables. The Second Law of Thermodynamics has disclosed a beautiful structure of the contributions of these sub-processes to those variables, summarised in the formulae (7.49), (7.50). According to these, the equilibrium contributions to the interaction forces, entropy, heat flux vector and CAUCHY stresses require non-vanishing derivatives of the production rate densities $\bar{\mathbf{\Phi}}_\alpha$, c_α and n_α in equilibrium with respect to the constituent velocities, time derivative of the temperature, temperature gradient and constituent stretching tensors (see (7.50)). Moreover, the weights by which these contributions affect $\bar{\mathbf{m}}^i_\beta\big|_{\mathrm{E}}$, $\eta\big|_{\mathrm{E}}$, $\mathbf{q}|_{\mathrm{E}}$ and $\bar{\mathbf{T}}_\alpha\big|_{\mathrm{E}}$, are fixed, once the HELMHOLTZ free energy is prescribed, see (7.49). The respective formulae (7.31), (7.36), (7.38), (7.41), (7.44) via (7.49), (7.50) make it very clear, which specializations of the constitutive behaviour are required to simplify the expressions of $\bar{\mathbf{m}}^i_\beta\big|_{\mathrm{E}}$, $\eta\big|_{\mathrm{E}}$, $\mathbf{q}|_{\mathrm{E}}$ and $\bar{\mathbf{T}}_\alpha\big|_{\mathrm{E}}$. This result indicates e. g. that possible transfer of certain terms of the constituent equilibrium interaction forces into the constituent equilibrium CAUCHY stresses should not be made, because they may destroy the structure of the formulae obtained via the Second Law of Thermodynamics.

The reduction of the general mixture theory to a binary, isothermal mixture model that accounts for two inert and density-preserving constituents of solid and fluid aggregation led to a rich structure of model equations. More specifically, the thermodynamic approach has led to expressions for the solid and fluid equilibrium and non-equilibrium constituent stresses and interaction momentum production rate densities, which are functionally determined. Depending on the specific choice of the former, elastic and hypo-plastic frictional effects are included in a functionally well defined form. Identification of the non-equilibrium pressures for the solid and fluid stresses and interaction force density has been implemented by following the methods of viscometry of viscous fluids with plastic yield and regularization procedures that guarantee vanishing of the non-equilibrium quantities in thermodynamic equilibrium. This demonstrated both the restrictions, which are imposed on the stress parameterizations by the simple shear configuration in standard viscometry, and the relative complexity of the identification of the parameters of this simple deformation.

These thermodynamic results also shed light on a popular *ad hoc* assumption, called 'pressure equilibrium'. The results indicate that this assumption must be rejected for the modelling of multiphase mixtures, because it prevents, first, the existence of configuration pressures which are essential for this kind of mixtures and second, eliminates all hypo-plastic effects. Following HUTTER et al. [63] a more restrictive assumption was introduced that avoids this collapse of the debris flow model.

Limitations and Outlook

Obvious limitations of the mixture model presented in this book are the many assumptions laid down as $[\mathbf{Aj}]$, $j = 1, 2, \ldots, 23$. Some of these affect the physical conditions to which the final model equations apply; some are made to simplify the mathematics or to streamline the deduction of explicit formulae which can better be interpreted in terms of physics. Perhaps, future research will show how some of these assumptions can be weakened or completely dispensed with. Such technical assumptions are e. g. [**A11**] through [**A15**]. Assumptions which restrict the physics are e. g. [**A2**] and [**A3**]. They restrict the model equations to continua with constituents having the same temperature and no changes of the aggregation states. These are limitations which exclude proper treatment of avalanching landslides from e. g. volcanoes in which melting and solidification of the solid or molten constituents can occur. A situation which is conceptually analogous, but not included in the theory, is the motion of a pyroclastic landslide along a glacier surface, in which the heat melts and evaporates the ice at the surface and incorporates these components via entrainment mechanisms into the avalanche.

The fact that aggregation changes are excluded in the model does not entirely exclude all mass changes between the constituents. If the solid constituents are characterized by disjoint particle size ranges, then fragmentation can, in principle, be handled as a process within this theory. However, the prerequisite for it to be applicable is that the energy which is dissipated into heat by the fracturing is small. Similarly, abrasion may be modelled by a small mass transfer from the abrading constituents to the component with the smallest size (clay, silt in a slurry).

Assumption [**A3**], stating that the constituent production rate of angular momentum in the model is given by the moment of the production of linear momentum of the same constituent, is technical in form – it makes all constituent CAUCHY stresses symmetric and thus limits considerations to non-polar theories – but is certainly physically doubtful. Rubbing friction between particles (boulders in the application to landslides); this is, between three-dimensional bodies transfers angular momentum between the bodies in contact. It follows in reality that encounters between particles redistribute the rotational and translational energies, which is not consistent with the form of [**A3**]. However, any deviation from [**A3**] requires polar formulations which are yet an order of magnitude more complex than the model in this book.

It can not be denied that incorporation of dry friction between the particles was introduced into the model only indirectly through symmetric tensor variables, for which evolution equations were postulated. This procedure was required because a thermodynamic setting of hypo-plasticity could only be achieved using such an 'indirect' approach. It is certainly a disadvantage, if the modeling of the hypo-plastic stress contribution is considered only in this somewhat awkward way. All derivations of the classical theories of hypo-plasticity postulate an evolution equation for the CAUCHY stress tensor. Future theories of elasto-visco-plastic solid-fluid mixtures should attempt to write down such evolution equations and incorporate these in the thermodynamic formulation. If such a formulation can be pursued with the MÜLLER

& LIU entropy approach, the resulting model equations will perhaps be less complex. In the past we have not been successful in any such attempt.

The above discussion indicates that considering the applicability of the derived model equations to common geophysical mass flow problems leads to a more realistic description of rapid granular fluid flows. From the present work it transpires that the derivation of thermodynamically consistent generalizations may be difficult, but nevertheless fairly straightforward. More urgent, however, is the application of the simplest form of these equations to a realistic situation of rapid flow down arbitrary topography. At this point we are still not able to estimate the adequacy and accuracy of the model, because we are still lacking explicit functional forms for some of the constitutive quantities such as Ψ_{sf}, $\mathbf{L}$, $\mathbf{N}$, m_{D}, κ_f, μ_f, λ_s and μ_s. The present work can only be regarded as a foundation or starting point for a mathematical debris flow model, because to deduce a numerical model that allows the reproduction of realistic debris flow events, more sophisticated mathematical refinements are necessary. First, the full field equations (with accelerations) have to be formulated in topography following coordinates, such that, second, a thin film (shallowness) approximation with respect to these coordinates can be performed. In a third step one can introduce a depth integration to reduce the, then, two-dimensional problem, to a one-dimensional problem. For the completion of the mathematical model we still have to face the challenge of finding appropriate boundary conditions, i. e. we have to deduce representations for the basal friction and the entrainment rate (e. g. erosion, water-infiltration etc.). Thus, for this type of model, there is still a long way to go to finally gain a deeper insight into the physics of debris flow initiation, evolution and deposition.

Appendix A
A Primer on Exterior Calculus

This Appendix summarizes the formal theory of Exterior Calculus and follows very closely the approximately first 130 pages of the book by EDELEN [35].

We assume that the reader is familiar with the basic concepts of (linear) algebra and follow in the subsequent formal analysis the approach taken by EDELEN in his book on 'Applied Exterior Calculus' [35]. So, let $\mathcal{E}_n$ be an n-dimensional Euclidean space and $\mathcal{S} \in \mathcal{E}_n$ a subset or region of $\mathcal{E}_n$. Let, moreover, $\mathcal{T}(\mathcal{E}_n)$ be the tangent space of $\mathcal{E}_n$ and $\{x^i\}$ its Cartesian coordinate cover.

Definition 2.1: A *vector field* V on a region $\mathcal{S}$ of $\mathcal{E}_n$ is a smooth (C^∞)-map

$$V:\ \mathcal{S} \to \mathcal{T}(\mathcal{S}) \ \mid\ v^i = v^i(x^j) \quad i = 1, \dots, n \tag{A.1}$$

such that evaluation at $(x^j) \in \mathcal{S}$ yields $\{v^i(x^j)\} \in \mathcal{T}(\mathcal{S})$. ◊

Let $\Lambda^0(\mathcal{E}_n)$ be the vector space of all (C^∞)-functions which form a commutative and associative algebra. This means that if $u^i(x^j)$, $v^i(x^j)$ are vector-fields as given in the above definition and $a(x^i)$ and $b(x^i)$ are (C^∞)-functions with $\mathcal{E}_n$ as domain, then

$$w^i(x^j) = a(x^i)u^i(x^j) + b(x^i)v^i(x^j) \tag{A.2}$$

is again an element of $\mathcal{T}(\mathcal{E}_n)$. It is seen that with this understanding $\Lambda^0(\mathcal{E}_n)$ is closed under vector addition and multiplication of vectors with scalars.

In what follows the operator representation of vector fields is of significance. To see this, recall that for n real-valued (C^∞)-functions $\gamma^i(t)$, $i = 1, \dots, n$, the assignment

$$\Gamma:\ \mathrm{J} \to \mathcal{E}_n \ \mid\ x^i = \gamma^i(t), \quad \gamma^i(0) = x^i_0, \quad \mathrm{J} \in \mathbb{R} \tag{A.3}$$

defines a smooth (C^∞)-function in $\mathcal{E}_n$. If, moreover, $\mathrm{F} : r = f(x^1, \dots, x^n) = f(x^i)$, where f is a smooth function, then $\hat{\mathrm{F}}(t) = f(\gamma^i(t))$ evaluates the

L. Schneider, K. Hutter, *Solid-Fluid Mixtures of Frictional Materials in Geophysical and Geotechnical Context*, Advances in Geophysical and Environmental Mechanics and Mathematics,
DOI 10.1007/978-3-642-02968-4_BM2,

function along the curve Γ; one may then also calculate its time rate of change,

$$\frac{d\hat{\mathrm{F}}}{dt} = \frac{d\gamma^i(t)}{dt} \frac{\partial f(x^j)}{\partial x^i}\bigg|_{x^i=\gamma^i(t)} . \tag{A.4}$$

Observe the occurrence of the coefficients $\{d\gamma^i(t)/dt\}$ on the right-hand side of (A.4), which is the velocity field in the direction of Γ. This suggests that vector fields can be defined in a manner similar to (A.4). To this end, we first simplify notation by introducing the abbreviation $\partial_i := \partial/\partial x^i$. Defining the operator H by

$$\mathrm{H} := h^i(x^j)\partial_i \tag{A.5}$$

its action on f is well defined,

$$\mathrm{H}\langle f\rangle\langle x^j\rangle = h^i(x^j)\partial_i f(x^j) = g(x^j) . \tag{A.6}$$

It follows from the definition (A.5) that if $f, g \in \Lambda^0(\mathcal{E}_n)$ and $a, b \in \mathbb{R}$,

$$\begin{aligned} \mathrm{H}\langle af + bg\rangle &= a\mathrm{H}\langle f\rangle + b\mathrm{H}\langle g\rangle , \\ \mathrm{H}\langle (f,g)\rangle &= (\mathrm{H}\langle f\rangle, g) + (f, \mathrm{H}\langle g\rangle) . \end{aligned} \tag{A.7}$$

Definition 2.2: An operation H on an algebra $\mathcal{A}$ is a *derivation* if and only if H maps $\mathcal{A}$ to $\mathcal{A}$ and satisfies (A.7). ◊

With these prerequisites we have

Definition 2.3: Let $\{v^i(x^j)\}$ be a (C^∞)-vector field on $\mathcal{E}_n$. Its *operator representation* is defined as the operator

$$\mathrm{V} = v^i(x^j)\partial_i \tag{A.8}$$

on the algebra $\Lambda^0(\mathcal{E}_n)$ of (C^∞)-functions on $\mathcal{E}_n$. (A.8) is called a derivation, and evaluation of V for f is given by

$$\mathrm{V}\langle f\rangle = v^i(x_j)\partial_i f . \tag{A.9}$$

◊

The simplest function to which (A.8) can be applied is x^k, for which $\partial_i x^k = \delta_i^k$ ($= 1$ for $k = i$, $= 0$ for $k \neq i$). It follows directly from (A.8) that

$$\mathrm{V}\langle x^k\rangle = v^i(x^j)\partial_i x^k = v^i(x^j)\delta_i^k = v^k(x^j) . \tag{A.10}$$

Clearly, any derivation $\mathrm{V} = v^i(x^j)\partial_i$ reproduces its coefficients $\{v^i(x^j)\}$ this way. This fact may also be stated as the following

Lemma 2.1: *Let* V *be an operator representation of a vector field* $\{v^i(x^j)\}$ *on* $\mathcal{E}_n$. *Then* V *is uniquely determined by the action on the* n *functions* $x^1, x^2, \ldots, x^n$,

$$\mathrm{V} = v^i(x^j)\,\partial_i = \mathrm{V}\langle x^j\rangle(x^j)\,\partial_i\ , \tag{A.11}$$

where $\mathrm{V}\langle x^j\rangle(x^j)$, *denotes the function that is obtained by allowing* V *to act on* x^i. *Thus,* $v^i(x^j) = \mathrm{V}\langle x^i\rangle(x^j)$, $i = 1, \ldots, n$, *and hence an operator representation of a vector field on* $\mathcal{E}_n$ *reproduces the vector field on* $\mathcal{E}_n$. □

This lemma makes it quite clear that any derivation on the algebra $\Lambda^0(\mathcal{E}_n)$ determines a vector field on $\mathcal{E}_n$. Furthermore, the tangent space $\mathcal{T}(\mathcal{E}_n)$ is the collection of all derivations on the algebra $\Lambda^0(\mathcal{E}_n)$. Moreover,

Lemma 2.2: *The* n *vector fields* $\partial_1, \partial_2, \ldots, \partial_n$ *constitute a basis for* $\mathcal{T}(\mathcal{E}_n)$. *This basis is called the natural basis of* $\mathcal{T}(\mathcal{E}_n)$ *with respect to the* (x^i) *coordinate cover of* $\mathcal{E}_n$. □

Proof: This is quite easy: Take $(\sum_{i=1}^n c_i\partial_i)\,\langle f\rangle = 0$ for any c_i and for $f = x^k$. Then, $(\sum_{i=1}^n c_i\partial_i)\,\langle x^k\rangle = c_k = 0$ for any $k \in (1, 2, \ldots, n)$, proving Lemma 2.2. □

Note that in view of (A.9) the equation

$$\mathrm{V}\langle f\rangle = 0 \quad \Rightarrow \quad v^i(x^j)\frac{\partial f}{\partial x^i} = 0 \tag{A.12}$$

can be viewed as a first order linear partial differential equation. With this interpretation $\mathrm{V} = v^i(x^j)\partial_i$ is then called the *characteristic equation* to the partial differential equation $\mathrm{V}\langle f\rangle = 0$.

The natural Lie algebra of $\mathcal{T}(\mathcal{E}_n)$**:** Recall that a vector space equipped with a binary operation $(\cdot,\cdot)$ becomes an algebra. The tangent space is the vector space whose elements are the derivations. Thus, the question arises as to whether a binary operation can be defined over $\mathcal{T}(\mathcal{E}_n) \times \mathcal{T}(\mathcal{E}_n)$ for which $\mathcal{T}(\mathcal{E}_n)$ becomes an algebra.

Given two derivations, U and V, one may compute

$$\begin{aligned} \mathrm{U}\langle \mathrm{V}\langle f\rangle\rangle &= \mathrm{U}\langle v^j\partial_j f\rangle = u^i\partial_i\left(v^j\partial_j f\right) \\ &= \underbrace{\underbrace{u^i\partial_i v^j}_{w^j}\partial_j f}_{=\text{derivation}} + \underbrace{u^i v^j\partial_i\partial_j f}_{\neq\text{derivation}} \end{aligned} \tag{A.13}$$

and thus sees, since the right-hand side is not expressible as a derivation, that $\mathrm{U}\langle \mathrm{V}\langle f\rangle\rangle$ does not define a binary operation on $\mathcal{T}(\mathcal{E}_n) \times \mathcal{T}(\mathcal{E}_n)$ to $\mathcal{T}(\mathcal{E}_n)$. However, it is easy to show that

$$\mathrm{W}\langle f\rangle := \mathrm{U}\langle \mathrm{V}\langle f\rangle\rangle - \mathrm{V}\langle \mathrm{U}\langle f\rangle\rangle = \underbrace{\left(u^i\partial_i v^j - v^i\partial_i u^j\right)\partial_j f}_{=\text{derivation}} + \underbrace{\left(u^i v^j - u^j v^i\right)\partial_i\partial_j f}_{=0}, \tag{A.14}$$

which *is* a derivation. The second term on the right-hand side vanishes since it is the contraction of a skew-symmetric with a symmetric tensor. One may conclude

Lemma 2.3: *The binary operation* $[\cdot,\cdot]$, *defined by*

$$[\mathrm{U},\mathrm{V}]\langle f\rangle = \mathrm{U}\langle \mathrm{V}\langle f\rangle\rangle - \mathrm{V}\langle \mathrm{U}\langle f\rangle\rangle \tag{A.15}$$

for all $f \in \Lambda^0(\mathcal{E}_n)$ *has the representation*

$$[\mathrm{U},\mathrm{V}] = (u^j\partial_j v^i - v^j\partial_j u^i)\partial_i \tag{A.16}$$

for every $\mathrm{U},\mathrm{V} \in \mathcal{T}(\mathcal{E}_n)$. *Thus,* $[\cdot,\cdot]$ *defines a binary map of* $\mathcal{T}(\mathcal{E}_n) \times \mathcal{T}(\mathcal{E}_n)$ *to* $\mathcal{T}(\mathcal{E}_n)$. □

The multiplication operation $[\cdot,\cdot]$ is called *Lie product.* The following facts on the LIE product hold:

Lemma 2.4: *The* LIE *product satisfies the following identities:*

(a) $$[\mathrm{U},\mathrm{V}] = -[\mathrm{V},\mathrm{U}], \tag{A.17}$$

(b) $$[\mathrm{U},a\mathrm{V}+b\mathrm{W}] = a[\mathrm{U},\mathrm{V}] + b[\mathrm{U},\mathrm{W}], \tag{A.18}$$

$$[a\mathrm{U}+b\mathrm{V},\mathrm{W}] = a[\mathrm{U},\mathrm{W}] + b[\mathrm{V},\mathrm{W}], \tag{A.19}$$

(c) $$[\mathrm{U},[\mathrm{V},\mathrm{W}]] + [\mathrm{V},[\mathrm{W},\mathrm{U}]] + [\mathrm{W},[\mathrm{U},\mathrm{V}]] = 0 \tag{A.20}$$

for all $\mathrm{U},\mathrm{V},\mathrm{W} \in \mathcal{T}(\mathcal{E}_n)$ *and all* $a, b \in \mathbb{R}$. *The property (A.20) is called Jacobi identity.* □

The *proof* of (a) follows directly from (A.15), (b) and (c) can be verified by performing the computations in long hand. The computation for the verification of (c) is straightforward but long. □

The algebra that is based on the binary operations in (a) and (c) of Lemma 2.4 is called a LIE algebra.

Exterior forms: Consider again, as before, the Euclidean space $\mathcal{E}_n$ with Cartesian coordinate cover $\{x^i\}$ and tangent space $\mathcal{T}(\mathcal{E}_n)$. We have made clear above that $\mathcal{T}(\mathcal{E}_n)$ is a LIE algebra of derivations on $\Lambda^0(\mathcal{E}_n)$ with the basis fields $\{\partial_i, i = 1,\ldots,n\}$. To define exterior forms the concept of the dual space $\mathcal{T}^*(\mathcal{E}_n)$ to $\mathcal{T}(\mathcal{E}_n)$ is needed.

Definition 2.4: The dual space $\mathcal{T}^*(\mathcal{E}_n)$ of $\mathcal{T}(\mathcal{E}_n)$ is the set of all functions $\langle\cdot,\cdot\rangle$ with the properties

$$\text{(a)} \qquad \omega : \mathcal{T}(\mathcal{E}_n) \to \mathbb{R} \mid r = \langle \omega, \mathrm{V} \rangle \ , \tag{A.21}$$

$$\text{(b)} \qquad \langle \omega, f\mathrm{U} + g\mathrm{V} \rangle = f\langle \omega, \mathrm{U} \rangle + g\langle \omega, \mathrm{V} \rangle \ , \tag{A.22}$$

$$\text{(c)} \qquad \langle f\omega + g\rho, \mathrm{V} \rangle = f\langle \omega, \mathrm{V} \rangle + g\langle \rho, \mathrm{V} \rangle \ , \tag{A.23}$$

for all $\omega, \rho \in \mathrm{T}^*(\mathcal{E}_n)$, all $\mathrm{U}, \mathrm{V} \in \mathcal{T}(\mathcal{E}_n)$ and all $f, g \in \Lambda^0(\mathcal{E}_n))$. ◊

In this definition $\omega \in \mathcal{T}^*(\mathcal{E}_n)$ and $\mathrm{V} \in \mathcal{T}(\mathcal{E}_n)$ are both fields defined on $\mathcal{E}_n$, a fact which may be expressed as $\omega(x^j)$ and $\mathrm{V}(x^j)$. Similarly, the value of $\omega(x^j)$ is given by $\langle \omega(x^j), \mathrm{V}(x^j) \rangle = h(x^j)$ with $h(x^j)$ an element of $\Lambda^0(\mathcal{E}_n)$. In this spirit one may define

$$\langle\cdot,\cdot\rangle : \mathcal{T}^*(\mathcal{E}_n) \times \mathcal{T}^(\mathcal{E}_n) \to \Lambda^0(\mathcal{E}_n) \mid \langle \omega(x^j, \mathrm{V}(x^j)) \rangle = h(x^j) \ . \tag{A.24}$$

Finally, the properties (b) and (c) in the above definition demonstrate closure under vector additions and multiplication with scalars so that $\mathcal{T}^*(\mathcal{E}_n)$ is indeed a vector space. One must equip $\mathcal{T}^*(\mathcal{E}_n)$ with a basis. This is established by

Definition 2.5: A collection of n fields $\{\theta^i, i = 1, 2, \ldots, n\}$ is a *natural basis* for $\mathcal{T}^*(\mathcal{E}_n)$ (with respect to the (x^i) coordinate cover) if and only if

$$\langle \theta, \partial_j \rangle = \delta^i_j \ . \tag{A.25}$$

◊

This definition of $\theta^i, i = 1, \ldots, n$ is abstract and does not disclose its meaning. EDELEN [35] writes: 'The fact that we can define something does not necessarily mean that we can use it'. On the other hand, if a solution to (A.25) exists, then it is unique. Indeed, let θ^i and ρ^i be two solutions; then, in view of (A.25),

$$\langle \theta^i, \partial_j \rangle - \langle \rho^i, \partial_j \rangle = \langle \theta^i - \rho^i, \partial_j \rangle = 0 \ . \tag{A.26}$$

Thus, for every choice of (C^∞)-functions $\{v^i(x^k)\}$ we have

$$v^j \langle \theta^i - \rho^i, \partial_j \rangle = \langle \theta^i - \rho^i, \mathrm{V} \rangle = 0, \quad \text{for all} \quad \mathrm{V} \in \mathcal{T}^*(\mathcal{E}_n); \tag{A.27}$$

thus, since the zero element $0^* \in \mathcal{T}(\mathcal{E}_n)$ is unique, we have $\theta^i - \rho^i = 0^*$, which establishes uniqueness.

To establish existence, let us construct a solution explicitly: For any $\mathrm{F} \in \Lambda^0(\mathcal{E}_n)$ and any $\mathrm{V} = v^i\partial_i \in \mathcal{T}(\mathcal{E}_n)$ we know that $\mathrm{V}\langle f \rangle = v^i \partial f / \partial x^i$ takes its values in $\mathbb{R}$. Thus, $\mathrm{V}\langle f\rangle$ may be viewed as a linear functional on $\mathcal{T}(\mathcal{E}_n)$ since it has the required linearity properties $\mathrm{V}\langle f+g\rangle = \mathrm{V}\langle f\rangle + \mathrm{V}\langle g\rangle$, $(\mathrm{U}+\mathrm{V})\langle f\rangle = \mathrm{U}\langle f\rangle + \mathrm{V}\langle f\rangle$.

Having shown existence of $\mathrm{V}\langle f\rangle$ as an element of $\mathcal{T}^*(\mathcal{E}_n)$ for all $\mathrm{V}\in\mathcal{T}(\mathcal{E}_n)$, let us now introduce a different notation for $\mathrm{V}\langle f\rangle$.

Definition 2.6: The symbol '*df*', defined by

$$\langle df, \mathrm{V}\rangle = \mathrm{V}\langle f\rangle \tag{A.28}$$

for all $\mathrm{V}\in\mathcal{T}(\mathcal{E}_n)$ is an element of $\mathcal{T}^*(\mathcal{E}_n)$ for each $f\in\Lambda^0(\mathcal{E}_n)$. ◊

This is an abstract definition of the symbol '*df*', saying only the outcome of it when it is applied to a vector field V through $\langle\cdot,\cdot\rangle$. More light into its meaning is brought by the following

Lemma 2.5: *The unique natural dual basis of $\mathcal{T}^*(\mathcal{E}_n)$ is given by the n elements $\{dx^i, i=1,\ldots,n\}$. Any element $\omega\in\mathcal{T}^*(\mathcal{E}_n)$ can be written uniquely as*

$$\omega = \omega_i(x^j)dx^i, \tag{A.29}$$

where the coefficients $\{\omega_i(x^j)\}$ are determined by

$$\omega_i = \langle\omega, \partial_i\rangle \quad i = 1,\ldots,n \tag{A.30}$$

and thus belongs to $\Lambda^\circ(\mathcal{E}_n)$. □

Proof: The quantities dx^i are defined via (A.28) by $\langle dx^i, \mathrm{V}\rangle = \mathrm{V}\langle x^i\rangle$ and therefore $\langle dx^i, \partial_j\rangle = \partial_j x^i = \delta^i_j$. This establishes the existence of the n quantities dx^i and identifies them with the θ_i's in (A.25), which have already been shown to be a basis for the n-dimensional space $\mathcal{T}^*(\mathcal{E}_n)$. Consequently, any $\omega\in\mathcal{T}^*(\mathcal{E}_n)$ can be expressed as (A.29) with components shown in (A.30). □

All this is formal, but does not yet establish a concrete realization of the abstract symbol *df*. If one applies *df* to the basis $\{\partial_i\}$, then $\langle df, \partial_i\rangle = \partial_i f$ is obtained. Alternatively, $df\in\mathcal{T}^*(\mathcal{E}_n)$ can also be written as $df = \gamma_i(x^k)dx^i$ with $\gamma_i(x^k) = \langle df, \partial_i\rangle = \partial_i f(x^k)$, implying that

$$df = \partial_i f(x^k)dx^i\ . \tag{A.31}$$

This is the familiar total differential of the function f of the n variables $x^1, x^2, \ldots, x^n$. This result is so important that we summarize it as

Lemma 2.6: *The abstract element $df\in\mathcal{T}^*(\mathcal{E}_n)$ that is defined by (A.28) for any $f\in\Lambda^0(\mathcal{E}_n)$ has the realization*

$$df = \partial_i f(x^k) dx^i \tag{A.32}$$

as the total differential of the function $f(x^k)$. □

In elementary calculus, if f and $\{x^i\}$ are viewed as functions of the variable t, (A.32) is shown in the form when both sides of (A.32) are divided by dt,

$$\frac{df}{dt} = \partial_i f(x^k) \frac{dx^i}{dt} \ . \tag{A.33}$$

In this form it appears as the chain rule of differentiation for $f : \mathcal{E}_n \to \mathbb{R}$ and $\{x^i(t)\} : \mathbb{R} \to \mathcal{E}_n$. This same view, when applied to an arbitrary element $\omega = \omega_i(x^k)dx^i \in \mathcal{T}^*(\mathcal{E}_n)$, takes the form

$$\frac{\omega}{dt} = \omega_i(x^j) \frac{dx^i}{dt} \ . \tag{A.34}$$

The right-hand side of this expression is well defined whenever the x's are functions of the variable t, but the left-hand side ω/dt makes no sense; it is not d/dt of any single function. One speaks of ω/dt as an *inexact differential.* It is simply an expression of the form

$$\omega = \omega_i(x^k) dx^i \tag{A.35}$$

that is not necessarily the derivative of a function $f(x^k)$ with respect to t. Quantities such as (A.35) are referred to as *differential forms* or *Pfaffian forms*; they appear as *exact* and are then expressible as the *total* differential of a function $f \in \Lambda^0(\mathcal{E}_n)$ or *inexact* and then cannot be written as the differential of a function. In association to $\Lambda^0(\mathcal{E}_n)$ as the set of $(\mathcal{C}^\infty)$-functions, $\mathcal{T}^*(\mathcal{E}_n)$ is now identified with $\Lambda^1(\mathcal{E}_n)$. We summarize this result as

Definition 2.7: $\Lambda^1(\mathcal{E}_n)$, the space of differential forms of degree one on $\mathcal{E}_n$, coincides with the dual space $\mathcal{T}^*(\mathcal{E}_n)$. The elements $\Lambda^1(\mathcal{E}_n)$ have the representation

$$\omega = \omega_i(x^j) dx^i \tag{A.36}$$

in terms of the natural basis $\{dx^i\}$ and are referred to as differential forms of degree one. $\Lambda^0(\mathcal{E}_n)$ is referred to as the *space of forms of degree zero.* ◊

Note as an application of (A.36) that

$$\begin{aligned} \langle \omega, \mathrm{V} \rangle &= \langle \omega_i dx^i, v^j \partial_j \rangle = \omega_i v^j \langle dx^i, \partial_j \rangle \\ &= \omega_i v^j \delta^i_j = \omega_i v^i \ . \end{aligned} \tag{A.37}$$

There still remains the following question: Given an inexact differential like (A.36), can it be transformed, e. g. by multiplication with $a \in \Lambda^0(\mathcal{E}_n)$, into a total differential; in other words, is

$$a\omega = a(x^j)\omega_i(x^j)dx^i \tag{A.38}$$

a total differential, and what are the conditions that functions $a(x^j) \in \Lambda^0(\mathcal{E}_n)$ can be found which make (A.38) a total differential? Or, what are the conditions that such an attempt is in vain? Towards this end we need

The exterior or 'veck' product: We start with

Definition 2.8: Let $\{dx^i\}$ be the natural basis for $\Lambda^1(\mathcal{E}_n)$ and $\alpha, \beta, \gamma \in \Lambda^1(\mathcal{E}_n)$ be differential forms. The *exterior* or *veck product* '$\wedge$' is defined on $\Lambda^1(\mathcal{E}_n) \times \Lambda^1(\mathcal{E}_n)$ by

$$\text{(a)} \qquad dx^i \wedge dx^j = -dx^j \wedge dx^i, \tag{A.39}$$

$$\text{(b)} \qquad dx^i \wedge f(x^k)dx^j = f(x^k)dx^i \wedge dx^j \tag{A.40}$$

for any function $f(x^k) \in \Lambda^0(\mathcal{E}_n)$. Furthermore,

$$\begin{aligned} \text{(c)} \qquad & \alpha \wedge (\beta + \gamma) = \alpha \wedge \beta + \alpha \wedge \gamma, \\ & \alpha \wedge \beta = -\beta \wedge \gamma \end{aligned} \tag{A.41}$$

for all $\alpha, \beta, \gamma \in \Lambda^1(\mathcal{E}_n)$. Property (a) implies at once

$$dx^i \wedge dx^i = \text{`0'} , \tag{A.42}$$

where '0' is the zero element of the 2-forms. ◊

With the aid of this definition the following properties are straightforward to prove:

$$\text{(1)} \qquad (\alpha + \beta) \wedge \gamma = \alpha \wedge \gamma + \beta \wedge \gamma , \tag{A.43}$$

$$\text{(2)} \qquad \alpha \wedge \beta = \alpha_i\beta_j dx^i \wedge dx^j = -\alpha_j\beta_i dx^i \wedge dx^j, \tag{A.44}$$

$$\text{(3)} \qquad 2\alpha \wedge \beta = (\alpha_i\beta_j - \alpha_j\beta_i)dx^i \wedge dx^j \quad . \tag{A.45}$$

Proof: For item (1) we use properties (a) and (c) of the above definition as follows:

$$\begin{aligned} (\alpha + \beta) \wedge \gamma &\overset{(a)}{=} -\gamma \wedge (\alpha + \beta) \overset{(c)}{=} -\gamma \wedge \alpha - \gamma \wedge \beta \\ &\overset{(a)}{=} \alpha \wedge \gamma + \beta \wedge \gamma . \end{aligned} \tag{A.46}$$

To prove item (2), let $\alpha = \alpha_i dx^i$ and $\beta = \beta_j dx^j$. Then evaluation of $\alpha \wedge \beta$ yields

$$\begin{aligned}
\alpha \wedge \beta &= (\alpha_i dx^i) \wedge (\beta_j dx^j) \\
&= (\alpha_i dx^j) \wedge (\beta_1 dx^1 + \cdots + \beta_n dx^n) \\
&\overset{\text{(c)}}{=} (\alpha_i dx^i) \wedge \beta_1 dx^1 + \cdots \\
&\quad + (\alpha_i dx^i) \wedge \beta_n dx^n \\
&\overset{\text{(b),(1)}}{=} \beta_1 (\alpha_i dx^i) \wedge dx^1 + \cdots \\
&\quad + \beta_n (\alpha_i dx^i) \wedge dx^n \\
&= \alpha_i \beta_j dx^i \wedge dx_j \ .
\end{aligned}$$

Property (3) is obtained by using the two evaluations of $\alpha \wedge \beta$ in item (2). □

If we form all linear combinations of quantities thus formed with coefficients from $\Lambda^0(\mathcal{E}_n)$, we obtain a vector space over $\Lambda^0(\mathcal{E}_n)$ that is denoted by $\Lambda^2(\mathcal{E}_n)$. Because of (A.39) its dimension is, however, not equal to n but $n(n-1)/2$. We state these properties as

Definition 2.9: The $n(n-1)/2$ dimensional vector space $\Lambda^2(\mathcal{E}_n)$ over $\Lambda^0(\mathcal{E}_n)$ with the basis

$$\{dx^i \wedge dx^j, \qquad i < j\} \tag{A.47}$$

is the vector space of the exterior 2-forms over $\mathcal{E}_n$. The elements of $\Lambda^2(\mathcal{E}_n)$ are referred to as *2-forms* or *exterior forms of degree 2*. The elements of $\Lambda^0(\mathcal{E}_n)$ and $\Lambda^1(\mathcal{E}_n)$ are now referred to as *exterior forms of degree* 0 *and degree* 1, respectively. ◊

Having defined exterior forms of degree 2 it is now clear how we can proceed to construct *exterior forms of degree 3* by forming all linear combinations of all veck products of the basis elements dx^i taken three at a time. In this process $\{dx^i \wedge dx^j \wedge dx^k\}$ is the natural basis for the vector space $\Lambda^3(\mathcal{E}_n)$, and for any $\alpha, \beta, \gamma \in \Lambda^3(\mathcal{E}_n)$ we request the multiplication to be associative,

$$\alpha \wedge (\beta \wedge \gamma) = (\alpha \wedge \beta) \wedge \gamma \ . \tag{A.48}$$

It is also easily shown that $dx^i \wedge dx^j \wedge dx^k$ does not change its value if the indices i, j, k are altered with an even permutation, but its value changes to its negative, if the permutation of i, j, k is odd; moreover and obviously, if any two indices are the same, then $dx^i \wedge dx^j \wedge dx^j = 0$. Explicitly,

$$\begin{aligned} dx^i \wedge dx^j \wedge dx^k &= dx^j \wedge dx^k \wedge dx^i \\ &= dx^k \wedge dx^i \wedge dx^j \\ &= -dx^i \wedge dx^k \wedge dx^j \\ &= -dx^j \wedge dx^i \wedge dx^k \\ &= -dx^k \wedge dx^j \wedge dx^i \ . \end{aligned} \tag{A.49}$$

Therefore, the dimension of the vector space $\Lambda^3(\mathcal{E}_n)$ is $\binom{n}{3} = [n(n-1)(n-2)/3!]$.

It is now obvious how one may proceed, but the construction must terminate at $\Lambda^n(\mathcal{E}_n)$ since this space only possesses one single independent element

$$dx^1 \wedge dx^2 \wedge \cdots \wedge dx^n \tag{A.50}$$

because any other such element in which the sequence of the dx's is altered is $\pm dx^1 \wedge dx^2 \wedge \cdots \wedge dx^n$. Moreover, vecking (A.50) with any dx^j, $j \leq n$ generates zero. So, $\Lambda^{n+m}(\mathcal{E}_n) = \{0\}$ for any $m > 0$. These considerations lead to the

Definition 2.10: The space $\Lambda^k(\mathcal{E}_n), 0 < k \leq n$, of *exterior forms of degree* k is the vector space of dimension

$$\binom{n}{k} = \frac{n!}{k!(n-k)!} \tag{A.51}$$

over $\Lambda^0(\mathcal{E}_n)$ with the natural basis

$$\{dx^{i_1} \wedge dx^{i_2} \wedge \ldots \wedge dx^{i_k}, \quad i_1 < i_2 < \ldots < i_k\} \ . \tag{A.52}$$

If $\alpha \in \Lambda^k(\mathcal{E}_n)$, then we write

$$\deg(\alpha) = k \ . \tag{A.53}$$

◊

The above analysis leads to the properties of the veck product in the following general form:

Lemma 2.7: *Let $\alpha \in \Lambda^r(\mathcal{E}_n)$ and $\beta, \gamma \in \Lambda^s(\mathcal{E}_n)$. The operation $\wedge$ of exterior multiplication generates a map $\wedge : \Lambda^r(x^j) \times \Lambda^s(x_j) \to \Lambda^{r+s}(x_j)$ with the following properties:*

$$(a) \qquad \alpha \wedge (\beta + \gamma) = \alpha \wedge \beta + \alpha \wedge \gamma \ , \tag{A.54}$$

$$(b) \qquad \alpha \wedge \beta = (-1)^{\deg(\alpha)\deg(\beta)} \beta \wedge \alpha \ , \tag{A.55}$$

$$(c) \qquad \alpha \wedge (\beta \wedge \gamma) = (\alpha \wedge \beta) \wedge \gamma \ . \tag{A.56}$$

□

The properties (a) and (c) need no further proof, for (b) the reader may test that (A.55) is correct for a selected case. However we state that

- elements from any $\Lambda^k(\mathcal{E}_n)$ may be multiplied with elements from $\Lambda^0(\mathcal{E}_n)$, but that elements from $\Lambda^k(\mathcal{E}_n)$ and $\Lambda^m(\mathcal{E}_n)$ with $k \neq m$ cannot be vecked;
- any element of $\Lambda^k(\mathcal{E}_n)$ can be vecked with any element from $\Lambda^m(\mathcal{E}_n)$, but the answer belongs to $\Lambda^{k+m}(\mathcal{E}_n)$. If $k + m > n$ this answer is zero.

A straightforward implication of (A.55) is the following:

- If $\omega \in \Lambda^k(\mathcal{E}_n)$ with $k = (2r + 1)$, then

$$\omega \wedge \omega = (-1)^{(2r+1)^2} \omega \wedge \omega \tag{A.57}$$

  and therefore $\omega \wedge \omega = 0$.
- However, if k is even, $k = 2r$, then $\omega \wedge \omega$ does not vanish in general.

These properties suggest the following definition for the space encompassing all exterior k-forms, $k = 0, 1, \ldots, n$.

Definition 2.11: The *graded exterior algebra* of differential forms over $\mathcal{E}_n$ is the direct sum

$$\Lambda(\mathcal{E}_n) = \Lambda^0(\mathcal{E}_n) \oplus \Lambda^1(\mathcal{E}_n) \oplus \cdots \oplus \Lambda^n(\mathcal{E}_n) \tag{A.58}$$

with the vector space operations of each $\Lambda^k(\mathcal{E}_n)$ and together with the exterior product $\wedge$ as a map from $\Lambda(\mathcal{E}_n) \times \Lambda(\mathcal{E}_n)$ to $\Lambda(\mathcal{E}_n)$. ◊

It is obvious that a basis for $\Lambda(\mathcal{E}_n)$ is

$$1 \oplus \{dx^i\} \oplus \{dx^i \wedge dx^j, i < j\} \oplus \cdots \oplus \{dx^1 \wedge dx^2 \wedge \cdots \wedge dx^n\} \tag{A.59}$$

and the dimension[1]

$$\dim(\wedge) = \sum_{k=0}^{n} \binom{n}{k} = 2^n \ . \tag{A.60}$$

We next perform a number of straightforward computations. To this end the following definition is helpful:

---

[1] Note that $(1+t)^n = \sum_{k=0}^{n} \binom{n}{k}$; so for $t = 2$ this yields $2^n = \sum_{k=0}^{n} \binom{n}{k}$.

Definition 2.12: An element $\omega \in \Lambda^k(\mathcal{E}_n)$ is said to be *simple* if and only if there exist k elements $\{\eta^1, \eta^2, \ldots, \eta^k\} = \{\eta^a\} \in \Lambda^1(\mathcal{E}_n)$ such that

$$\omega = \eta^1 \wedge \eta^2 \wedge \cdots \wedge \eta^k \; . \tag{A.61}$$

◊

- We have already encountered simple elements of $\Lambda^k(\mathcal{E}_n)$, namely the basis elements of $\Lambda^k(\mathcal{E}_n)$. Now, the representation

$$\omega = \sum_{i_1 < i_2 < \cdots < i_k} \omega_{i_1, i2 \cdots ik}(x^m) dx^{i_1} \wedge dx^{i_2} \wedge \cdots \wedge dx^{i_k} \tag{A.62}$$

  shows that *every element of $\Lambda^k(\mathcal{E}_n)$ consists of a linear combination of simple elements of $\Lambda^k(\mathcal{E}_n)$ with coefficients from $\Lambda^0(\mathcal{E}_n)$.*
- Every $\omega \in \Lambda^n(\mathcal{E}_n)$ is simple. This follows from the above by choosing $k = n$. Or we may recall that $dx^1 \wedge dx^2 \wedge \cdots \wedge dx^n$ is the natural basis for $\Lambda^n(\mathcal{E}_n)$. Thus, any element of $\Lambda^n(\mathcal{E}_n)$ is simple.
- Let $\omega^1, \omega^2, \ldots, \omega^k$ be k given elements of $\Lambda^1(\mathcal{E}_n)$ and construct the simple k-form

$$\Omega = \omega^1 \wedge \omega^2 \wedge \cdots \wedge \omega^k \; . \tag{A.63}$$

  Then:

  - A necessary and sufficient condition that the k given 1-forms are *linearly dependent* is

$$\Omega = 0 \; . \tag{A.64}$$

  - A necessary and sufficient condition that the k given 1-forms are *linearly independent* is

$$\Omega \neq 0 \; . \tag{A.65}$$

  We leave the proof to the reader (see EDELEN [35], page 89).
- If $\eta^1, \eta^2, \ldots, \eta^k$ are k elements of $\Lambda^1(\mathcal{E}_n)$ and another collection of k elements of $\Lambda^l(\mathcal{E}_n)$ is defined by

$$\omega^a = K^a_b \eta^b \; , \tag{A.66}$$

  then

$$\omega^1 \wedge \omega^2 \wedge \cdots \wedge \omega^k = \det(K^a_b) \eta^1 \wedge \eta^2 \wedge \cdots \wedge \eta^k \; . \tag{A.67}$$

- Let $\omega^1, \ldots, \omega^k \in \Lambda^l(\mathcal{E}_n)$ be linearly independent 1-forms; suppose, moreover, that the 1-forms $\gamma_1, \ldots, \gamma_r$ satisfy

$$\sum_{i=1}^{r} \omega^i \wedge \gamma_i = 0 \; . \tag{A.68}$$

Then, each of the r 1-forms $\gamma_1, \ldots, \gamma_r$ belongs to the subspace spanned by $\omega^1, \ldots, \omega^r$. Thus, there exists a matrix (A_{ab}) such that $\gamma_a = A_{ab}\omega^b$ with $A_{ab} = A_{ba}, a, b = 1, \ldots, r$.

Proof: This is quite easy. Since $\omega^1, \ldots, \omega^r$ are linearly independent $\Omega = \omega^1 \wedge \omega^2 \wedge \cdots \wedge \omega^r \neq 0$. Thus, vecking (A.68) with $\Omega^j_{r-1} := \omega^1 \wedge \cdots \wedge \omega^{j-1} \wedge \omega^{j+1} \wedge \cdots \wedge \omega^r$ (ω^j is missing), we obtain

$$\sum_{i=1}^{r} \gamma^i \wedge \Omega^j_{r-1} = \gamma^i \wedge \Omega = 0, \qquad j = 1, \ldots, r\ . \tag{A.69}$$

So, each of the γ's is linearly dependent on the ω's. This establishes the representation $\gamma_a = A_{ab}\omega^b$ with $A_{ab} = A_{ba}$.

Inner multiplication. The exterior product $\wedge : \Lambda^k(\mathcal{E}_n) \times \Lambda^m(\mathcal{E}_n) \to \Lambda^{k+m}(\mathcal{E}_n)$ is an ascending operation. Hence, once reaching the topmost collection $\Lambda^n(\mathcal{E}_n)$ a return to elements of $\Lambda^k(\mathcal{E}_n)$ with $k < n$ cannot be achieved. However, this backward operation down the ladder is provided by the pull down operation of

Definition 2.13: *Inner multiplication* or the *pull down* is a map

$$\vee : \mathcal{T}(\mathcal{E}_n) \times \Lambda^k(\mathcal{E}_n) \ \to \ \Lambda^{k-l}(\mathcal{E}_n) \tag{A.70}$$

with the following properties:

$$(a) \quad \mathrm{V} \vee f = 0 \quad \text{for all} \quad \mathrm{V} \in \mathcal{T}(\mathcal{E}_n) \text{ and all } f \in \Lambda^0(\mathcal{E}_n)\ , \tag{A.71}$$

$$(b) \quad \mathrm{V} \vee \omega = \langle \omega, \mathrm{V} \rangle \text{ for all } \mathrm{V} \in \mathcal{T}(\mathcal{E}_n) \text{ and all } \omega \in \Lambda^1(\mathcal{E}_n)\ , \tag{A.72}$$

$$\begin{aligned}(c) \quad & \mathrm{V} \vee (\alpha + \beta) = \mathrm{V} \vee \alpha + \mathrm{V} \vee \beta \text{ for all } \alpha, \beta \in \Lambda^k(\mathcal{E}_n), \\ & k = 1, 2, \ldots, n \text{ and all } \mathrm{V} \in \mathcal{T}(\mathcal{E}_n)\end{aligned} \tag{A.73}$$

$$\begin{aligned}(d) \quad & \mathrm{V} \vee (\alpha \wedge \beta) = (\mathrm{V} \vee \alpha) \wedge \beta + (-1)^{\deg(\alpha)} \alpha \wedge (\mathrm{V} \vee \beta) \\ & \text{for all } \alpha, \beta \in \Lambda(\mathcal{E}_n) \text{ and all } \mathrm{V} \in \mathcal{T}(\mathcal{E}_n)\ .\end{aligned} \tag{A.74}$$

◊

These properties also imply

$$(e) \quad (f\mathrm{U} + g\mathrm{V}) \vee \omega = f(\mathrm{U} \vee \omega) + g(\mathrm{V} \vee \omega) \tag{A.75}$$

$$(f) \quad \mathrm{U} \vee (\mathrm{V} \vee \omega) = -\mathrm{V} \vee (\mathrm{U} \vee \omega) \quad \to \quad \mathrm{V} \vee (\mathrm{V} \vee \omega) = 0\ . \tag{A.76}$$

The *proofs* of (e) and (f) are facilitated by recognizing that every k-form is expressible as a linear combination of simple k-forms. So, for (e),

$$
\begin{aligned}
&\mathrm{V} \vee (\omega^1 \wedge \omega^2 \wedge \cdots \wedge \omega^k) \\
&= (\mathrm{V} \vee \omega^1) \wedge \omega^2 \wedge \cdots \wedge \omega^k - \omega^1 \wedge (\mathrm{V} \vee \omega^2 \wedge \cdots \wedge \omega^k) \\
&= \langle \omega^1, \mathrm{V} \rangle \, (\omega^2 \wedge \cdots \wedge \omega^k) - \omega^1 \wedge \{(\mathrm{V} \vee \omega^2) \wedge \omega^3 \wedge \cdots \wedge \omega^k \\
&\qquad - \omega^2 \wedge (\mathrm{V} \vee (\omega^3 \wedge \cdots \wedge \omega^k))\} \\
&= \langle \omega^1, \mathrm{V} \rangle \, \omega^2 \wedge \cdots \wedge \omega^k - \langle \omega^2, \mathrm{V} \rangle \, \omega^1 \wedge \omega^3 \wedge \cdots \wedge \omega^k \\
&\qquad + \langle \omega^3, \mathrm{V} \rangle \, \omega^1 \wedge \omega^2 \wedge \cdots \wedge \omega^k + \cdots \\
&\qquad + (-1)^{k-1} \langle \omega^k, \mathrm{V} \rangle \, \omega^1 \wedge \cdots \omega^{k-2} \wedge \omega^{k-1} \\
&= \sum_{j=1}^{k-1} (-1)^{j-1} \langle \omega^j, \mathrm{V} \rangle \, \Omega^{(j)}_{k-1} \, ,
\end{aligned}
\tag{A.77}
$$

where

$$
\Omega^{(j)}_{k-1} = \omega^1 \wedge \cdots \wedge \omega^{j-1} \wedge \omega^{j+1} \wedge \cdots \wedge \omega^{k-1} \in \Lambda^{k-1}(\mathcal{E}_n), \tag{A.78}
$$

which implies that (A.77) belongs to $\Lambda^{k-1}(\mathcal{E}_n)$. To prove (f), we start again with (A.77) and allow $\mathrm{U}\vee$ to act on both sides of (A.77), reverse the roles of U and V and then add the results. For basis elements of $\Lambda^2(\mathcal{E}_n)$ this yields

$$
\begin{aligned}
\mathrm{V} \vee (dx^i \wedge dx^j) \quad &= (\mathrm{V} \vee dx^i) \wedge dx^j - dx^i \wedge (\mathrm{V} \vee dx^j) \\
&= \langle dx^i, \mathrm{V} \rangle \, dx^j - \langle dx^j, \mathrm{V} \rangle dx^i \\
&= v^i dx^j - v^j dx^i \, ,
\end{aligned}
\tag{A.79}
$$

$$
\begin{aligned}
\mathrm{U} \vee (\mathrm{V} \vee (dx^i \wedge dx^j)) \quad &= v^i \mathrm{U} \vee dx^j - v^j \mathrm{U} \vee dx^i \\
&= v^i u^j - v^j u^i \, ,
\end{aligned}
\tag{A.80}
$$

$$
\mathrm{V} \vee (\mathrm{U} \vee (dx^i \wedge dx^j)) \quad = u^i v^j - u^j v^i \, , \tag{A.81}
$$

which corroborates (A.76).

Exterior derivatives: In the preceding analysis the abstract operator 'd' was introduced as an element of $\Lambda^1(\mathcal{E}_n)$ (see Definition 2.6 and following discussion), and it was shown that it has realizations as exact and inexact differentials (see Lemma 2.6 and following discussion), i. e., forms of degree zero. We ask whether an abstract operator 'd' can also be defined for forms other than degree zero.

This is indeed so, and we begin by noting that any element of $\Lambda^k(\mathcal{E}_n)$ is the sum of terms of the form $f(x^j)dx^{i_1} \wedge \cdots \wedge dx^{i_k}$ and that we know the meaning of $df(x^j)$. For any $\alpha, \beta \in \Lambda(\mathcal{E}_n)$ we now propose to define 'd' via the computational rules

$$d(\alpha + \beta) = d\alpha + d\beta \ , \tag{A.82}$$

$$d(\alpha \wedge \beta) = d\alpha \wedge \beta + (-1)^{\deg(\alpha)} \alpha \wedge d\beta \ . \tag{A.83}$$

(A.82) makes d a linear operator, whilst (A.83) says that d is an *anti-derivation* because of the factor $(-1)^{\deg(\alpha)}$. For any $f \in \Lambda^0(\mathcal{E}_n)$ we have

$$\begin{aligned}
df &= \frac{\partial f}{\partial x^j} dx^j \ , \\
ddf &= d\left(\frac{\partial f}{\partial x^j} dx^j\right) = d\left(\frac{\partial f}{\partial x^j}\right) \wedge dx^j + \frac{\partial f}{\partial x^j} ddx^j \\
&= \frac{\partial^2 f}{\partial dx^k \partial dx^j} dx^k \wedge dx^j + \frac{\partial f}{\partial x^j} ddx^j \\
&\overset{(A.83)}{=} \frac{1}{2}\left(\frac{\partial^2 f}{\partial x^k \partial x^j} - \frac{\partial^2 f}{\partial x^j \partial x^k}\right) dx^k \wedge dx^j + \frac{\partial f}{\partial x^j} ddx^j \\
&= \frac{\partial f}{\partial x^j} ddx^j \ .
\end{aligned} \tag{A.84}$$

Thus, the operator will provide a statement of necessary symmetry, namely $\partial^2 f/(\partial x^k \partial x^j) = \partial^2 f/(\partial x^j \partial x^k)$ if

$$ddx^i = 0 \qquad i = 1, \ldots, n \ . \tag{A.85}$$

We must now show that an operator d with the above properties is defined for all elements of $\Lambda(\mathcal{E}_n)$. If $\beta \in \Lambda^k(\mathcal{E}_n)$, it is composed of terms of the form

$$\gamma = f(x^m) dx^{i_1} \wedge dx^{i_2} \wedge \cdots \wedge dx^{i_k} \ . \tag{A.86}$$

So, if we apply d to both sides of (A.86) and use (A.83), we obtain

$$\begin{aligned}
d\gamma &= df \wedge dx^{i_1} \wedge \cdots \wedge dx^{i_k} + f d\{dx^{i_1} \wedge \cdots \wedge dx^{i_k}\} \\
&\overset{(A.83,A.85)}{=} \frac{\partial f}{\partial x^j} dx^j \wedge dx^{i_1} \wedge \cdots \wedge dx^{i_k} \ ,
\end{aligned} \tag{A.87}$$

which proves the statement that d is defined for every element of $\Lambda(\mathcal{E}_n)$; in particular, $d : \Lambda^k(\mathcal{E}_n) \to \Lambda^{k+1}(\mathcal{E}_n)$. If we now apply d to both sides of (A.87), we obtain

$$dd\gamma = 0 \tag{A.88}$$

for all $\gamma \in \Lambda(\mathcal{E}_n)$. Uniqueness is secured through the final requirement that $df = (\partial f/\partial dx^i)dx^i$ holds for all $f \in \Lambda^0(\mathcal{E}_n)$.

In summary we have shown:

Theorem 2.1: *There is one and only one operator d on $\Lambda(\mathcal{E}_n)$ with the following properties:*

$$(a) \qquad d(\alpha + \beta) = d\alpha + d\beta \ , \tag{A.89}$$

$$(b) \qquad d(\alpha \wedge \beta) = d\alpha \wedge \beta + (-1)^{\deg(\alpha)} d\beta \ , \tag{A.90}$$

$$(c) \qquad df = \frac{\partial f}{\partial x^i} dx^i, \quad f \in \Lambda^0(\mathcal{E}_n) \ , \tag{A.91}$$

$$(d) \qquad dd\alpha = 0 \ . \tag{A.92}$$

If $\alpha \in \Lambda^k(\mathcal{E}_n)$, then $d\alpha \in \Lambda^{k+1}(\mathcal{E}_n)$ and hence d may be viewed as the map

$$d : \Lambda^k(\mathcal{E}_n) \ \rightarrow \ \Lambda^{k+1}(\mathcal{E}_n) \ . \tag{A.93}$$

which satisfies properties (a)-(d). d is called exterior differentiation. □

The actual calculation of the exterior derivative of any exterior form follows from the properties (a) through (d) of the above theorem.

- For example, if $\alpha = \alpha_i(x^k)dx^i$ is a 1-form, then

$$d\alpha = d\alpha_i \wedge dx^i = \frac{\partial \alpha_i}{\partial dx^j} dx^j \wedge dx^i = \tfrac{1}{2}(\partial_j \alpha_i - \partial_i \alpha_j) dx^j \wedge dx^i \ , \tag{A.94}$$

  because $dx^j \wedge dx^i = -dx^i \wedge dx^j$.
- If $\beta = \frac{1}{2}\beta_{ij} dx^i \wedge dx^j$, $\beta_{ij} = -\beta_{ji}$, then

$$\begin{aligned} d\beta &= \tfrac{1}{2} d\beta_{ij} \wedge dx^i \wedge dx^j = \tfrac{1}{2}(\partial_k \beta_{ij}) dx^k \wedge dx^i \wedge dx^j \\ &= \tfrac{1}{3!}(\partial_k \beta_{ij} + \partial_i \beta_{jk} + \partial_j \beta_{ki}) dx^k \wedge dx^i \wedge dx^j \ . \end{aligned} \tag{A.95}$$

- If $\omega \in \Lambda^n(\mathcal{E}_n)$, then $d\omega = 0$. Indeed, since d maps $\Lambda^n(\mathcal{E}_n)$ into $\Lambda^{n+1}(\mathcal{E}_n)$ and $\Lambda^{n+1}(\mathcal{E}_n)$ only contains the element 0 we have $d\omega = 0$.

Definition 2.14

(1) An element $\alpha \in \Lambda(\mathcal{E}_n)$ is said to be *closed*, if and only if α is in the kernel of d:

$$d\alpha = 0 \ . \tag{A.96}$$

(2) An element $\alpha \in \Lambda(\mathcal{E}_n)$ is said to be *exact*, if α is in the range of d:

$$\alpha = d\beta \ . \tag{A.97}$$

◊

Theorem 2.2 *The collection of all closed elements of $\Lambda(\mathcal{E}_n)$ forms a subspace $\mathcal{C}(\mathcal{E}_n)$ of $\Lambda(\mathcal{E}_n)$ over $\mathbb{R}$, but not over $\Lambda^0(\mathcal{E}_n)$. The collection of all exact elements of $\Lambda(\mathcal{E}_n)$ forms a subspace $\mathcal{D}(\mathcal{E}_n)$ of $\Lambda(\mathcal{E}_n)$ over $\mathbb{R}$, but not over $\Lambda^0(\mathcal{E}_n)$. Moreover,*

$$\mathcal{D}(\mathcal{E}_n) \subset \mathcal{C}(\mathcal{E}_n) \ . \tag{A.98}$$

□

Proof: If α and β are closed elements, then $d\alpha = 0$ and $d\beta = 0$. For a linear combination $f\alpha + g\beta$ we have

$$\begin{aligned} d(f\alpha + g\beta) &= df \wedge \alpha + dg \wedge \beta + (-1)^0 f d\alpha + (-1)^0 g d\beta \\ &= df \wedge \alpha + dg \wedge \beta \ . \end{aligned} \tag{A.99}$$

Thus, if $df = 0$, $dg = 0$, that is $f =$ constant and $g =$ constant, then $d(f\alpha + g\beta) = 0$ and $\mathcal{C}(\mathcal{E}_n)$ is a subspace of $\Lambda(\mathcal{E}_n)$ over $\mathbb{R}$ but not over $\Lambda^0(E_n)$, since f, g must be constant.

Analogously, if α and β belong to $\mathcal{D}(\mathcal{E}_n)$, there exist elements ρ and η such that $\alpha = d\rho$ and $\beta = d\eta$. Then, $f\alpha + g\beta = f d\rho + g d\eta = d(f\rho) + d(g\eta)$ $-df \wedge \rho - dg \wedge \eta$. But this only reduces to $d(f\rho + g\eta)$ if $df = 0$ and $dg = 0$, hence $f =$ constant and $g =$ constant. So, $\mathcal{D}(\mathcal{E}_n)$ is a vector subspace of $\Lambda(\mathcal{E}_n)$ over $\mathbb{R}$, but not over $\Lambda^0(\mathcal{E}_n)$. Furthermore, for any element $\alpha \in \mathcal{D}(\mathcal{E}_n)$ we automatically have $\alpha = d\beta$ and, a forteriori, $d\alpha = dd\beta = 0$. So, every $\alpha \in \mathcal{D}(\mathcal{E}_n)$ is also element of $\mathcal{C}(\mathcal{E}_n)$, but not vice versa, proving (A.98). □

The inclusion $\mathcal{D}(\mathcal{E}_n) \subset \mathcal{C}(\mathcal{E}_n)$ is the basis by which a large number of significant problems are solved. For example, suppose that we are given a 1-form

$$F = F_i(x^j) dx^i \ , \tag{A.100}$$

and we would like to find an element $\eta \in \Lambda^0(\mathcal{E}_n)$ such that

$$F = d\eta = \partial_i \eta dx^i \ . \tag{A.101}$$

Then, comparison shows that this is only the case if

$$F_i(x^j) = \partial_i \eta(x^j) \ . \tag{A.102}$$

In other words, the 'force' with components $F_i(x^j)$ admits the potential function η. Such a function can exist if and only if F is exact, namely

$0 = dF = \frac{1}{2}(\partial_i F_j - \partial_j F_i) dx^i \wedge dx^j$, or

$$\partial_i F_j = \partial_j F_i \ . \tag{A.103}$$

Theorem 2.2, however, says quite a bit more. Suppose that $\alpha \in \Lambda^k(\mathcal{E}_n)$ is a k-form, and we ask the question under which conditions can we find a $(k-1)$-form β such that $\alpha = d\beta$. Since this says that α is exact if there exists such a β, then such an α must likewise be closed: $d\alpha = 0$.

Suppose that we satisfy these necessary conditions for the existence of solutions: $\alpha = d\beta$ only if $d\alpha = 0$. The question then arises as to whether we can actually find a β that makes the equation $\alpha = d\beta$ true for given α. This would indeed be the case if we could show that every closed form is an exact form. However, this is just wishful thinking because such a result can not be true in general. For instance, suppose that we are in two dimensions and $\mathbf{F}$ satisfies $\nabla \times \mathbf{F} = \mathbf{0}$ on a region with a hole in it. We then know that $\mathbf{F} = \nabla\eta$ can not necessarily be satisfied by a single-valued function η because of the hole. This is only so if there is no hole, that is if the region is simply connected. A similar situation also exists here: under restricted properties of the region $\mathcal{S} \in \mathcal{E}_n$ there is a partial converse to (A.98) expressed in the following

Lemma 2.8: (POINCARÉ *theorem*) *If $\mathcal{S}$ is a region of $\mathcal{E}_n$ that can be shrunk to a point in a smooth way ($\mathcal{S}$ is star shaped with respect to one of its points), then*

$$\mathcal{C}(\mathcal{S}) \subset \mathcal{D}(\mathcal{S}) \ . \tag{A.104}$$

That is, if $d\alpha = 0$ on $\mathcal{S}$ then there exists a β on $\mathcal{S}$ such that $\alpha = d\beta$. □

The proof of this theorem is given in Chap. 5 of EDELEN [35].

For the further developments we need to recall the definition of an *ideal.* To this end recall that a vector space $\mathcal{V}$ over $\mathbb{R}$ together with a binary operation $(\cdot \mid \cdot)$ is an algebra if

$$(a\mathrm{U} + b\mathrm{V} \mid \mathrm{W}) = a(\mathrm{U} \mid \mathrm{W}) + b(\mathrm{V} \mid \mathrm{W}) \ , \tag{A.105}$$

$$(\mathrm{U} \mid a\mathrm{V} + b\mathrm{W}) = a(\mathrm{U} \mid \mathrm{V}) + b(\mathrm{U} \mid \mathrm{W}) \tag{A.106}$$

for $\mathrm{U}, \mathrm{V}, \mathrm{W} \in \mathcal{V}$ and $a, b \in \mathbb{R}$. An ideal is a subspace $\mathcal{U}$ of an algebra $\{\mathcal{V}, (\cdot \mid \cdot)\}$ such that

$$(\mathrm{W} \mid \mathrm{V}) \text{ belong to } \mathcal{U} \text{ for all } \mathrm{W} \in \mathcal{U} \text{ and all } \mathrm{V} \in \mathcal{V} \ , \tag{A.107}$$

$$(\mathrm{V} \mid \mathrm{W}) \text{ belong to } \mathcal{U} \text{ for all } \mathrm{W} \in \mathcal{U} \text{ and all } \mathrm{V} \in \mathcal{V} \ . \tag{A.108}$$

Definition 2.15: An ideal I of $\Lambda(\mathcal{E}_n)$ is said to be *closed* if and only if $d\rho \in \mathrm{I}$ for every $\rho \in \mathrm{I}$ in which case we write

$$dI \subset I \,. \tag{A.109}$$

◊

Obviously, any finitely generated ideal $I(\omega^1, \omega^2, \ldots, \omega^k)$ can be closed by formation of a new ideal $\bar{I}(\omega^1, \omega^2, \ldots, \omega^k, d\omega^1, d\omega^2, \ldots, d\omega^k)$, called the closure of $I(\omega^1, \omega^2, \ldots, \omega^k)$. For if $\rho \in \bar{I}$, then

$$\rho = \gamma_a \wedge \omega^a + \Gamma_a \wedge d\omega^a \tag{A.110}$$

and hence

$$d\rho = d\gamma_a \wedge \omega^a + \{(-1)^{\deg(\gamma_a)}\gamma_a + d\Gamma_a\} \wedge d\omega^a \tag{A.111}$$

which belongs to $\bar{I}$.

The important question to answer is: When is a given ideal $I(\omega^1, \omega^2, \ldots, \omega^k)$ a closed ideal? The following theorems answer this question.

Theorem 2.3: *Let* $I(\omega^1, \omega^2, \ldots, \omega^k) \overset{def}{=} I(\omega^a)$ *be an ideal of* $\Lambda(\mathcal{E}_n)$ *such that each of its generators is of the same degree. Then,* $I(\omega^a)$ *is a closed ideal if and only if there exist* k^2 *1-forms* $\{\Gamma^a_b\}$ *such that the generators satisfy*

$$d\omega^a = \Gamma^a_b \wedge \omega^b \,. \tag{A.112}$$

□

Proof: Since $\rho \in I\{\omega^a\}$ it possesses the representation $\rho = \gamma_a \wedge \omega^a$ for some k-tuple of forms $\{\gamma_a\}$ with common $\deg(\gamma_a) = b$. Therefore,

$$\begin{aligned} d\rho &= \underbrace{d\gamma_a \wedge \omega^a}_{\subset I\{\omega^a\}} + (-1)^b \gamma_a \wedge d\omega^a \\ &= (-1)^b \gamma_a \wedge d\omega^a \quad \mathrm{mod} \quad I\{\omega^k\} \,. \end{aligned} \tag{A.113}$$

Now, $d\rho \in I\{\omega^a\}$ for every $\rho \in I\{\omega^a\}$ if and only if each of the forms $d\omega^a$ belongs to $I\{\omega^a\}$. Thus, since $d\omega^a$ has one degree greater than $\omega^a, d\omega^a \in I\{\omega^a\}$ for $a = 1, \ldots, k$, if and only if there exist k^2 1-forms such that the generators satisfy (A.112). □

Note that the Γ^a_b's in (A.112) are not entirely arbitrary, since $dd\omega^a = 0$. Indeed

$$dd\omega^a = (d\Gamma^a_b - \Gamma^a_c \Gamma^c_b) \wedge \omega^b = 0 \,. \tag{A.114}$$

It seems quite plausible that not every ideal allows solutions satisfying (A.112), (A.114); *not every ideal generated by forms of the same degree is closed.*

Of particular interest is the situation when the forms generating an ideal are of degree 1. What are the conditions that such ideals are closed? The answer to this question is given by the following

Theorem 2.4: *Let* $\mathrm{I}\{\omega^a\}$ *be an ideal of* $\Lambda(\mathcal{E}_n)$ *whose generators* $\omega^1, \dots, \omega^k$ *are linearly independent 1-forms such that*

$$\omega^1 \wedge \omega^2 \wedge \cdots \omega^k = \Omega(\mathrm{I}) \neq 0 \tag{A.115}$$

and $k < n-1$. *Then* $d\mathrm{I}\{\omega^a\} \subset \mathrm{I}\{\omega^a\}$ *if and only if*

$$d\omega^a \wedge \Omega(\mathrm{I}) = 0, \qquad a = 1, \dots, k\ . \tag{A.116}$$

□

Proof: If the generators satisfy (A.112), then they also satisfy (A.116),

$$d\omega^a \wedge \Omega(\mathrm{I}) = \Gamma^a_b \wedge \omega^b \wedge \Omega(\mathrm{I}) = 0\ , \tag{A.117}$$

since ω^b arises twice in $\omega^b \wedge \Omega(\mathrm{I})$. To establish the converse, let $\omega^1, \dots, \omega^k$, $\omega^{k+1}, \dots, \omega^n$ be a basis of $\Lambda^1(\mathcal{E}_n)$ and label the additional 1-forms with the indices $\alpha, \beta, \dots$. Since $d\omega^a \in \Lambda^2(\mathcal{E}_n)$ we may always write

$$d\omega^a = \xi^a_{ij}\omega^i \wedge \omega^j = \xi^a_{bc}\omega^b \wedge \omega^c + 2\xi^a_{\alpha c}\omega^\alpha \wedge \omega^c + \xi^a_{\alpha\beta}\omega^\alpha \wedge \omega^\beta\ , \tag{A.118}$$

and hence

$$\Omega(\mathrm{I}) \wedge d\omega^a = \xi^a_{\alpha\beta}\Omega(\mathrm{I}) \wedge \omega^\alpha \wedge \omega^\beta\ . \tag{A.119}$$

Since $k < n-1$, we have $\deg\left(\Omega(\mathrm{I}) \wedge \omega^\alpha \wedge \omega^\beta\right) \leq n$; consequently, $\Omega(\mathrm{I}) \wedge \omega^\alpha \wedge \omega^\beta$ is a simple nonzero $(k+2)$-form, because ω^α and ω^β are independent and not members of the subspace spanned by $\omega^1, \dots, \omega^k$. So, (A.116) is only satisfied when $\xi^a_{\alpha\beta} = 0$, $\alpha, \beta = k+1, \dots, n$, $a = 1, \dots, k$; that is (A.118) reduces to

$$d\omega^a = \underbrace{\left(\xi^a_{bc}\omega^b + 2\xi^a_{\alpha c}\omega^\alpha\right)}_{\Gamma^a_c} \wedge \omega^c = \Gamma^a_c \wedge \omega^c\ . \tag{A.120}$$

□

This proof excludes the case $k \geq n-1$ because $\Omega(\mathrm{I}) \wedge d\omega^a$ would have degree greater than n and hence vanish identically. This case is covered in

Theorem 2.5: *If the ideal* I *is generated by either* $n-1$ *or* n *linearly independent 1-forms, then* $d\mathrm{I} \subset \mathrm{I}$. □

Proof:

(1) For $k = n$, $\{\omega^c\}$ is a basis for $\Lambda^1(\mathrm{E}_n)$ and therefore $d\omega^a = (\xi^a_{bc}\omega^b) \wedge \omega^c = \Gamma^a_b \wedge \omega^c$ which agrees with (A.112).

(2) For the case $k = n-1$, let γ be an additional 1-form such that $\{\omega^1, \ldots, \omega^{n-1}, \gamma\}$ is a basis for $\Lambda^1(\mathcal{E}_n)$. We then have

$$\begin{aligned} d\omega^a &= \xi^a_{bc}\omega^b \wedge \omega^c + \xi^a_c \gamma \wedge \omega^c \\ &= (\xi^a_{bc}\omega^b + \xi^a_c \gamma^c) \wedge \omega^c = \Gamma^a_b \wedge \omega^c \ , \end{aligned} \tag{A.121}$$

again agreeing with (A.112). □

A simple example illustrates the above theorems. Consider the 1-form ω on $\mathcal{E}_n$ that is defined by

$$\omega = f(x^j)dg(x^j) = f\partial_i g dx^i, \qquad f \neq 0 \ . \tag{A.122}$$

Exterior differentiation yields

$$d\omega = df \wedge dg \tag{A.123}$$

implying that

$$\omega \wedge d\omega = f dg \wedge df \wedge dg = 0 \ . \tag{A.124}$$

The result (A.124) coincides with Theorem 2.4 and agrees with (A.116). It is here formulated for the ideal I$\{\omega\}$ generated by ω, which is closed: dI $\subset$ I. Indeed since $f \neq 0$, we can rewrite (A.123) in the equivalent form

$$d\omega = \frac{1}{f}df \wedge \underbrace{fdg}_{\omega} \overset{(A.122)}{=} \frac{1}{f}df \wedge \omega \ , \tag{A.125}$$

which is of the form (A.112) in Theorem 2.3.

This idea is now generalized to the case of more than a single 1-form, namely

$$\omega^a = K^a_b(x^j)dg^b(x^j), \qquad K^a_b(x^j) \in \Lambda^0(\mathcal{E}_n) \ . \tag{A.126}$$

Its answer is the so-called *Frobenius Theorem.* We restrict attention to a system of 1-forms which can without loss of generality be assumed to be linearly independent and to span $\Lambda^1(\mathcal{E}_n)$.

Definition 2.16: We call a collection of r linearly independent 1-forms, $\{\omega^a, a = 1, \ldots, r\}$ an *exterior system of dimension* r and use the symbol D_r to identify it. The ideal that is generated by $\{\omega^a\}$ is denoted by I$\{D_r\}$. ◊

Definition 2.17: An exterior system D_r with 1-forms $\{\omega^a\}$ is said to be *completely integrable* if and only if there exist r independent functions $\{g^a(x^j)\}$ such that each of the r 1-forms $\{\omega^a\}$ vanishes on each of the r-parameter family of $(n-r)$-dimensional surfaces

$$\{g^a(x^j)\} = c^a, \qquad a = 1, \ldots, r \ , \tag{A.127}$$

generated by allowing the r constants $\{c^a\}$ to range over all r-tuples of real numbers. ◊

Theorem 2.6: *An exterior system D_r with 1-forms $\{\omega^a\}$ is completely integrable if and only if there exists a non-singular $(r \times r)$-matrix of functions $\left(A^a_b(x^j)\right)$ and r independent functions $\{g^b(x^j)\}$ such that*

$$\omega^a = A^a_b dg^b \ . \tag{A.128}$$

□

Proof: Assume that (A.128) holds true with the non-singular matrix $A^a_b(x^j)$ of functions. Thus, since $\det(A^a_b) \neq 0$, we have

$$\begin{aligned} d\omega^a &= dA^a_b \wedge dg^b = (dA^a_c)(A^{-1})^e_c \wedge \underbrace{A^c_b dg^b}_{\omega^c} \\ &= (dA^a_e)(A^{-1})^e_c \wedge \omega^c = \Gamma^a_c \wedge \omega^c \ , \end{aligned} \tag{A.129}$$

which agrees with formula (A.112) in Theorem 2.3. Conversely, suppose that $\{\omega^a\}$ is completely integrable, so that r linearly independent integrable functions $g^a(x^j)$ exist. We can then construct the ideal $\mathrm{I}\{dg^a\}$, and this ideal is the largest closed ideal such that every of its elements vanishes on the surfaces $\{g^a(x^j) = \text{constant},\ a = 1, \ldots, r\}$. However, complete integrability of $\{\omega^a\}$ says that each ω^a vanishes on the surfaces $\{g^a(x^j) = \text{constant},\ a = 1, \ldots, r\}$, and hence $\mathrm{I}\{D_r\} \subset \mathrm{I}\{dg^a\}$. Since both the ω's and dg's are 1-forms, the ideal inclusion is satisfied only if there exists a matrix (A^a_b) such that $\omega^a = A^a_b dg^b$. Since the ω's and dg's are both linearly independent, $\omega^1 \wedge \cdots \wedge \omega^r \neq 0$, and $dg^1 \wedge \cdots \wedge dg^r \neq 0$. Therefore, since $\omega^1 \wedge \cdots \wedge \omega^r = \det(A^a_b) dg^1 \wedge \cdots \wedge dg^r$, we have $\det(A^a_b) \neq 0$. □

Theorem 2.3 shows that the ideal $\mathrm{I}\{D_r\}$ is closed if D_r is completely integrable. The converse is the famous

Theorem 2.7: (Frobenius) *An exterior system D_r that is defined by 1-forms $\{\omega^a\}$ is completely integrable if and only if the ideal $\mathrm{I}\{D_r\}$ is closed, that is $d\mathrm{I}\{D_r\} \subset \mathrm{I}\{D_r\}$ which is equivalent to*

$$d\omega^a = \Gamma^a_b \wedge d\omega^b \tag{A.130}$$

or

$$\omega^1 \wedge \omega^2 \wedge \cdots \wedge \omega^r \wedge d\omega^a = 0, \qquad a = 1, \ldots, r \ . \tag{A.131}$$

□

The proof of this theorem involves deeper methods of exterior calculus and is given in EDELEN [35].

Appendix B
Auxiliary Results

B.1 Manipulation of the Entropy Inequality

We repeat (5.12):

$$
\begin{aligned}
\pi^{\rho\eta} = {} & \partial\left(\rho\eta\right) - \nabla\cdot\left(\boldsymbol{\phi}^{\rho\eta} - \rho\eta\mathbf{v}\right) && \text{I}\\
& -\sum\lambda_\alpha^\rho\left\{\partial\bar{\rho}_\alpha + \nabla\cdot\left(\bar{\rho}_\alpha\mathbf{v}_\alpha\right) - \bar{\rho}_\alpha c_\alpha\right\} && \text{II}\\
& -\sum\boldsymbol{\lambda}_\alpha^v\cdot\left\{\partial\left(\bar{\rho}_\alpha\mathbf{v}_\alpha\right) - \nabla\cdot\left(\bar{\mathbf{T}}_\alpha - \bar{\rho}_\alpha\mathbf{v}_\alpha\otimes\mathbf{v}_\alpha\right) - \bar{\mathbf{m}}_\alpha\right\} && \text{III}\\
& -\lambda^\varepsilon\left\{\partial\left(\rho\varepsilon\right) + \nabla\cdot\left(\mathbf{q} + \rho\varepsilon\mathbf{v}\right) - \mathbf{T}\cdot\left(\nabla\mathbf{v}\right)\right\} && \text{IV}\\
& -\sum_{\alpha=1}^{n-1}\lambda_\alpha^\nu\left\{\partial\nu_\alpha + \nabla\cdot\left(\nu_\alpha\mathbf{v}_\alpha\right) - \bar{n}_\alpha\right\} && \text{V}\\
& -\lambda_n^\nu\Big\{-\partial\Big(\sum_{\beta=1}^{n-1}\nu_\beta\Big) + \nabla\cdot\mathbf{v}_n - \nabla\cdot\Big(\sum_{\beta=1}^{n-1}\nu_\beta\mathbf{v}_n\Big) - \bar{n}_n\Big\} && \text{VI}\\
& -\sum\boldsymbol{\lambda}_\alpha^Z\cdot\left\{\acute{\bar{\mathbf{Z}}}_\alpha - \left[\boldsymbol{\Omega}_\alpha, \bar{\mathbf{Z}}_\alpha\right] - \bar{\boldsymbol{\Phi}}_\alpha\right\} && \text{VII}\\
\geqslant 0\,, &
\end{aligned}
\tag{B.1}
$$

where Latin identifiers are assigned to each row.
First, let us focus on the mixture quantities, namely lines I and IV. The partial time derivatives and the second terms in the brackets of the divergence terms are combined to give

$$
\begin{aligned}
& \partial\left(\rho\eta\right) + \nabla\cdot\left(\rho\eta\mathbf{v}\right) - \lambda^\varepsilon\partial\left(\rho\varepsilon\right) - \lambda^\varepsilon\nabla\cdot\left(\rho\varepsilon\mathbf{v}\right)\\
&\quad = \partial\left(\rho\eta\right) - \lambda^\varepsilon\partial\left(\rho\varepsilon\right)\\
&\qquad + \left\{\nabla\left(\rho\eta\right) - \lambda^\varepsilon\nabla\left(\rho\varepsilon\right)\right\}\cdot\mathbf{v} + \left(\rho\eta - \lambda^\varepsilon\rho\varepsilon\right)\nabla\cdot\mathbf{v}\,.
\end{aligned}
\tag{B.2}
$$

The last term in (B.2) can be transformed into

$$(\rho\eta - \lambda^{\varepsilon}\rho\varepsilon)\,\nabla\cdot\mathbf{v} = \\ (\eta - \lambda^{\varepsilon}\varepsilon)\sum\left\{\bar{\rho}_\alpha\nabla\cdot\mathbf{v}_\alpha + \nu_\alpha\mathbf{u}_\alpha\cdot\nabla\rho_\alpha + \rho_\alpha\mathbf{u}_\alpha\cdot\nabla\nu_\alpha\right\}\ , \tag{B.3}$$

where (3.55), (3.56) and the product rule of differentiation have been used. The divergence terms in lines I, III, IV containing the constitutive quantities $\boldsymbol{\phi}^{\rho\eta}$, $\bar{\mathbf{T}}_\alpha$ and $\mathbf{q}$ are simply added to yield

$$-\left\{\nabla\cdot\boldsymbol{\phi}^{\rho\eta} + \lambda^{\varepsilon}\nabla\cdot\mathbf{q} - \sum_{\alpha=1}^{n}\left(\boldsymbol{\lambda}^{v}_{\alpha}\cdot(\nabla\cdot\bar{\mathbf{T}}_\alpha)\right)\right\}\ . \tag{B.4}$$

(B.4) and the first two terms in (B.2) have the desired structure to introduce $\mathcal{P}$ and $\boldsymbol{\mathcal{F}}$.

In a second step, line II is split into the sums for the m compressible constituents and the $(n-m)$ density-preserving constituents. The result reads

$$\begin{aligned}
&-\sum_{\alpha=1}^{n}\lambda^{\rho}_{\alpha}\left\{\ \partial\bar{\rho}_\alpha + \nabla\cdot(\bar{\rho}_\alpha\mathbf{v}_\alpha) - \bar{\rho}_\alpha c_\alpha\right\} \\
&= -\sum_{\alpha=1}^{m}\lambda^{\rho}_{\alpha}\left\{\nu_\alpha(\partial\rho_\alpha) + \rho_\alpha(\partial\nu_\alpha) + \nu_\alpha\mathbf{v}_\alpha\cdot\nabla\rho_\alpha\right. \\
&\qquad\qquad \left. +\rho_\alpha\mathbf{v}_\alpha\cdot\nabla\nu_\alpha + \bar{\rho}_\alpha\nabla\cdot\mathbf{v}_\alpha - \bar{\rho}_\alpha c_\alpha\right\} \\
&\quad -\sum_{\alpha=m+1}^{n}\lambda^{\rho}_{\alpha}\left\{\rho_\alpha(\partial\nu_\alpha) + \rho_\alpha\mathbf{v}_\alpha\cdot\nabla\nu_\alpha + \bar{\rho}_\alpha\nabla\cdot\mathbf{v}_\alpha - \bar{\rho}_\alpha c_\alpha\right\}\ ,
\end{aligned} \tag{B.5}$$

where $\rho_\alpha = \text{const.}$ for $\alpha = m+1,\ldots,n$ and again the product rule of differentiation has been used. The remainder of line III can be written as

$$\begin{aligned}
\mathrm{III} - \sum_{\alpha=1}^{n}\left(\boldsymbol{\lambda}^{v}_{\alpha}\cdot(\nabla\cdot\bar{\mathbf{T}}_\alpha)\right) = -\sum_{\alpha=1}^{n}\boldsymbol{\lambda}^{v}_{\alpha}\cdot\left\{\ \nu_\alpha\mathbf{v}_\alpha(\partial\rho_\alpha) + \rho_\alpha\mathbf{v}_\alpha(\partial\nu_\alpha) + \bar{\rho}_\alpha(\partial\mathbf{v}_\alpha)\right. \\
+\nu_\alpha\mathbf{v}_\alpha\otimes\mathbf{v}_\alpha\nabla\rho_\alpha + \rho_\alpha\mathbf{v}_\alpha\otimes\mathbf{v}_\alpha\nabla\nu_\alpha \\
\left. +\bar{\rho}_\alpha\nabla\mathbf{v}_\alpha\mathbf{v}_\alpha + \bar{\rho}_\alpha\mathbf{v}_\alpha\nabla\cdot\mathbf{v}_\alpha - \bar{\mathbf{m}}_\alpha\right\}\ ,
\end{aligned} \tag{B.6}$$

where

$$\begin{aligned}\nabla\cdot(\bar\rho_\alpha\mathbf{v}_\alpha\otimes\mathbf{v}_\alpha)\cdot\boldsymbol{\lambda}^v_\alpha &= (\boldsymbol{\lambda}^v_\alpha)_i\ (\bar\rho_\alpha\ v^\alpha_i\ v^\alpha_j)_{,j}\\ &= (\boldsymbol{\lambda}^v_\alpha)_i\ (\bar\rho_\alpha)_{,j}\ v^\alpha_i\ v^\alpha_j\\ &\quad +(\boldsymbol{\lambda}^v_\alpha)_i\ \bar\rho_\alpha\ (v^\alpha_i)_{,j}\ v^\alpha_j+(\boldsymbol{\lambda}^v_\alpha)_i\ \bar\rho_\alpha\ v^\alpha_i\ (v^\alpha_j)_{,j}\\ &= \boldsymbol{\lambda}^v_\alpha\cdot(\mathbf{v}_\alpha\otimes\mathbf{v}_\alpha)(\nabla\bar\rho_\alpha)\\ &\quad +\bar\rho_\alpha\boldsymbol{\lambda}^v_\alpha\cdot(\nabla\mathbf{v}_\alpha\mathbf{v}_\alpha)+\bar\rho_\alpha\boldsymbol{\lambda}^v_\alpha\cdot\mathbf{v}_\alpha\nabla\cdot\mathbf{v}_\alpha \end{aligned}\tag{B.7}$$

was applied. Later we will use the transformation

$$\boldsymbol{\lambda}^v_\alpha\cdot(\nabla\mathbf{v}_\alpha\mathbf{v}_\alpha)=(\boldsymbol{\lambda}^v_\alpha)_i\ (v^\alpha_i)_{,j}\ v^\alpha_j=(\boldsymbol{\lambda}^v_\alpha\otimes\mathbf{v}_\alpha)\cdot\nabla\mathbf{v}_\alpha \tag{B.8}$$

for the first term in the last line of (B.7).

The remainder of line IV can be written as

$$\begin{aligned}&\lambda^\varepsilon\ \mathbf{T}\cdot(\nabla\mathbf{v})\\ &=\lambda^\varepsilon\ \rho^{-1}\ \mathbf{T}\cdot\sum\Big\{\bar\rho_\alpha(\nabla\mathbf{v}_\alpha)+\rho_\alpha(\mathbf{u}_\alpha\otimes\nabla\nu_\alpha)+\nu_\alpha(\mathbf{u}_\alpha\otimes\nabla\rho_\alpha)\Big\}\ ,\end{aligned}\tag{B.9}$$

where the right-hand side of (B.9) is deduced from (3.56) and

$$\begin{aligned}\mathbf{T}\cdot(\rho\nabla\mathbf{v}) &= \mathbf{T}\cdot\nabla(\rho\mathbf{v})-\mathbf{T}\cdot(\mathbf{v}\otimes\nabla\rho)\\ &= \mathbf{T}\cdot\sum\nabla(\bar\rho_\alpha\mathbf{v}_\alpha)-\mathbf{T}\cdot\sum\mathbf{v}\otimes(\nabla\bar\rho_\alpha)\\ &\ \ \vdots\\ &= \mathbf{T}\cdot\sum\bar\rho_\alpha\nabla\mathbf{v}_\alpha+\mathbf{T}\cdot\sum\mathbf{u}_\alpha\otimes(\nabla\bar\rho_\alpha)\ .\end{aligned}\tag{B.10}$$

Leaving lines V and VI unaltered, we transform line VII into

$$\begin{aligned}&-\sum\boldsymbol{\lambda}^Z_\alpha\cdot\Big\{\acute{\bar{\mathbf{Z}}}_\alpha-[\boldsymbol{\Omega}_\alpha\ ,\ \bar{\mathbf{Z}}_\alpha]-\bar{\boldsymbol{\Phi}}_\alpha\Big\}\\ &\qquad=-\sum\boldsymbol{\lambda}^Z_\alpha\cdot\{\partial\bar{\mathbf{Z}}_\alpha+\nabla\bar{\mathbf{Z}}_\alpha\mathbf{v}_\alpha-[\boldsymbol{\Omega}_\alpha,\bar{\mathbf{Z}}_\alpha]-\bar{\boldsymbol{\Phi}}_\alpha\}\ ,\end{aligned}$$

where definition (3.8) has been applied. With these modifications and the saturation condition, we can collect the coefficients belonging to $\partial\rho_\alpha$, $\nabla\rho_\alpha$, $\partial\nu_\alpha$, $\nabla\nu_\alpha$, $\partial\mathbf{v}_\alpha$, $\nabla\mathbf{v}_\alpha$, $\partial\bar{\mathbf{Z}}_\alpha$ and $\nabla\bar{\mathbf{Z}}_\alpha$ to obtain

$$
\begin{aligned}
\pi^{\rho\eta} = \; & \partial(\rho\eta) - \lambda^{\varepsilon}\partial(\rho\varepsilon) + \{\nabla(\rho\eta) - \lambda^{\varepsilon}\nabla(\rho\varepsilon)\}\cdot\mathbf{v} \\
& -\Big\{\nabla\cdot\boldsymbol{\phi}^{\rho\eta} + \lambda^{\varepsilon}\nabla\cdot\mathbf{q} - \sum_{\alpha=1}^{n}\left(\boldsymbol{\lambda}^{v}_{\alpha}\cdot(\nabla\cdot\bar{\mathbf{T}}_{\alpha})\right)\Big\} \\
& -\sum_{\alpha=1}^{m}\{\lambda^{\rho}_{\alpha}\nu_{\alpha} + \nu_{\alpha}\boldsymbol{\lambda}^{v}_{\alpha}\cdot\mathbf{v}_{\alpha}\}(\partial\rho_{\alpha}) \\
& -\sum_{\alpha=1}^{m}\Big\{\lambda^{\rho}_{\alpha}\nu_{\alpha}\mathbf{v}_{\alpha} + \nu_{\alpha}\left(\boldsymbol{\lambda}^{v}_{\alpha}\cdot\mathbf{v}_{\alpha}\right)\mathbf{v}_{\alpha} \\
& \qquad -\nu_{\alpha}\lambda^{\varepsilon}\rho^{-1}\mathbf{T}\,\mathbf{u}_{\alpha} - \nu_{\alpha}(\eta - \lambda^{\varepsilon}\varepsilon)\,\mathbf{u}_{\alpha}\Big\}\cdot(\nabla\rho_{\alpha}) \\
& -\sum_{\alpha=1}^{n-1}\{\lambda^{\rho}_{\alpha}\rho_{\alpha} - \lambda^{\rho}_{n}\rho_{n} + \rho_{\alpha}\boldsymbol{\lambda}^{v}_{\alpha}\cdot\mathbf{v}_{\alpha} - \rho_{n}\boldsymbol{\lambda}^{v}_{n}\cdot\mathbf{v}_{n} + \lambda^{\nu}_{\alpha} - \lambda^{\nu}_{n}\}(\partial\nu_{\alpha}) \\
& -\sum_{\alpha=1}^{n-1}\Big\{\underbrace{(\eta - \lambda^{\varepsilon}\,\varepsilon)(-\rho_{\alpha}\mathbf{u}_{\alpha} + \rho_{n}\mathbf{u}_{n})}_{\text{energy \& entropy}} + \underbrace{\lambda^{\rho}_{\alpha}\,\rho_{\alpha}\,\mathbf{v}_{\alpha} - \lambda^{\rho}_{n}\,\rho_{n}\,\mathbf{v}_{n}}_{\text{mass}} \\
& \quad + \underbrace{\rho_{\alpha}(\boldsymbol{\lambda}^{v}_{\alpha}\cdot\mathbf{v}_{\alpha})\mathbf{v}_{\alpha} - \rho_{n}(\boldsymbol{\lambda}^{v}_{n}\cdot\mathbf{v}_{n})\mathbf{v}_{n}}_{\text{momentum}} \\
& \quad - \underbrace{\lambda^{\varepsilon}\rho^{-1}\rho_{\alpha}(\mathbf{T}\,\mathbf{u}_{\alpha}) + \lambda^{\varepsilon}\rho^{-1}\rho_{n}(\mathbf{T}\,\mathbf{u}_{n})}_{\text{stress-energy}} + \underbrace{\lambda^{\nu}_{\alpha}\mathbf{v}_{\alpha} - \lambda^{\nu}_{n}\mathbf{v}_{n}}_{\text{volume fraction}}\Big\}\cdot(\nabla\nu_{\alpha}) \\
& -\sum_{\alpha=1}^{n}\bar{\rho}_{\alpha}\boldsymbol{\lambda}^{v}_{\alpha}\cdot(\partial\mathbf{v}_{\alpha}) \\
& -\sum_{\alpha=1}^{n}\Big\{\left(\bar{\rho}_{\alpha}\lambda^{\rho}_{\alpha} + \bar{\rho}_{\alpha}\boldsymbol{\lambda}^{v}_{\alpha}\cdot\mathbf{v}_{\alpha} + \nu_{\alpha}\lambda^{\nu}_{\alpha}\right)\mathbf{I} \\
& \qquad -\bar{\rho}_{\alpha}\rho^{-1}\lambda^{\varepsilon}\mathbf{T} - \bar{\rho}_{\alpha}(\eta - \lambda^{\varepsilon}\varepsilon)\,\mathbf{I} + \bar{\rho}_{\alpha}\boldsymbol{\lambda}^{v}_{\alpha}\otimes\mathbf{v}_{\alpha}\Big\}\cdot(\nabla\mathbf{v}_{\alpha}) \\
& -\sum_{\alpha=1}^{n}\Big\{\boldsymbol{\lambda}^{Z}_{\alpha}\cdot(\partial\bar{\mathbf{Z}}_{\alpha}) + \left(\boldsymbol{\lambda}^{Z}_{\alpha}\otimes\mathbf{v}_{\alpha}\right)\cdot(\nabla\bar{\mathbf{Z}}_{\alpha})\Big\} \\
& -\sum_{\alpha=1}^{n}\left[\bar{\mathbf{Z}}_{\alpha},\,\boldsymbol{\lambda}^{Z}_{\alpha}\right]\cdot\boldsymbol{\Omega}_{\alpha} + \sum_{\alpha=1}^{n}\bar{\boldsymbol{\Phi}}_{\alpha}\cdot\boldsymbol{\lambda}^{Z}_{\alpha} \\
& +\sum_{\alpha=1}^{n}\{\boldsymbol{\lambda}^{v}_{\alpha}\cdot(\bar{\mathbf{m}}_{\alpha} - \mathbf{v}_{\alpha}\bar{\rho}_{\alpha}c_{\alpha}) + \boldsymbol{\lambda}^{v}_{\alpha}\cdot\mathbf{v}_{\alpha}\,\bar{\rho}_{\alpha}c_{\alpha} + \lambda^{\rho}_{\alpha}\bar{\rho}_{\alpha}c_{\alpha} + \lambda^{\nu}_{\alpha}\bar{n}_{\alpha}\} \\
\geqslant\; & 0\,. \qquad \text{(B.11)}
\end{aligned}
$$

In the eighth sum in inequality (B.11) the identity

$$\begin{aligned} \boldsymbol{\lambda}^Z_\alpha \cdot \nabla\bar{\mathbf{Z}}_\alpha \, \mathbf{v}_\alpha = \lambda^{Z\alpha}_{ij} \, \bar{Z}^\alpha_{ij,k} \, v^\alpha_k = \lambda^{Z\alpha}_{ij} \, v^\alpha_k \, \bar{Z}^\alpha_{ij,k} \\ = \left(\boldsymbol{\lambda}^Z_\alpha \otimes \mathbf{v}_\alpha\right) \cdot \nabla\bar{\mathbf{Z}}_\alpha \, , \end{aligned} \tag{B.12}$$

and in the ninth sum

$$\begin{aligned} -\boldsymbol{\lambda}^Z_\alpha \cdot \left[\boldsymbol{\Omega}_\alpha, \bar{\mathbf{Z}}_\alpha\right] &= \lambda^{Z\alpha}_{ij} \, \bar{Z}^\alpha_{ik} \, \Omega^\alpha_{kj} - \lambda^{Z\alpha}_{ij} \, \bar{Z}^\alpha_{kj} \, \Omega^\alpha_{ik} \\ &= \bar{Z}^\alpha_{ki} \, \lambda^{Z\alpha}_{ij} \, \Omega^\alpha_{kj} \; - \lambda^{Z\alpha}_{ij} \, \bar{Z}^\alpha_{jk} \, \Omega^\alpha_{ik} \\ &= \bar{Z}^\alpha_{kj} \, \lambda^{Z\alpha}_{ji} \, \Omega^\alpha_{ki} \; - \lambda^{Z\alpha}_{ij} \, \bar{Z}^\alpha_{jk} \, \Omega^\alpha_{ik} \\ &= \left(\bar{\mathbf{Z}}_\alpha \, \boldsymbol{\lambda}^Z_\alpha\right)_{ki} \, \Omega^\alpha_{ki} \; - \left(\boldsymbol{\lambda}^Z_\alpha \, \bar{\mathbf{Z}}_\alpha\right)_{ik} \, \Omega^\alpha_{ik} \\ &= \left[\bar{\mathbf{Z}}_\alpha, \boldsymbol{\lambda}^Z_\alpha\right] \cdot \boldsymbol{\Omega}_\alpha \end{aligned} \tag{B.13}$$

were applied. Moreover, at several places the fact was used that $\bar{\mathbf{T}}_\alpha$, $\mathbf{T}$, $\boldsymbol{\lambda}^Z_\alpha$, $\bar{\mathbf{Z}}_\alpha$ are symmetric second order tensors. If we now substitute abbreviations of the form

$$\begin{aligned} l^\rho_\alpha &:= \lambda^\rho_\alpha + \boldsymbol{\lambda}^v_\alpha \cdot \mathbf{v}_\alpha, & \alpha = 1, ..., m, \\ l^\nu_\alpha &:= \rho_\alpha \left(l^\rho_\alpha\right) + \lambda^\nu_\alpha, & \alpha = 1, ..., n, \\ \boldsymbol{\Gamma} &:= \lambda^\varepsilon \rho^{-1} \mathbf{T} + (\eta - \lambda^\varepsilon \varepsilon) \, \mathbf{I} = \boldsymbol{\Gamma}^{\mathrm{T}} \, , \\ s &:= - \, l^\nu_n, \\ \mathbf{s} &:= s \mathbf{v}_n + \rho_n \boldsymbol{\Gamma} \mathbf{u}_n \, , \end{aligned} \tag{B.14}$$

into (B.11) and recall (4.6) we immediately recover inequality (5.18). □

B.2 Other Auxiliary Results from Section 5.2

Identity (5.28) is deduced in the following way: First, we use the definition of the mixture total derivative, $\dot{(\cdot)} = \partial(\cdot) + \nabla(\cdot)\mathbf{v}$, to merge the two brackets in the first line of (5.28), i.e.

$$\begin{aligned} \left\{\mathcal{P}_\theta \, \partial\theta + \mathcal{P}_{\dot\theta} \, \partial\dot\theta + \mathcal{P}_{\nabla\theta} \cdot \partial(\nabla\theta)\right\} + \left\{\mathcal{P}_\theta \, \nabla\theta + \mathcal{P}_{\dot\theta} \, \nabla\dot\theta + \mathcal{P}_{\nabla\theta} \nabla(\nabla\theta)\right\} \cdot \mathbf{v} \\ = \mathcal{P}_\theta \dot\theta + \mathcal{P}_{\nabla\theta} \cdot (\nabla\theta)^{\cdot} + \mathcal{P}_{\dot\theta} \, \ddot\theta \, . \end{aligned} \tag{B.15}$$

Obviously, $(\nabla\theta)^{\cdot}$ satisfies the identity

$$\begin{aligned}\nabla\dot{\theta} &= (\partial\theta + \theta_{,k}\, v_k)_{,j}\, \mathbf{e}_j \\ &= (\partial\theta_{,j} + \theta_{,kj}\, v_k + \theta_{,k}\, v_{k,j})\mathbf{e}_j = (\nabla\theta)^{\cdot} + \mathbf{L}^{\mathrm{T}}\nabla\theta \ ,\end{aligned} \tag{B.16}$$

and from the sum relation for $\mathbf{L}$ (see (3.58)) we obtain

$$\begin{aligned}(\nabla\theta)^{\cdot} &= \nabla\dot{\theta} - \rho^{-1}\sum \left\{\bar{\rho}_\alpha \mathbf{L}_\alpha^{\mathrm{T}}\nabla\theta + (\nabla\bar{\rho}_\alpha \otimes \mathbf{u}_\alpha)\nabla\theta\right\} \\ &= \nabla\dot{\theta} - \rho^{-1}\Big\{\sum \left(\bar{\rho}_\alpha \mathbf{L}_\alpha^{\mathrm{T}}\nabla\theta\right) + \sum_{\alpha=1}^{m}\left(\nu_\alpha(\nabla\rho_\alpha \otimes \mathbf{u}_\alpha)\nabla\theta\right) \\ &\quad - \sum_{\alpha=1}^{n-1}\left(\rho_\alpha(\nabla\nu_\alpha \otimes \mathbf{u}_\alpha)\nabla\theta\right) + \sum_{\alpha=1}^{n-1}\left(\rho_n(\nabla\nu_\alpha \otimes \mathbf{u}_n)\nabla\theta\right)\Big\} \ .\end{aligned} \tag{B.17}$$

If we use the decomposition of $\mathbf{L}_\alpha$ (see (3.9)) and the identities

$$\begin{aligned}-\rho^{-1}\mathcal{P}_{\nabla\theta}\cdot\sum \bar{\rho}_\alpha\, \mathbf{L}_\alpha^{\mathrm{T}}\nabla\theta &= -\rho^{-1}\sum \bar{\rho}_\alpha \mathcal{P}_{\nabla\theta}\cdot(\mathbf{D}_\alpha - \mathbf{W}_\alpha)\,\nabla\theta \\ &= -\rho^{-1}\sum \bar{\rho}_\alpha\, (\mathcal{P}_{\nabla\theta})_j\, (\mathbf{D}_\alpha - \mathbf{W}_\alpha)_{ji}\, \theta_{,i} \\ &= -\rho^{-1}\sum \bar{\rho}_\alpha\, (\mathcal{P}_{\nabla\theta}\otimes\nabla\theta)\cdot(\mathbf{D}_\alpha - \mathbf{W}_\alpha) \ ,\end{aligned} \tag{B.18}$$

$$\begin{aligned}-\rho^{-1}\mathcal{P}_{\nabla\theta}\cdot\sum_{\alpha=1}^{n-1}\rho_\alpha\, (\nabla\nu_\alpha\otimes\mathbf{u}_\alpha)\nabla\theta &= -\rho^{-1}\sum_{\alpha=1}^{n-1}\rho_\alpha\,(\mathbf{u}_\alpha\cdot\nabla\theta)\ \mathcal{P}_{\nabla\theta}\cdot\nabla\nu_\alpha \\ &= -\rho^{-1}\sum_{\alpha=1}^{n-1}\rho_\alpha\,\{(\mathcal{P}_{\nabla\theta}\otimes\nabla\theta)\,\mathbf{u}_\alpha\}\cdot\nabla\nu_\alpha \ ,\end{aligned} \tag{B.19}$$

$$\begin{aligned}-\rho^{-1}\mathcal{P}_{\nabla\theta}\cdot\sum_{\alpha=1}^{m}\nu_\alpha\, (\nabla\rho_\alpha\otimes\mathbf{u}_\alpha)\nabla\theta &= -\rho^{-1}\sum_{\alpha=1}^{m}\nu_\alpha\,(\mathbf{u}_\alpha\cdot\nabla\theta)\ \mathcal{P}_{\nabla\theta}\cdot\nabla\rho_\alpha \\ &= -\rho^{-1}\sum_{\alpha=1}^{m}\nu_\alpha\,\{(\mathcal{P}_{\nabla\theta}\otimes\nabla\theta)\,\mathbf{u}_\alpha\}\cdot\nabla\rho_\alpha \ ,\end{aligned} \tag{B.20}$$

we immediately obtain relation (5.28). □

To prove (5.31), that is

$$\begin{aligned}\mathcal{P}_{\mathbf{B}_\alpha}\cdot(\partial\mathbf{B}_\alpha) + \left(\mathcal{P}_{\mathbf{B}_\alpha}\otimes\mathbf{v}\right)\cdot(\nabla\mathbf{B}_\alpha) &= \langle\mathbf{B}_\alpha\ ,\ \mathcal{P}_{\mathbf{B}_\alpha}\rangle\cdot\mathbf{D}_\alpha \\ &\quad - [\mathbf{B}_\alpha\ ,\ \mathcal{P}_{\mathbf{B}_\alpha}]\cdot\mathbf{W}_\alpha - \left(\mathcal{P}_{\mathbf{B}_\alpha}\otimes\mathbf{u}_\alpha\right)\cdot(\nabla\mathbf{B}_\alpha) \ ,\end{aligned} \tag{B.21}$$

we first take (3.56) to transform the left-hand side of (B.21) into

$$\begin{aligned}&\mathcal{P}_{\mathbf{B}_\alpha}\cdot(\partial\mathbf{B}_\alpha)+\left(\mathcal{P}_{\mathbf{B}_\alpha}\otimes\mathbf{v}\right)\cdot(\nabla\mathbf{B}_\alpha)\\&\quad=\underbrace{\mathcal{P}_{\mathbf{B}_\alpha}\cdot(\partial\mathbf{B}_\alpha)+(\mathcal{P}_{\mathbf{B}_\alpha}\otimes\mathbf{v}_\alpha)\cdot(\nabla\mathbf{B}_\alpha)}_{(1)}-(\mathcal{P}_{\mathbf{B}_\alpha}\otimes\mathbf{u}_\alpha)\cdot(\nabla\mathbf{B}_\alpha)\ ,\end{aligned}\tag{B.22}$$

where

$$\begin{aligned}(1)&=(\mathcal{P}_{\mathbf{B}_\alpha})_{ij}\,(\partial\mathbf{B}_\alpha)_{ij}+(\mathcal{P}_{\mathbf{B}_\alpha})_{ij}\,B^\alpha_{ij,k}v^\alpha_k\\&=(\mathcal{P}_{\mathbf{B}_\alpha})_{ij}\left\{(\partial\mathbf{B}_\alpha)_{ij}+(\nabla\mathbf{B}_\alpha)_{ijk}\,v^\alpha_k\right\}\\&=(\mathcal{P}_{\mathbf{B}_\alpha})_{ij}\left(\acute{\mathbf{B}}_\alpha\right)_{ij}\ .\end{aligned}\tag{B.23}$$

The last term in (B.23), which denotes the material derivative of $\mathbf{B}_\alpha$ following constituent K_α, can be transformed into

$$\begin{aligned}\acute{\mathbf{B}}_\alpha&=\acute{\mathbf{F}}_\alpha\mathbf{F}^{\mathrm{T}}_\alpha+\mathbf{F}_\alpha\acute{\mathbf{F}}^{\mathrm{T}}_\alpha\\&=\mathbf{L}_\alpha\mathbf{F}_\alpha\mathbf{F}^{\mathrm{T}}_\alpha+\mathbf{F}_\alpha\mathbf{F}^{\mathrm{T}}_\alpha\mathbf{L}^{\mathrm{T}}_\alpha=\mathbf{L}_\alpha\mathbf{B}_\alpha+\mathbf{B}_\alpha\mathbf{L}^{\mathrm{T}}_\alpha\ ,\end{aligned}\tag{B.24}$$

where $\acute{\mathbf{F}}_\alpha=\mathbf{L}_\alpha\mathbf{F}_\alpha$ has been used. Now, (1) has the form

$$(1)=\mathcal{P}_{\mathbf{B}_\alpha}\cdot\left(\mathbf{L}_\alpha\mathbf{B}_\alpha+\mathbf{B}_\alpha\mathbf{L}^{\mathrm{T}}_\alpha\right)\tag{B.25}$$

which, with the decomposition $(3.9)_1$ and the definitions of the Lie- and Jacobi-brackets (see $(2.10)_{3,4}$), can be written as

$$\begin{aligned}(1)&=\langle\mathcal{P}_{\mathbf{B}_\alpha},\mathbf{B}_\alpha\rangle\cdot\mathbf{D}_\alpha+[\mathcal{P}_{\mathbf{B}_\alpha},\mathbf{B}_\alpha]\cdot\mathbf{W}_\alpha\\&=\langle\mathbf{B}_\alpha,\mathcal{P}_{\mathbf{B}_\alpha}\rangle\cdot\mathbf{D}_\alpha-[\mathbf{B}_\alpha,\mathcal{P}_{\mathbf{B}_\alpha}]\cdot\mathbf{W}_\alpha\ .\end{aligned}\tag{B.26}$$

The expression (B.26) together with (B.22) yields (B.21). □

B.3 Deduction of the Liu Identities $(6.7)_{2,3}$

Let $\mathbf{A}_I$ be a symmetric second rank tensor representing elements of either $\{\vec{\bar{\mathbf{Z}}}\}$ or $\{\vec{\mathbf{B}}\}$, and let $\mathcal{P}_{\mathbf{A}_I}$ be one corresponding element of the set of one-forms $\{\mathcal{P}_{\mathbf{B}_1},\ \cdots,\ \mathcal{P}_{\mathbf{B}_n},\mathcal{P}_{\bar{\mathbf{Z}}_1},\ \cdots,\mathcal{P}_{\bar{\mathbf{Z}}_n}\}$. We then write equations $(6.2)_{7,8}$ in the form

$$\left\{(\boldsymbol{\mathcal{F}}_{\mathbf{A}_I})^{\mathrm{T}}+(\mathcal{P}_{\mathbf{A}_I}\otimes\mathbf{u}_\alpha)\right\}\cdot(\nabla\mathbf{A}_I)=0\qquad\forall\ \nabla\mathbf{A}_I\ ,\tag{B.27}$$

where the constituent-index of $\mathbf{u}_\alpha$ has to coincide with that of $\mathbf{A}_I$, when for $\mathbf{A}_I$ an element of $\{\vec{\bar{\mathbf{Z}}},\ \vec{\mathbf{B}}\}$ is chosen. In index notation (B.27) reads

$$\left\{(\boldsymbol{\mathcal{F}}_{\mathbf{A}_I})^{\mathrm{T}} + (\boldsymbol{\mathcal{P}}_{\mathbf{A}_I} \otimes \mathbf{u}_\alpha)\right\}_{ijk} (\nabla \mathbf{A}_I)_{ijk} = 0 \qquad \forall\, (\nabla \mathbf{A}_I)_{ijk} \ . \tag{B.28}$$

As $\mathbf{A}_I$ is symmetric and (B.28) must be true for all choices of $\nabla \mathbf{A}_I$ (I fixed), it can only be satisfied if $\left\{(\boldsymbol{\mathcal{F}}_{\mathbf{A}_I})^{\mathrm{T}} + (\boldsymbol{\mathcal{P}}_{\mathbf{A}_I} \otimes \mathbf{u}_\alpha)\right\}_{ijk}$ is skew-symmetric in the first two indices. Using the latter result and the symmetry of $\mathbf{A}_I$ in $\boldsymbol{\mathcal{F}}_{\mathbf{A}_I}$ and $\boldsymbol{\mathcal{P}}_{\mathbf{A}_I}$ yields

$$\begin{aligned} \mathcal{F}_{iA^I_{jk}} + \mathcal{P}_{A^I_{ij}} u^\alpha_k &= \mathcal{F}_{jA^I_{ki}} - \mathcal{P}_{A^I_{ij}} u^\alpha_k \\ &= \mathcal{F}_{jA^I_{ik}} + \mathcal{P}_{A^I_{ji}} u^\alpha_k \ , \end{aligned} \tag{B.29}$$

which proves that $\left\{(\boldsymbol{\mathcal{F}}_{\mathbf{A}_I})^{\mathrm{T}} + (\boldsymbol{\mathcal{P}}_{\mathbf{A}_I} \otimes \mathbf{u}_\alpha)\right\}_{ijk}$ is also symmetric with respect to the first two inidices. This implies that

$$2\left\{(\boldsymbol{\mathcal{F}}_{\mathbf{A}_I})^{\mathrm{T}} + (\boldsymbol{\mathcal{P}}_{\mathbf{A}_I} \otimes \mathbf{u}_\alpha)\right\}_{ijk} = 0 \ , \tag{B.30}$$

or

$$(\boldsymbol{\mathcal{F}}_{\mathbf{A}_I})^{\mathrm{T}} = -\left(\boldsymbol{\mathcal{P}}_{\mathbf{A}_I} \otimes \mathbf{u}_\alpha\right) \ , \tag{B.31}$$

which can also be written as

$$\boldsymbol{\mathcal{F}}_{\mathbf{A}_I} = -\left(\mathbf{u}_\alpha \otimes \boldsymbol{\mathcal{P}}_{\mathbf{A}_I}\right) \ . \tag{B.32}$$

Identifying $\mathbf{A}_I$ with the elements of $\{\vec{\bar{\mathbf{Z}}}, \vec{\mathbf{B}}\}$, we obtain relations $(6.7)_{1,2}$. □

B.4 Deduction of an Isotropic Representation of the Mixture Flux Density

Let us first recall relation (6.56):

$$\begin{aligned} (\delta_{ij} k_k - \delta_{ik} k_j) - \sum_{\alpha=1}^{n} \left((\boldsymbol{\lambda}^v_\alpha)_j (\bar{\mathbf{T}}_\alpha)_{ki} - (\boldsymbol{\lambda}^v_\alpha)_k (\bar{\mathbf{T}}_\alpha)_{ji} \right) & \\ = \sum_{\alpha=1}^{n} \left(\mathcal{F}^{\mathbf{v}_\alpha}_{ij} v^\alpha_k - \mathcal{F}^{\mathbf{v}_\alpha}_{ik} v^\alpha_j \right) . & \end{aligned} \tag{B.33}$$

If we now use assumption [**A11**], the principle of objectivity for $\sum \boldsymbol{\mathcal{F}}_{\mathbf{v}_\alpha}$, i.e.,

$$\mathbf{v} \otimes \sum \boldsymbol{\mathcal{F}}_{\mathbf{v}_\alpha} = \mathbf{0} \ ,$$

and definition (3.56), we can write (B.33) in the form

$$(\delta_{ij}k_k - \delta_{ik}k_j) + \lambda^\varepsilon \sum_{\alpha=1}^{n} \Big(u_j^\alpha (\bar{\mathbf{T}}_\alpha)_{ki} - u_k^\alpha (\bar{\mathbf{T}}_\alpha)_{ji} \Big)$$

$$= \sum_{\alpha=1}^{n} \left(\mathcal{F}_{ij}^{\mathbf{v}_\alpha} u_k^\alpha - \mathcal{F}_{ik}^{\mathbf{v}_\alpha} u_j^\alpha \right) , \qquad \text{(B.34)}$$

which with the aid of [**A12**] can be transformed into

$$(\delta_{ij}k_k - \delta_{ik}k_j) = -\lambda^\varepsilon \sum_{\alpha=1}^{n} \Big(u_j^\alpha \underbrace{(\bar{\mathbf{T}}_\alpha + \tfrac{1}{\lambda^\varepsilon}\boldsymbol{\mathcal{F}}_{\mathbf{v}_\alpha})_{ki}}_{=:(\mathbf{G}_\alpha)_{ki}}$$

$$- u_k^\alpha \underbrace{(\bar{\mathbf{T}}_\alpha + \tfrac{1}{\lambda^\varepsilon}\boldsymbol{\mathcal{F}}_{\mathbf{v}_\alpha})_{ji}}_{=:(\mathbf{G}_\alpha)_{ji}} \Big) . \qquad \text{(B.35)}$$

Here, the abbreviation $\mathbf{G}_\alpha$ is a symmetric second rank tensor. For the special choices of indices, (6.57), we obtain from (B.35):

- $(i,\ j) = (1,\ 1)$ $\quad k = 2: \quad k_2 = -\lambda^\varepsilon \sum \left\{ u_1^\alpha G_{21}^\alpha - u_2^\alpha G_{11}^\alpha \right\}$,

  $k = 3: \quad k_3 = -\lambda^\varepsilon \sum \left\{ u_1^\alpha G_{31}^\alpha - u_3^\alpha G_{11}^\alpha \right\}$,
- $(i,\ j) = (2,\ 2)$ $\quad k = 1: \quad k_1 = -\lambda^\varepsilon \sum \left\{ u_2^\alpha G_{12}^\alpha - u_1^\alpha G_{22}^\alpha \right\}$,

  $k = 3: \quad k_3 = -\lambda^\varepsilon \sum \left\{ u_2^\alpha G_{32}^\alpha - u_3^\alpha G_{22}^\alpha \right\}$,
- $(i,\ j) = (3,\ 3)$ $\quad k = 1: \quad k_1 = -\lambda^\varepsilon \sum \left\{ u_3^\alpha G_{13}^\alpha - u_1^\alpha G_{33}^\alpha \right\}$,

  $k = 2: \quad k_2 = -\lambda^\varepsilon \sum \left\{ u_3^\alpha G_{23}^\alpha - u_2^\alpha G_{33}^\alpha \right\}$. (B.36)

Next, we add the expressions for k_i $(i = 1, 2, 3)$, and obtain

$$2k_2 = \lambda^\varepsilon \sum \Big\{ - u_1^\alpha G_{21}^\alpha + u_2^\alpha G_{11}^\alpha - u_3^\alpha G_{23}^\alpha + u_2^\alpha G_{33}^\alpha \Big\} . \qquad \text{(B.37)}$$

If we now add and subtract $\lambda^\varepsilon \sum (u_2^\alpha G_{22}^\alpha)$ on the right hand-side of (B.37), we obtain

$$2k_2 = \lambda^\varepsilon \sum \Big\{ - u_1^\alpha G_{21}^\alpha - u_2^\alpha G_{22}^\alpha - u_3^\alpha G_{23}^\alpha + u_2^\alpha \operatorname{tr}(\mathbf{G}_\alpha) \Big\} . \qquad \text{(B.38)}$$

A similar procedure for k_1 and k_3 leads to

$$2k_1 = \lambda^\varepsilon \sum \Big\{ - u_1^\alpha G_{11}^\alpha - u_2^\alpha G_{12}^\alpha - u_3^\alpha G_{13}^\alpha + u_1^\alpha \operatorname{tr}(\mathbf{G}_\alpha) \Big\} \tag{B.39}$$

and

$$2k_3 = \lambda^\varepsilon \sum \Big\{ - u_1^\alpha G_{31}^\alpha - u_2^\alpha G_{32}^\alpha - u_3^\alpha G_{33}^\alpha - u_3^\alpha \operatorname{tr}(\mathbf{G}_\alpha) \Big\} . \tag{B.40}$$

In symbolic notation (B.38) to (B.40) read

$$2\mathbf{k} = -\lambda^\varepsilon \sum \{\mathbf{G}_\alpha - \operatorname{tr}(\mathbf{G}_\alpha)\,\mathbf{I}\}\mathbf{u}_\alpha \,, \tag{B.41}$$

and if we replace $\mathbf{G}_\alpha$ by its definition in (B.35), (B.41) can be written as

$$\begin{aligned} \mathbf{k} = -\tfrac{1}{2}\lambda^\varepsilon \sum \Big\{ &\{\bar{\mathbf{T}}_\alpha - \operatorname{tr}(\bar{\mathbf{T}}_\alpha)\mathbf{I}\}\mathbf{u}_\alpha \\ &+ (\lambda^\varepsilon)^{-1}\{\boldsymbol{\mathcal{F}}_{\mathbf{v}_\alpha} - \operatorname{tr}(\boldsymbol{\mathcal{F}}_{\mathbf{v}_\alpha})\mathbf{I}\}\mathbf{u}_\alpha \Big\} \,, \end{aligned} \tag{B.42}$$

which is the desired relation (6.58). □

B.5 Auxiliary Results for Section 7.1

Equation $(7.3)_1$ is derived in the following way:

$$\begin{aligned} (\mathbf{u}_\gamma),_{\rho_\alpha} &\overset{(3.56)}{=} \underbrace{(\mathbf{v}_\gamma),_{\rho_\alpha}}_{\mathbf{0}} - \Big(\rho^{-1}\sum_{\beta=1}^{n} \bar{\rho}_\beta \mathbf{v}_\beta\Big),_{\rho_\alpha} \\ &= -\rho^{-2}\sum_{\beta=1}^{n}\big(\rho\,\nu_\alpha\,\delta_{\beta\alpha}\,\mathbf{v}_\beta - \bar{\rho}_\beta\,\nu_\alpha\,\mathbf{v}_\beta\big) \\ &= -\rho^{-1}\nu_\alpha\,\mathbf{v}_\alpha + \nu_\alpha\,\rho^{-2}\underbrace{\sum_{\beta=1}^{n}\bar{\rho}_\beta\mathbf{v}_\beta}_{\rho\mathbf{v}} \\ &= -\rho^{-1}\nu_\alpha\mathbf{u}_\alpha \,. \end{aligned} \tag{B.43}$$

□

Equation $(7.3)_2$ is obtained via

$$
\begin{aligned}
(\mathbf{u}_\alpha),_{\nu_\beta} &\overset{(3.56)}{=} \underbrace{(\mathbf{v}_\alpha),_{\nu_\beta}}_{\mathbf{0}} - \big(\rho^{-1}\sum_{\gamma=1}^{n}\bar{\rho}_\gamma \mathbf{v}_\gamma\big),_{\nu_\beta} \\
&= -\Big\{(\rho^{-1}),_{\nu_\beta}\sum_{\gamma=1}^{n}\bar{\rho}_\gamma \mathbf{v}_\gamma + \rho^{-1}\big(\sum_{\gamma=1}^{n}\bar{\rho}_\gamma \mathbf{v}_\gamma\big),_{\nu_\beta}\Big\} \\
&= -\Big\{-\rho^{-2}\rho,_{\nu_\beta}\sum_{\gamma=1}^{n}\bar{\rho}_\gamma \mathbf{v}_\gamma \\
&\qquad + \rho^{-1}\big(\sum_{\gamma=1}^{n-1}\bar{\rho}_\gamma \mathbf{v}_\gamma\big),_{\nu_\beta} + \rho^{-1}\left(\nu_n\rho_n\mathbf{v}_n\right),_{\nu_\beta}\Big\} \\
&= -\Big\{-\rho^{-2}\big\{\sum_{\gamma=1}^{n-1}(\nu_\gamma\rho_\gamma),_{\nu_\beta} + (\nu_n\rho_n),_{\nu_\beta}\big\}\sum_{\gamma=1}^{n}\bar{\rho}_\gamma \mathbf{v}_\gamma \\
&\qquad + \rho^{-1}(\rho_\beta\mathbf{v}_\beta) + \rho^{-1}\rho_n\mathbf{v}_n\big(1-\sum_{\gamma=1}^{n-1}\nu_\gamma\big),_{\nu_\beta}\Big\} \\
&= -\Big\{\rho^{-2}(\rho_n-\rho_\beta)\underbrace{\sum_{\gamma=1}^{n}\bar{\rho}_\gamma \mathbf{v}_\gamma}_{\rho\mathbf{v}} + \rho^{-1}\big(\rho_\beta\mathbf{v}_\beta - \rho_n\mathbf{v}_n\big)\Big\} \\
&= -\xi_\beta\mathbf{u}_\beta + \xi_n\mathbf{u}_n\ .
\end{aligned}
\tag{B.44}
$$

□

B.6 Derivation of Residual Inequality (7.5)

We recall inequality (7.4),

$$
\begin{aligned}
\pi^{\rho\eta} = \quad & \mathcal{P}_\theta\,(\dot{\theta}) - \Big\{\mathbf{k},_\theta - (\lambda^\varepsilon),_\theta \{\mathbf{q} + \sum \bar{\mathbf{T}}_\alpha \mathbf{u}_\alpha\}\Big\} \cdot (\nabla\theta) & \text{(I)}\\
& + \sum_{\alpha=1}^{m} \Big\{\nu_\alpha\big(\boldsymbol{\Gamma}^* - l_\alpha^\rho \mathbf{I}\big)\mathbf{u}_\alpha - \mathbf{k},_{\rho_\alpha} - \nu_\alpha \lambda^\varepsilon \rho^{-1} \mathbf{T}_I\, \mathbf{u}_\alpha \\
& \qquad\qquad - \rho^{-1}\nu_\alpha \,\mathrm{skw}\,(\mathcal{P}_{\nabla\theta} \otimes \nabla\theta)\,\mathbf{u}_\alpha\Big\} \cdot (\nabla\rho_\alpha) & \text{(II)}\\
& + \sum_{\alpha=1}^{n-1} \Big\{\big(\rho_\alpha \boldsymbol{\Gamma}^* - l_\alpha^\nu \mathbf{I}\big)\mathbf{u}_\alpha - \mathbf{k},_{\nu_\alpha} - \lambda^\varepsilon \mathbf{T}_I \big(\xi_\alpha \mathbf{u}_\alpha - \xi_n \mathbf{u}_n\big)\\
& \qquad + s\mathbf{v} - \mathbf{s}^* - \mathrm{skw}\,(\mathcal{P}_{\nabla\theta} \otimes \nabla\theta)\,\big(\xi_\alpha \mathbf{u}_\alpha - \xi_n \mathbf{u}_n\big)\Big\} \cdot (\nabla\nu_\alpha) & \text{(III)}\\
& + \sum_{\alpha=1}^{n} \Big\{\nu_\alpha\big(\rho_\alpha \boldsymbol{\Gamma}^* - l_\alpha^\nu \mathbf{I}\big) + \bar{\rho}_\alpha \lambda^\varepsilon (\mathbf{u}_\alpha \otimes \mathbf{u}_\alpha) + \langle \mathbf{B}_\alpha\ ,\ \mathcal{P}_{\mathbf{B}_\alpha}\rangle \\
& \qquad\qquad - \mathbf{k},_{\mathbf{v}_\alpha} + \lambda^\varepsilon \big(\bar{\mathbf{T}}_\alpha - \bar{\xi}_\alpha \mathbf{T}_I\ \big)\Big\} \cdot (\mathbf{D}_\alpha) & \text{(IV)}\\
& + \sum_{\alpha=1}^{n} \boldsymbol{\lambda}_\alpha^Z \cdot \bar{\boldsymbol{\Phi}}_\alpha & \text{(V)}\\
& + \sum_{\alpha=1}^{n} \{\boldsymbol{\lambda}_\alpha^v \cdot \bar{\mathbf{m}}_\alpha^i + l_\alpha^\rho \bar{\rho}_\alpha c_\alpha + \lambda_\alpha^\nu \bar{n}_\alpha\} & \text{(VI)}\\
\geqslant 0\ . &
\end{aligned}
\tag{B.45}
$$

If we write $\boldsymbol{\Gamma}^*$ in the form

$$
\begin{aligned}
\boldsymbol{\Gamma}^* &\overset{(6.10)}{=} \boldsymbol{\Gamma} - \rho^{-1}\,\mathrm{sym}\,(\mathcal{P}_{\nabla\theta} \otimes \nabla\theta)\\
&\overset{(5.21)}{=} \lambda^\varepsilon \rho^{-1}\mathbf{T} + (\eta - \lambda^\varepsilon \varepsilon)\mathbf{I} - \rho^{-1}\,\mathrm{sym}\,(\mathcal{P}_{\nabla\theta} \otimes \nabla\theta)\\
&= \lambda^\varepsilon \rho^{-1}\,(\mathbf{T}_I + \mathbf{T}_D) - \lambda^\varepsilon \Psi^G \mathbf{I} - \rho^{-1}\,\mathrm{sym}\,(\mathcal{P}_{\nabla\theta} \otimes \nabla\theta)\ ,
\end{aligned}
\tag{B.46}
$$

line (II) of inequality (B.45) can be transformed into

$$
\begin{aligned}
(\mathrm{II}) = \sum_{\alpha=1}^{m} \Big\{ & \{\nu_\alpha \rho^{-1} \lambda^\varepsilon \mathbf{T}_I + \nu_\alpha \rho^{-1} \lambda^\varepsilon \mathbf{T}_D - \nu_\alpha \lambda^\varepsilon \Psi^G \mathbf{I} \\
& \qquad - \nu_\alpha \rho^{-1} \operatorname{sym}(\mathcal{P}_{\nabla\theta} \otimes \nabla\theta) \,\} \mathbf{u}_\alpha \\
& \underbrace{- \bar{l}^{\rho}_{\alpha} \mathbf{u}_\alpha - \tfrac{1}{2} \lambda^\varepsilon \nu_\alpha (\mathbf{u}_\alpha \cdot \mathbf{u}_\alpha) \mathbf{u}_\alpha}_{-\bar{l}^{\rho}_{\alpha \mathrm{I}} \, \mathbf{u}_\alpha} + \tfrac{1}{2} \lambda^\varepsilon \nu_\alpha (\mathbf{u}_\alpha \cdot \mathbf{u}_\alpha) \mathbf{u}_\alpha - \mathbf{k}_{,\rho_\alpha} \\
& - \nu_\alpha \rho^{-1} \lambda^\varepsilon \mathbf{T}_I \mathbf{u}_\alpha - \nu_\alpha \, \rho^{-1} \operatorname{skw}(\mathcal{P}_{\nabla\theta} \otimes \nabla\theta) \, \mathbf{u}_\alpha \Big\} \cdot (\nabla \rho_\alpha)
\end{aligned}
$$

$$
\begin{aligned}
= \sum_{\alpha=1}^{m} \Big\{ & \nu_\alpha \lambda^\varepsilon \{ \rho^{-1} \mathbf{T}_D - \Psi^G_D \mathbf{I} + \tfrac{1}{2} (\mathbf{u}_\alpha \cdot \mathbf{u}_\alpha) \mathbf{I} \\
& \qquad - (\lambda^\varepsilon \rho)^{-1} \operatorname{sym}(\mathcal{P}_{\nabla\theta} \otimes \nabla\theta) \,\} \mathbf{u}_\alpha \\
& - \lambda^\varepsilon \{ \nu_\alpha \Psi^G_I + (\lambda^\varepsilon)^{-1} \bar{l}^{\rho}_{\alpha \mathrm{I}} \} \mathbf{u}_\alpha \\
& - \mathbf{k}_{,\rho_\alpha} - \nu_\alpha \, \rho^{-1} \operatorname{skw}(\mathcal{P}_{\nabla\theta} \otimes \nabla\theta) \, \mathbf{u}_\alpha \Big\} \cdot (\nabla \rho_\alpha) \,, \qquad \text{(B.47)}
\end{aligned}
$$

in which $\Psi^G = \Psi^G_I + \Psi^G_D$, $\mathbf{T} = \mathbf{T}_I + \mathbf{T}_D$ were used. From (6.104) we deduce

$$
-\lambda^\varepsilon \nu_\alpha \{ \Psi^G_I + (\lambda^\varepsilon)^{-1} l^{\rho}_{\alpha \mathrm{I}} \} = \lambda^\varepsilon \nu_\alpha (\rho_\alpha)^{-1} p^G_\alpha \,, \qquad \text{(B.48)}
$$

from which, together with the definition

$$
\boldsymbol{\Delta}^{*\alpha}_D := \rho^{-1} \Big(\mathbf{T}_D - (\lambda^\varepsilon)^{-1} \operatorname{sym}(\mathcal{P}_{\nabla\theta} \otimes \nabla\theta) \Big) - \Psi^G_D \, \mathbf{I} + \tfrac{1}{2} (\mathbf{u}_\alpha \cdot \mathbf{u}_\alpha) \, \mathbf{I} \,, \qquad \text{(B.49)}
$$

we obtain

$$
\begin{aligned}
(\mathrm{II}) = \lambda^\varepsilon \sum_{\alpha=1}^{m} \Big\{ & \nu_\alpha \{ \boldsymbol{\Delta}^{*\alpha}_D + (\rho_\alpha)^{-1} p^G_\alpha \mathbf{I} \} \mathbf{u}_\alpha - (\lambda^\varepsilon)^{-1} \mathbf{k}_{,\rho_\alpha} \\
& - \nu_\alpha (\lambda^\varepsilon \rho)^{-1} \operatorname{skw}(\mathcal{P}_{\nabla\theta} \otimes \nabla\theta) \, \mathbf{u}_\alpha \Big\} \cdot (\nabla \rho_\alpha) \,. \qquad \text{(B.50)}
\end{aligned}
$$

In line (III), we again apply (B.46) to derive

$$
\begin{aligned}
\text{(III)} = \sum_{\alpha=1}^{n-1} \Big\{ & \lambda^{\varepsilon}\rho_\alpha\{\rho^{-1}(\mathbf{T}_I + \mathbf{T}_D) - \Psi_D^G\mathbf{I} + \tfrac{1}{2}(\mathbf{u}_\alpha\cdot\mathbf{u}_\alpha)\mathbf{I} \\
& \qquad -(\rho\lambda^{\varepsilon})^{-1}\,\mathrm{sym}\,(\mathcal{P}_{\nabla\theta}\otimes\nabla\theta)\,\}\mathbf{u}_\alpha \\
& -\tfrac{1}{2}\lambda^{\varepsilon}\rho_\alpha(\mathbf{u}_\alpha\cdot\mathbf{u}_\alpha)\mathbf{u}_\alpha - \lambda^{\varepsilon}\rho_\alpha\Psi_I^G\mathbf{u}_\alpha - l_\alpha^\nu\,\mathbf{u}_\alpha + s\mathbf{v} - \mathbf{s}^* - \mathbf{k}_{,\nu_\alpha} \\
& -\lambda^{\varepsilon}\mathbf{T}_I(\xi_\alpha\mathbf{u}_\alpha - \xi_n\mathbf{u}_n) - \mathrm{skw}\,(\mathcal{P}_{\nabla\theta}\otimes\nabla\theta)\,(\xi_\alpha\mathbf{u}_\alpha - \xi_n\mathbf{u}_n)\Big\}\cdot(\nabla\nu_\alpha) \\
= \sum_{\alpha=1}^{n-1} \Big\{ & \lambda^{\varepsilon}\rho_\alpha\boldsymbol{\Delta}_D^{*\alpha}\mathbf{u}_\alpha - \lambda^{\varepsilon}\rho_\alpha\Psi_I^G\mathbf{u}_\alpha - \underbrace{\big(l_\alpha^\nu + \tfrac{1}{2}\lambda^{\varepsilon}\rho_\alpha(\mathbf{u}_\alpha\cdot\mathbf{u}_\alpha) - \mathsf{k}\big)}_{\underset{\alpha}{l}{}_{\mathrm{I}}^{\nu}}\mathbf{u}_\alpha \\
& -\mathsf{k}\mathbf{u}_\alpha + s\mathbf{v} - \mathbf{s}^* \\
& -\mathbf{k}_{,\nu_\alpha} + \lambda^{\varepsilon}\xi_n\mathbf{T}_I\mathbf{u}_n - \mathrm{skw}\,(\mathcal{P}_{\nabla\theta}\otimes\nabla\theta)\,(\xi_\alpha\mathbf{u}_\alpha - \xi_n\mathbf{u}_n)\Big\}\cdot(\nabla\nu_\alpha) \\
= \sum_{\alpha=1}^{n-1} \Big\{ & \lambda^{\varepsilon}\rho_\alpha\boldsymbol{\Delta}_D^{*\alpha}\mathbf{u}_\alpha - \lambda^{\varepsilon}\underbrace{\Big((\rho_\alpha - \rho_n)\Psi_I^G + (\lambda^{\varepsilon})^{-1}\underset{\alpha}{l}{}_{\mathrm{I}}^{\nu} + \varsigma\Big)}_{-\beta_\alpha^G}\mathbf{u}_\alpha \\
& -\lambda^{\varepsilon}\rho_n\Psi_I^G\mathbf{u}_\alpha + \lambda^{\varepsilon}\varsigma\mathbf{u}_\alpha - \mathsf{k}\mathbf{u}_\alpha + s\mathbf{v} - \mathbf{s}^* - \mathbf{k}_{,\nu_\alpha} \\
& +\lambda^{\varepsilon}\xi_n\mathbf{T}_I\mathbf{u}_n - \mathrm{skw}\,(\mathcal{P}_{\nabla\theta}\otimes\nabla\theta)\,(\xi_\alpha\mathbf{u}_\alpha - \xi_n\mathbf{u}_n)\Big\}\cdot(\nabla\nu_\alpha),
\end{aligned} \tag{B.51}
$$

where the definitions of $\boldsymbol{\Delta}_D^{*\alpha}$, (B.49), and of $\underset{\alpha}{l}{}_{\mathrm{I}}^{\nu}$, (6.93), have been used. If we now consider relation (6.105), we are able to write

$$
\begin{aligned}
\text{(III)} = \sum_{\alpha=1}^{n-1} \Big\{ & \lambda^{\varepsilon}\rho_\alpha\boldsymbol{\Delta}_D^{*\alpha}\mathbf{u}_\alpha + \lambda^{\varepsilon}\Big(\beta_\alpha^G - \rho_n\Psi_I^G + \varsigma\Big)\mathbf{u}_\alpha - \mathsf{k}\mathbf{u}_\alpha + s\mathbf{v} - \mathbf{s}^* - \mathbf{k}_{,\nu_\alpha} \\
& + \lambda^{\varepsilon}\xi_n\mathbf{T}_I\mathbf{u}_n - \mathrm{skw}\,(\mathcal{P}_{\nabla\theta}\otimes\nabla\theta)\,(\xi_\alpha\mathbf{u}_\alpha - \xi_n\mathbf{u}_n)\Big\}\cdot(\nabla\nu_\alpha)\;.
\end{aligned} \tag{B.52}
$$

Furthermore, by defining

$$
\zeta_\alpha := \begin{cases} \beta_\alpha^G - \rho_n\Psi_I^G + \varsigma & \alpha = 1,\ \ldots, n-1 \\ -\rho_n\Psi_I^G + \varsigma & \alpha = n\ , \end{cases} \tag{B.53}
$$

and recalling the definition of $\mathbf{s}^*$ in (6.11), viz.,

$$
\mathbf{s}^* = s\mathbf{v}_n + \rho_n\boldsymbol{\Gamma}^*\mathbf{u}_n\ , \tag{B.54}
$$

and the representation of $\boldsymbol{\Gamma}^*$ in (B.46), (B.52) can be transformed into

$$
\begin{aligned}
(\mathrm{III}) = \sum_{\alpha=1}^{n-1} \Big\{ & \lambda^\varepsilon \rho_\alpha \boldsymbol{\Delta}_D^{*\alpha} \mathbf{u}_\alpha + \lambda^\varepsilon \zeta_\alpha \mathbf{u}_\alpha - \mathsf{k}\mathbf{u}_\alpha + s\mathbf{v} - s\mathbf{v}_n \\
& -\rho_n \big\{ \lambda^\varepsilon \rho^{-1} (\mathbf{T}_I + \mathbf{T}_D) - \lambda^\varepsilon (\Psi_I^G + \Psi_D^G)\mathbf{I} \\
& \qquad - \rho^{-1} \operatorname{sym}(\mathcal{P}_{\nabla\theta} \otimes \nabla\theta) \big\} \mathbf{u}_n \\
& -\mathbf{k}_{,\nu_\alpha} + \lambda^\varepsilon \xi_n \mathbf{T}_I \mathbf{u}_n - \operatorname{skw}(\mathcal{P}_{\nabla\theta} \otimes \nabla\theta)(\xi_\alpha \mathbf{u}_\alpha - \xi_n \mathbf{u}_n) \Big\} \cdot (\nabla \nu_\alpha) \\
= \sum_{\alpha=1}^{n-1} \Big\{ & \{\lambda^\varepsilon \rho_\alpha \boldsymbol{\Delta}_D^{*\alpha} + \lambda^\varepsilon \zeta_\alpha \mathbf{I} - \tfrac{1}{2} \lambda^\varepsilon \rho_n (\mathbf{u}_n \cdot \mathbf{u}_n)\mathbf{I}\} \mathbf{u}_\alpha - \lambda^\varepsilon \rho_n \boldsymbol{\Delta}_D^{*n} \, \mathbf{u}_n \\
& + \lambda^\varepsilon \big(\rho_n \Psi_I^G - \varsigma\big) \mathbf{u}_n - \mathbf{k}_{,\nu_\alpha} \\
& - \operatorname{skw}(\mathcal{P}_{\nabla\theta} \otimes \nabla\theta)(\xi_\alpha \mathbf{u}_\alpha - \xi_n \mathbf{u}_n) \Big\} \cdot (\nabla \nu_\alpha) \, .
\end{aligned}
\tag{B.55}
$$

In the last manipulation, we also used the definitions for k, $(6.93)_3$, $\boldsymbol{\Delta}_D^{*n}$, (B.49), and that of ς, (6.103). If we take into account ζ_n in (B.53) and define

$$
\mathbf{c} = \big(\rho_n \boldsymbol{\Delta}_D^{*n} + \zeta_n\big) \mathbf{u}_n \ , \tag{B.56}
$$

we obtain from (B.55)

$$
\begin{aligned}
(\mathrm{III}) = \lambda^\varepsilon \sum_{\alpha=1}^{n-1} \Big\{ & \{\rho_\alpha \boldsymbol{\Delta}_D^{*\alpha} + \zeta_\alpha \mathbf{I} - \tfrac{1}{2} \rho_n (\mathbf{u}_n \cdot \mathbf{u}_n)\mathbf{I}\} \mathbf{u}_\alpha - \mathbf{c} - (\lambda^\varepsilon)^{-1} \mathbf{k}_{,\nu_\alpha} \\
& - (\lambda^\varepsilon)^{-1} \operatorname{skw}(\mathcal{P}_{\nabla\theta} \otimes \nabla\theta)(\xi_\alpha \mathbf{u}_\alpha - \xi_n \mathbf{u}_n) \Big\} \cdot (\nabla \nu_\alpha).
\end{aligned}
\tag{B.57}
$$

With the same definitions and the same manipulations as for lines (II) and (III), line (IV) is transformed into

$$
\begin{aligned}
(\mathrm{IV}) = \lambda^\varepsilon \sum_{\alpha=1}^{n} \Big\{ & \bar{\rho}_\alpha \big(\boldsymbol{\Delta}_D^{*\alpha} + \mathbf{u}_\alpha \otimes \mathbf{u}_\alpha\big) + \bar{\zeta}_\alpha \mathbf{I} - \tfrac{1}{2} \nu_\alpha \rho_n (\mathbf{u}_n \cdot \mathbf{u}_n)\mathbf{I} \\
& + (\lambda^\varepsilon)^{-1} \langle \mathbf{B}_\alpha \, , \, \mathcal{P}_{\mathbf{B}_\alpha} \rangle - (\lambda^\varepsilon)^{-1} \mathbf{k}_{,\mathbf{v}_\alpha} + \bar{\mathbf{T}}_\alpha \Big\} \cdot \mathbf{D}_\alpha \, .
\end{aligned}
\tag{B.58}
$$

The term $\langle \mathbf{B}_\alpha \, , \, \mathcal{P}_{\mathbf{B}_\alpha} \rangle \cdot \mathbf{D}_\alpha$ is modified in the following way:

$$
\begin{aligned}
\langle \mathbf{B}_\alpha \ , \ \mathcal{P}_{\mathbf{B}_\alpha} \rangle \cdot \mathbf{D}_\alpha &= \\
\langle \mathcal{P}_{\mathbf{B}_\alpha} \ , \ \mathbf{B}_\alpha \rangle \cdot \mathbf{D}_\alpha &= \Big\{ \big(\mathcal{P}_{\mathbf{B}_\alpha} \mathbf{B}_\alpha \big) + \big(\mathbf{B}_\alpha \mathcal{P}_{\mathbf{B}_\alpha} \big) \Big\} \cdot \mathbf{D}_\alpha \\
&= (\mathcal{P}_{\mathbf{B}_\alpha})_{ij} \ (\mathbf{B}_\alpha)_{jk} \ (\mathbf{D}_\alpha)_{ik} + (\mathbf{B}_\alpha)_{ij} \ (\mathcal{P}_{\mathbf{B}_\alpha})_{jk} \ (\mathbf{D}_\alpha)_{ik} \\
&= (\mathcal{P}_{\mathbf{B}_\alpha})_{ij} \ \big\{ (\mathbf{B}_\alpha)_{jk} \ (\mathbf{D}_\alpha)_{ik} \big\} + (\mathcal{P}_{\mathbf{B}_\alpha})_{jk} \ \big\{ (\mathbf{B}_\alpha)_{ji} \ (\mathbf{D}_\alpha)_{ik} \big\} \\
&= (\mathcal{P}_{\mathbf{B}_\alpha})_{ij} \ \big\{ (\mathbf{B}_\alpha)_{jk} \ (\mathbf{D}_\alpha)_{ik} \big\} + (\mathcal{P}_{\mathbf{B}_\alpha})_{ji} \ \big\{ (\mathbf{B}_\alpha)_{jk} \ (\mathbf{D}_\alpha)_{ki} \big\} \\
&= \big\{ (\mathcal{P}_{\mathbf{B}_\alpha})_{ij} + (\mathcal{P}_{\mathbf{B}_\alpha})_{ji} \big\} (\mathbf{B}_\alpha)_{jk} \ (\mathbf{D}_\alpha)_{ik} \\
&= \big\{ 2 \, \mathrm{sym}(\mathcal{P}_{\mathbf{B}_\alpha}) \mathbf{B}_\alpha \big\} \cdot \mathbf{D}_\alpha \ . \qquad \text{(B.59)}
\end{aligned}
$$

If we further replace $(\mathcal{P}_{\mathbf{B}_\alpha})$ by $\big(-\lambda^\varepsilon \, \rho \, (\Psi_I^G)_{,\mathbf{B}_\alpha} \big)$ (see relation $(6.96)_7$) we obtain

$$
\begin{aligned}
\text{(IV)} = \lambda^\varepsilon \sum_{\alpha=1}^{n} \Big\{ & \bar{\rho}_\alpha \big(\boldsymbol{\Delta}_D^{*\alpha} + \mathbf{u}_\alpha \otimes \mathbf{u}_\alpha \big) + \bar{\zeta}_\alpha \mathbf{I} - \tfrac{1}{2} \nu_\alpha \rho_n (\mathbf{u}_n \cdot \mathbf{u}_n) \mathbf{I} \\
& - 2\rho \ \mathrm{sym} \big((\Psi_I^G)_{,\mathbf{B}_\alpha} \big) \mathbf{B}_\alpha - (\lambda^\varepsilon)^{-1} \mathbf{k}_{,\mathbf{v}_\alpha} + \bar{\mathbf{T}}_\alpha \Big\} \cdot \mathbf{D}_\alpha \ .
\end{aligned}
\qquad \text{(B.60)}
$$

In line (V) we simply replace $(\boldsymbol{\lambda}_\alpha^Z)$ by $\big(-\lambda^\varepsilon \rho (\Psi_I^G)_{,\bar{\mathbf{Z}}_\alpha} \big)$ (see relation $(6.96)_8$) and in line (VI) we apply [**A11**] to obtain

$$
\text{(VI)} = \lambda^\varepsilon \sum_{\alpha=1}^{n} \Big\{ - \mathbf{u}_\alpha \cdot \bar{\mathbf{m}}_\alpha^i + \frac{l_\alpha^\rho}{\lambda^\varepsilon} \, \bar{\rho}_\alpha \, c_\alpha + \frac{\lambda_\alpha^\nu}{\lambda^\varepsilon} \, \bar{n}_\alpha \Big\} \ . \qquad \text{(B.61)}
$$

If we further use the definitions of the GIBBS free energies (see (6.107))

$$
\underset{\alpha}{\mu}{}_{\mathrm{I}}^{G} := -(\lambda^\varepsilon)^{-1} \, \underset{\alpha}{l}{}_{\mathrm{I}}^{\rho} \qquad \text{(B.62)}
$$

and that of $\boldsymbol{\iota}_\alpha$

$$
\boldsymbol{\iota}_\alpha := (\lambda^\varepsilon)^{-1} \lambda_\alpha^\nu \ , \qquad \text{(B.63)}
$$

we obtain

$$
\text{(VI)} = -\lambda^\varepsilon \sum_{\alpha=1}^{n} \Big\{ \mathbf{u}_\alpha \cdot \bar{\mathbf{m}}_\alpha^i + \bar{\rho}_\alpha \ \big(\underset{\alpha}{\mu}{}_{\mathrm{I}}^{G} + \tfrac{1}{2} \mathbf{u}_\alpha \cdot \mathbf{u}_\alpha \big) c_\alpha - \boldsymbol{\iota}_\alpha \bar{n}_\alpha \Big\} \ . \qquad \text{(B.64)}
$$

If all these results are substituted in (B.45) we obtain inequality (7.5). □

References

[1] Alts, T., Hutter, K.: Continuum description of the dynamics and thermodynamics of phase boundaries between ice and water. 1. Surface balance laws and their interpretation in terms of 3-dimensional balance laws averged over the phase-change boundary-layer. Journal of Non-Equilibrium Thermodynamics **13**, 221–257 (1988)

[2] Alts, T., Hutter, K.: Continuum description of the dynamics and thermodynamics of phase boundaries between ice and water. 2. Thermodynamics. Journal of Non-Equilibrium Thermodynamics **13**, 259–280 (1988)

[3] Alts, T., Hutter, K.: Continuum description of the dynamics and thermodynamics of phase boundaries between ice and water. 3. Thermodynamics and its consequences. Journal of Non-Equilibrium Thermodynamics **13**, 301–329 (1988)

[4] Alts, T., Hutter, K.: Continuum description of the dynamics and thermodynamics of phase boundaries between ice and water. 4. On thermostatic stability and well-posedness. Journal of Non-Equilibrium Thermodynamics **14**, 1–22 (1989)

[5] Ancey, C.: Plasticity and geophysical flows: A review. J. Non-Newtonian Fluid Mech. **142**, 4–35 (2007)

[6] Anderson, T., Jackson, R.: A fluid mechanical description of fluidized beds: equations of motion. Ind. Eng. Chem. Fundam. **6**, 527–539 (1967)

[7] Batchelor, G., Green, J.: The determination of the bulk stress in a suspension of spherical-particles to order c^2. J. Fluid Mech. **56**, 401–427 (1972)

[8] Bauer, B.: Modelling of critical stress in hypoplasticity. In: A. Balkema (ed.) NUMOG V, pp. 15–20 (1995)

[9] Bauer, B.: Calibration of a comprehensive constitutive equation for granular materials. Soils and Foundations **36**(1), 13–26 (1996)

[10] Bauer, G.: Thermodynamische Betrachtung einer gesättigten Mischung. Ph.D. Thesis, Institut für Mechanik AG III, Technische Universität Darmstadt, Germany (1997). (German)

[11] Bauer, H.: Measure and integration theory. de Gruyter (2001)
[12] Bingham, E.: Fluidity and Plasticity. McGraw-Hill, New York (1922)
[13] Bluhm, J.: A consistent model for saturated and empty porous media. Tech. Rep. 74, Habilitationsschrift, Fachbereich Bauwesen, Universität Essen (1997)
[14] Bowen, R.: Theory of Mixtures, part I. In: A. Eringen (ed.) Continuum Physics III, pp. 2–127. Academic Press (1976)
[15] Bowen, R., Wang, C.C.: Introduction to Vectors and Tensors, *Mathematical Concepts and Methods in Science and Engineering*, vol. 1. Plenum Press (1976)
[16] Bowen, R., Wang, C.C.: Introduction to Vectors and Tensors, *Mathematical Concepts and Methods in Science and Engineering*, vol. 2. Plenum Press (1976)
[17] Caratheodory, C.: Untersuchung über die Grundlagen der Thermodynamik. Math. Ann. **67**, 355–386 (1909). (German)
[18] Cartan, H.: Differnential Forms. Houghton Mifflin, Boston (1971)
[19] Cartan, H.: Differentialformen. BI-Wissenschaftsverlag, Mannheim-Wien-Zürich (1974)
[20] Casey, J.: On infinitesimal deformation measures. Journal of Elasticity **28**, 257–269 (1992)
[21] Casey, J.: On elastic-thermo-plastic materials at finite deformation. International Journal of Plasticity **14**, 173–191 (1998)
[22] Chadwick, P.: Continuum Mechanics, Concise Theory and Problems. Dover Publications (1999)
[23] Chambon, R.: Une classe de lois de comportement incrémentalement nonlinéaires pour les sols non visqueux, résolution de quelques problèmes de cohérence. C.R. Acad. Sci. **308(II)**, 1571–1576 (1989). (French)
[24] Chambon, R.: Discussion of paper "shear and objective stress rates in hypoplasticity" by D. Kolymbas and I. Herle. International Journal for Numerical and Analytical Methods in Geomechanics **28**, 365–372 (2004)
[25] Chambon, R., Desrues, J.: Bifurcation par localisation et non linéarité incrémentale: un example heuristique d'analyse complete. In: Plastic instability, pp. 101–113. Presses ENPC (1985). (French)
[26] Chambon, R., Desrues, J., Hammad, W., Charlier, R.: CloE, a new rate-type constitutive model for geomaterials. Theoretical basis and implementation. International Journal for Numerical and Analytical Methods in Geomechanics **18**, 253–278 (1994)
[27] Chen, K.C., Tai, Y.C.: Volume-weighted mixture theory for granular materials. Cont. Mech. Thermodyn. **19**(7), 457–474 (2008)
[28] Coleman, B., Noll, W.: The thermodynamics of elastic materials with heat conduction and viscosity. Arch. Rational. Mech. Anal. **13**, 167–178 (1963)

[29] Darve, F.: Une formulation incrémentale non-linéaire des lois rhéologique; application aux sols. Ph.D. Thesis, Université Scientifique et Médicale de Grenoble (1978). (French)
[30] Darve, F.: The expression of rheological laws in incremental form and the main classes of constitutive equations. In: F. Darve (ed.) Geomaterials: Constitutive Equations and Modelling, pp. 123–148. Elsevier (1990)
[31] Davis, R., Mullenger, G.: A rate-type constitutive model for soil with a critical state. Int. J. Numer. Anal. Methods Geomech. **2**, 255–282 (1978)
[32] De Kee, D., Turcotte, G.: Viscosity of biomaterials. Chem. Eng. Commun. **6**, 273–282 (1980)
[33] Denlinger, R., Iverson, R.: Flow of variably fluidized granular masses across three-dimensional terrain 2. Numerical predictions and experimental tests. J. Geophys. Res. solid earth **106**, 553–566 (2001)
[34] Drew, D., Passman, S.: Theory of Multicomponent Fluids, *Applied Mathematical Sciences*, vol. 135. Springer (1999)
[35] Edelen, D.: Applied Exterior Calculus. Dover Publications, Inc. (2005)
[36] Einstein, A.: Eine neue Bestimmung der Molekueldimensionen. Annalen der Physik **19**, 289 (1906). (German)
[37] Einstein, A.: Berichtigung zu meiner Arbeit "Eine neue Bestimmung der Molekueldimensionen". Annalen der Physik **34**, 591 (1911). (German)
[38] Eringen, A., Ingram, J.: A continuum theory of chemically reacting media-I. Int. J. Eng. Sci. **3**, 197–212 (1965)
[39] Fang, C.: A thermo-mechanical continuum theory with internal length of cohesionless granular materials. Ph.D. Thesis, Institut für Mechanik AG III, Technische Universität Darmstadt, Germany (2005)
[40] Fang, C., Lee, C.H.: A unified evolution equation for the Cauchy stress tensor of an isotropic elasto-visco-plastic material, Part II. Normal stress difference in a viscometric flow, and an unsteady flow with a moving boundary. Cont. Mech. Thermodyn. **19**(7), 441–455 (2008)
[41] Fang, C., Wang, Y., Hutter, K.: A thermo-mechanical continuum theory with internal length for cohesionless granular materials, Part I. A class of constitutive models. Cont. Mech. Thermodyn. **17**(8), 545–576 (2006)
[42] Fang, C., Wang, Y., Hutter, K.: A thermo-mechanical continuum theory with internal length for cohesionless granular materials, Part II. Non-equilibrium postulates and numerical simulations of simple shear, plane Poiseuille and gravity driven problems. Cont. Mech. Thermodyn. **17**(8), 577–607 (2006)
[43] Fang, C., Wang, Y., Hutter, K.: A unified evolution equation for the Cauchy stress tensor of an isotropic elasto-visco-plastic material, Part I. On thermodynamically consistent evolution. Cont. Mech. Thermodyn. **19**(7), 423–440 (2008)

[44] Farkas, G.: A Fourier-féle mechanikai elv alkalmazái. [On the applications of the mechanical principle of Fourier.]. Mathematikai ès Termèszettundaomànyi Értesitö **12**, 457–472 (1894). (Hungarian)
[45] Grauel, A.: Feldtheoretische Beschreibung von Grenyflächen, *Lecture Notes in Physics*, vol. 326. Springer (1989)
[46] Greve, R.: Kontinuumsmechanik - Ein Grundkurs. Springer (2003). (German)
[47] Gudehus, G.: A comparison of some constitutive laws for soils under radically symmetric loading and unloading. In: W. Wittke (ed.) Proc 3rd Int. Conf. on Numerical Methods in Geomechanics, Aachen, pp. B09–B23 (1979)
[48] Gudehus, G.: On the physical background of soil strength. In: D. Kolymbas (ed.) Constitutive Modelling of Granular Materials, pp. 291–301. Springer (1999)
[49] Gurtin, M.: An Introduction to Continuum Mechanics, *Mathematics in Science and Engineering*, vol. 158. Elsvier Science (2003)
[50] Harbitz, C.: Snow avalanche modeling, mapping and warning in Europe (SAME). Report of the fourth European Network Programme: Environment and Climate. Norwegian, Geotechnical Institute, Norway (1999)
[51] Haupt, P.: Continuum Mechanics and Theory of Materials. Springer (2000)
[52] Hauser, R., Kirchner, N.: A historical note on the entropy principle of Müller and Liu. Cont. Mech. Thermodyn. **14**, 223–226 (2002)
[53] Heil, E.: Differentialformen (und Anwendungen auf Vektoranalysis, Differentialgleichungen, Geometrie). BI-Wissenschaftsverlag, Mannheim-Wien-Zürich (1974)
[54] Herle, I., Gudehus, G.: Determination of parameters of a hypoplastic constitutive model from properties of grain assemblies. Mechanics of Cohesive-Frictional Materials **4**(5), 461–486 (1999)
[55] Herschel, W., Bulkley, R.: Konsistenzmessungen von Gummi-Benzol-Lösungen. Kolloid Z. **39**, 291–300 (1926). (German)
[56] Huang, W., Bauer, E.: Numerical investigations of shear localization in micro-polar hypoplastic material. International Journal for Numerical and Analytical Methods in Geomechanics **27**(4), 325–352 (2003)
[57] Husband, D., Aksel, N., Geissle, W.: The existence of static yield stresses in suspensions containing non-colloidal particles. J. Rheol. **37**, 215–235 (1993)
[58] Hutter, K.: The foundation of thermodynamcis, its basic postulates and implications. A review of modern thermodynamics. Acta Mechanica **27**, 1–54 (1977)
[59] Hutter, K.: Avalanche dynamics. In: V. Singh (ed.) Hydrology of disasters, pp. 317–394. Kluwer Academic Publishers, Dordrecht etc. (1996)
[60] Hutter, K.: Fluid und Thermodynamik, eine Einführung, 2nd edn. Springer (2002). (German)

[61] Hutter, K.: Geophysical granular and particle laden flows: review of the field. Phil. Trans. R. Soc. L. A **363**, 1497–1505 (2005)

[62] Hutter, K., Jöhnk, K.: Continuum Methods of Physical Modeling. Springer (2004)

[63] Hutter, K., Laloui, L., Vulliet, L.: Thermodynamically based mixture models of saturated and unsaturated soils. Mechanics of Cohesive-Frictional Materials **4**, 295–338 (1999)

[64] Hutter, K., Rajagopal, K.: On flows of granular materials. Continuum Mechanics and Thermodynamics **6**, 81–139 (1994)

[65] Hutter, K., Wang, Y.: Phenomenological thermodynamics and entropy principles. In: A. Greven, G. Keller, G. Warnecke (eds.) Entropy, pp. 57–77. Princeton University Press (2003)

[66] Iverson, R.: The physics of debris flows. Rev. Geophys. **35**, 245–296 (1997)

[67] Iverson, R.: Forecasting runout of rock and debris avalanches. In: S.G. Evans, G.S. Mugnozza, A. Strom, R. Hermanns (eds.) Landslides from Massive Rock Slope Failure, *NATO Science Series*, vol. 49, pp. 197–209. Springer (2003)

[68] Iverson, R.: Regulation of landslide motion by dilatancy and pore pressure feedback. J. Geophys. Res. Solid earth **110** (2005). F02015

[69] Iverson, R., Denlinger, R.: Flow of variably fluidized granular masses across three-dimensional terrain 1. Coulomb mixture theory. J. Geophys. Res. Solid Earth **106**, 537–552 (2001)

[70] Jackson, R.: The Dynamics of Fluidized Particles. Cambridge University Press (2000)

[71] Kirchner, N.: Thermodynamics of Structured Granular Materials. Ph.D. Thesis, Institut für Mechanik AG III, Technische Universität Darmstadt, Germany (2001)

[72] Kirchner, N.: Thermodynamically consistent modelling of abrasive granular materials. I Non-equilibrium theory. Proc. R. Soc. Lond. **458**, 2153–2176 (2002)

[73] Klingbeil, E.: Tensorrechnung für Ingenieure. BI-Wiss.-Verlag, Mannheim (1993). (German)

[74] Kolymbas, D.: Eine Konstitutive Theorie für Böden und andere Körnige Stoffe. Ph.D. Thesis, University of Karlsruhe, Germany (1978). (German)

[75] Kolymbas, D.: An outline of hypoplasticity. Archive of Applied Mechanics **61**, 143–151 (1991)

[76] Kolymbas, D.: Introduction to Hypoplasticity. No. 1 in Advances in Geotechnical Engineering und Tunneling. A.A.Balkema, Rotterdam (2000)

[77] Kovalevskaya, S.: Zur Theorie der partiellen Differentialgleichungen. [On the theory of partial differential equations]. Journal für die reine und angewandte Mathematik **80**, 1–32 (1875). (German)

[78] Liu, I.S.: Method of Lagrange Mulitpliers for Exploitation of the Entropy Principle. Arch. Rational Mech. Anal. **46**, 131–148 (1972)
[79] Liu, I.S.: On the entropy supply in a classical and relativistic fluid. Arch. Rational. Mech. Anal. **50**, 111–117 (1973)
[80] Liu, I.S.: On entropy flux - heat flux relation in thermodynamics with Lagrange multipliers. Continuum Mech. Thermodyn. **8**, 247–256 (1996)
[81] Liu, I.S.: Continuum Mechanics. Springer (2002)
[82] Luca, I., Hutter, K., Tai, Y.C., Kuo, C.Y.: A hierachy of avalanche models on arbitrary topography. Acta Mech. **205**, 121–149 (2009)
[83] Marsden, J., Hughes, T.: Mathematical Foundations of Elasticity. Dover Publications (1983)
[84] Masin, D.: Hypo-plastic models for fine-grained soils. Ph.D. Thesis, Charles University Prague, Institute of Hydrology, Engineering Geology and Applied Geophysics (2006)
[85] McDougall, S., Hungr, O.: Dynamic modelling of entrainment in rapid landslides. Canadian Geotechnical Journal **42**, 1437–1448 (2005)
[86] McDougall, S., Hungr, O.: A model for the analysis of rapid landslide motion across three-dimensional terrain. Canadian Geotechnical Journal **41**, 1084–1097 (2005)
[87] Mendes, P., Dutra, E.: Viscosity function for Yield-Stress Liquids. Appl. Rheol. **14**, 296–302 (2004)
[88] Minkowski, H.: Geometrie der Zahlen. [Geometry of numbers.], 1st edn. Teubner, Leipzig (1896). (German)
[89] Morland, L.: Simple constitutive theory for a fluid-saturated porous solid. Journal of Geophysical research **77**, 890–900 (1972)
[90] Morland, L.: Theory of slow fluid-flow through porous thermoelastic matrix. Geophys. J. Roy. Astr. Soc. **55**, 393–410 (1978)
[91] Morland, L.: Flow of viscous fluids through a porous deformable matrix. Surveys in Geophysics **13**, 209–268 (1992)
[92] Morland, L.: Flow in a Porous Matrix with Anisotropic Structure. Transport in Porous Media (2009). DOI 10.1007/s11242-009-9392-3
[93] Morland, L., Sellers, S.: Multiphase mixtures and singular surfaces. International Journal of non-linear mechanics **36**, 131–146 (2001)
[94] Müller, I.: Thermodynamic Theory of Mixtures of Fluids. Arch. Rational. Mech. Anal. **28**, 1–39 (1968)
[95] Müller, I.: The coldness, a universal function in thermoelastic bodies. Arch. Rational. Mech. Anal. **41**, 319–332 (1971)
[96] Müller, I.: Die Kältefunktion, eine universelle Funktion in der Thermodynamik viskoser wärmeleitender Flüssigkeiten. Arch. Rational. Mech. Anal. **40**, 1–36 (1971). (German)
[97] Müller, I.: Thermodynamics. Pitman, Boston (1985)
[98] Muschik, W., Papenfuss, C., Ehrentraut, H.: A sketch of continuum thermodynamics. J. Non-Newtonian Fluid Mech. **96**, 255–299 (2001)

[99] Niemunis, A.: A visco-plastic model for clay and its FE implementation. In: E. Dembicki, W. Cichy, L. Balachowski (eds.) Recent results in mechanics of soils and rocks, pp. 151–162. TU Gdansk (1996)

[100] Norem, H., Irgens, F., Schieldrop, B.: Continuum model for calculating snow avalanches. In: B. Salm, H.U. Gubler (eds.) Avalanche Formation, Movements and Effects, pp. 263–279. IAHS Publ. (1986)

[101] Ogden, R.: Non-linear Elastic Deformations. Dover (1987)

[102] Papanastasiou, T.: Flows of materials with yield. J. Rheol. **31**, 385–404 (1987)

[103] Passman, S., Nunziato, J., Walsh, E.: A theory of multiphase mixtures. In: C. Truesdell (ed.) Rational Thermodynamics, 2nd ed., pp. 286–325. Springer (1984)

[104] Pitman, E., Le, L.: A two-fluid model for avalanche and debris flows. Phil. Trans. R. Soc. Lond. A - Mathematical Physical and Engineering Sciences **363**, 1573–1601 (2005)

[105] Pudasaini, S., Hutter, K.: Avalanche Dynamics. Dynamics of Rapid Flows of Dense Granular Avalanches. Springer (2006)

[106] Pudasaini, S., Wang, Y., Hutter, K.: Modelling debris flows down general channels. Natural Hazards and Earth System Sciences **5**, 799–819 (2005)

[107] Rajagopal, K., Tao, L.: Mechanics of Mixtures, *Series of Advances of Applied Sciences*, vol. 35. World Scientific Singapor etc. (1995)

[108] Rankine, W.: On the stability of loose earth. Phil. Trans. R. Soc. Lond. **147**, 9–27 (1857)

[109] Rivlin, R., Saunders, D.: Large Elastic Deformations of Isotropic Materials VII. Experiments on the Deformation of Rubber. Phil. Trans. R. Soc. Lond. **243 A**, 252–288 (1951)

[110] Rockafellar, R.: Conjugate convex functions in optimal control and the calculus of variations. J. Math. Analysis Appl. **32**, 174–222 (1970)

[111] Romano, M.: A continuum theory for granular media with a critical state. Arch. Mech. **20**, 1011–1028 (1974)

[112] Roos, H., Hermann, A. (eds.): MAX PLANCK Vorträge Reden Erinnerungen. Springer (2001). (German)

[113] Schneider, L.: A non-classical debris flow model - based on a concise thermodynamic analyis. Master's Thesis, Technische Universtität Darmstadt (2005)

[114] Stutz, A.: Foundations of Plasticity, chap. Comportement elastplastique des milieux granulaires. Noordhoff (1972). (French)

[115] Svendsen, B., Hutter, K.: On the thermodynamics of a mixture of isotropic materials with constraints. Int. J. Eng. Sci. **33**, 2021–2054 (1995)

[116] Svendsen, B., Hutter, K., Laloui, L.: Constitutive models for granular materials including quasi-static frictional behaviour: Toward a thermodynamic theory of plasticity. Continuum Mech. Thermodyn. **4**, 263–275 (1999)

[117] Teufel, A.: Simple flow configurations in hypoplastic abrasive materials. Master's Thesis, Institut für Mechanik AG III, Technische Universität Darmstadt, Germany (2001)

[118] Truesdell, C.: Sulle basi della termodinamica delle miscele. Rend. Accad. Naz. Lincei. **44**(8), 381–383 (1957). (Italian)

[119] Truesdell, C.: Rational Thermodynamics, 2nd edn. Springer (1984)

[120] Truesdell, C., Muncaster, R.: Fundamentals of Maxwells's Kinetic Theory of a Simple Monatonic Gas, treated as a Branch of Rational Mechanics. Academic Press (1980)

[121] Truesdell, C., Noll, W.: The Non-linear Field Theories of Mechanics, 3nd edn. Springer (2004). Editor: Antman, S.S.

[122] Wildemuth, C., Wiliams, M.: A new interpretation of viscosity and yield stress in dense slurries; coal and other irregular particles. Rheol. Acta **24**, 75–91 (1985)

[123] Wilmański, K.: Lagrange Model of Two-Phase Porous Materials. J. Non-Equilib. Thermodyn. **20**, 50–77 (1995)

[124] Wilmański, K.: The thermodynamical model of compressible porous materials with the balance equation of porosity. Tech. rep., Weierstrass-Institut für Angewandte Analysis und Stochastik (1997). 310 WIAS

[125] Wilmański, K.: Mass exchange, diffusion and large deformations of poroelastic materials. Tech. rep., Weierstrass-Institut für Angewandte Analysis und Stochastik (2001). 628 WIAS

[126] von Wolffersdorff, P.: A hypoplastic relation for granular materials with a predefined limit state surface. Mechanics of Cohesive-Frictional Materials **1**, 251–271 (1996)

[127] Wu, W.: Hypoplastizität als mathematisches Modell zum mechanischen Verhalten granularer Stoffe. Ph.D. Thesis, Karlsruhe University (1992). (German)

[128] Wu, W., Bauer, E., Kolymbas, D.: Hypoplastic constitutive model with critical state for granular materials. Mechnanics of Materials **23**, 45–69 (1996)

[129] Wu, W., Kolymbas, D.: Hypoplasticity then and now. In: D. Kolymbas (ed.) Constitutive Modelling of Granular Materials, pp. 57–105. Springer (1999)

[130] Zhu, H., Kim, Y., Kee, D.: Non-Newtonian fluids with a yield stress. J. Non-Newtonian Fluid Mech. **129**, 177–181 (2005)

Index

MIX
Papier aus verantwortungsvollen Quellen
Paper from responsible sources
FSC® C105338

If you have any concerns about our products,
you can contact us on
ProductSafety@springernature.com

In case Publisher is established outside the EU,
the EU authorized representative is:
Springer Nature Customer Service Center GmbH
Europaplatz 3, 69115 Heidelberg, Germany

Printed by Libri Plureos GmbH
in Hamburg, Germany